Signals and Processes
a foundation course

John D Martin

School of Electrical Engineering, University of Bath

Pitman

PITMAN PUBLISHING
128 Long Acre, London WC2E 9AN
A Division of Longman Group UK Limited

© J. D. Martin 1991

First published in Great Britain in 1991

British Library Cataloguing in Publication Data
Martin, John D.
  Signal and processes.
  1. Signals. Processing
  I. Title
  621.3822

ISBN 0-273-03256-9

ISBN 0 273 03256 9

Printed and bound in Great Britain

Contents

Preface

Who is it for?

This book is designed for the potential *user* of signal processing. A veritable galaxy of digital hardware and software algorithms is available for the practice of signal processing, but the major problem for the user is to understand what the results mean and how items should be chosen.

Having followed this course, the student should be able to appreciate the operation of most modern measuring instruments and communication subsystems, and be able to specify the kind of process to be used for a particular application. Digital signal processing occurs frequently in preference to analog processing, and the student should be able to appreciate why.

Many engineers will progress to the *design* of signal processing systems, so would need this course to be supplemented by further study. The designer of a signal process algorithm or instrument must have a deeper understanding of the limitations and sources of error in such a process, and so must understand the mathematical background more thoroughly than a user, and be aware of approximations and short-cuts in deriving common algorithms. There is a wealth of textbooks which cover this kind of detail.

Only a few engineers will become *innovators* of signal processing systems, but these will require an additional detailed course in signal processing of the more traditional variety. The innovator must have an intimate and detailed knowledge of the theorems and development of the subject of signal processing, and must study the mathematical background in a rigorous fashion in order to be able to extend the subject and devise new processes. Much of this material appears in professional papers.

How is it presented?

This teaching material is for signal processing in its engineering context, and is presented under the conviction that

a) in the early stages of learning a subject, concepts matter more than mathematical completeness;

b) if concepts are properly understood, then mathematics will be applied correctly;

c) there is nothing intrinsically difficult about signal processing.

This is a foundation course in the sense that it is suitable for a first encounter with the subject. The aim is to provide the firm foundation of a physical and conceptual insight into signals and signal processing, enabling the student later to benefit from the many excellent rigorous mathematical texts, if so desired. Having understood the basic principles, many students have no need to delve further into the detail of signal processing.

A *top-down* development is attempted, starting from the information that signals represent, and proceeding to various simple processing applications. Concepts and principles are introduced systematically, taking a natural sequence rather than the traditional one. Sufficient detail enables the student to *use* each concept, either in analysis or in actual signal processing. This seems important, since many digital processes are packaged, so that the *doing* is not difficult, but problems occur in practice due to mis-application or mis-interpretation of results.

Acknowledgements

This book has been in gestation for a long time. Signal processing was first introduced by the author into the final year of the electrical engineering course at Bath in 1967–68, and has evolved steadily to become a key topic in the first-year course since 1980–81. Course notes have taken many different turnings during this time, driven by the current state of the signal processing art.

The late Dr. A. W. Keen and Professor W. Gosling particularly encouraged me in this development. Discussion with one's colleagues is a key benefit in any university department, and I gratefully acknowledge many such fruitful interactions. A succession of students too have helped hone these ideas into something of a coherent story.

My initial lessons in signal processing were learnt from the clear writing of W. W. Harman in his book *Principles of the Statistical Theory of Communication*, while inspiration for developing such a course at first-year level came from K. Steiglitz, *An Introduction to Discrete Systems*.

John Cushion and Roger Hill at Pitman Publishing have been most supportive in the onerous task of bringing the book into print. The diagrams have been expertly transcribed and enhanced by Ian Daisley at Alpha Graphics.

John Martin

Introduction to Information

OBJECTIVES

To introduce information as a measurable quantity, and the ultimate basis of signal processing. In particular

  a) To define information.
  b) To introduce the operations on information—transmission, storage and processing.
  c) To examine some examples of transmission, storage and processing of information.

COVERAGE

Although largely qualitative, this material establishes the context of and reasons for signal processing. Applications are chosen which may be grasped without any appreciable engineering experience, and orders of magnitude and scale are emphasised.

Signals represent information; so in order to understand why signals need to be processed, and to discover what the best way to process them is, we need to start by examining *information*. What is information, where does it occur, and what can be done with it? In this chapter we aim to establish the end product of signal processing, so that as later we delve into the fine detail we can keep the ultimate objectives firmly in view. This is the destination to which our journey of study will take us.

I.1 Information technology

Definition:

> *The acquisition, transmission, processing and presentation of information in all forms—audio, video, text and graphics.*

1

The implications of this definition range from carrier pigeons to supercomputers!

Information technology is dependent on the ability to process electrical signals at low cost, high speed and with great reliability. The *enabling technology* which has brought about the current information explosion includes microelectronics, computation, signal processing and communications.

About half the UK's working population could be described as *information workers*. Information and its distribution and interpretation form the basis of rapid changes taking place throughout the world. The Industrial Revolution brought immense changes to society through the replacement of human muscle and effort by energy, harnessed in machines powered by steam or hydraulic pressure. The Information Revolution is bringing changes of an even greater kind, by supplementing and even replacing human thought by systems operating on information.

I.2 What is information?

I.2.1 An engineering approach

Information, as understood by human beings, is qualitative. It is related to ideas, to concepts and creativity. People differ enormously in their assessment of a lecture, a book or documentary film. We judge a presentation in terms of its *content*.

Engineering measures of information must be quantitative, since engineering designs are to be based on our assessment of information. In this use of the word *information*, we are concerned more with the *quantity* of information, rather than its *content*, more aligned to *syntax* than *semantics*. In engineering systems, it is the amount of information which must be stored or transmitted which is important, rather than what it means or represents.

A simple example is the posting of a letter, the contents of which are of extreme value to the recipient, but are of no interest at all to the postmen, sorters and van drivers who handle it on its journey. Those who transport the letter form an information communication system, while the sender and recipient are the *users* of the system. A further example is that of a book, where the cost of producing a copy is proportional to its length, irrespective of whether it is a gripping novel or a treatise on electrical circuits.

Let us now begin to put some precision to these ideas. Consider a message which consists of *n symbols*, each of which has been chosen from an *alphabet* of α possible symbols. Then the number of different messages of this type which may be formed is *N*, where

$$N = \alpha^n \tag{I.1}$$

However, the information content of the message is proportional to its length n. Hence an information measure I may be given by

$$I = \log_\alpha N \tag{I.2}$$

Such a measure of information content is *logarithmic*, and depends upon the alphabet size α. So, if we are considering written English then $\alpha = 27$ (including the 'space'), while if we were considering decimal numbers then $\alpha = 10$. Clearly this is not a satisfactory measurement, where the units change from problem to problem. So we take the logarithm to base 2 and call the resulting unit, a *bit*. All messages can be expressed in binary form, so this is a convenient common denominator for measuring information. Hence

$$I = \log_2 N \quad \text{bits} \tag{I.3}$$

Note that this use of the *bit* implies more than just a single binary digit, it has become a precise measure of information content. A binary word of, say, 8 bits has the *potential* to store 8 bits of information, but in practice may represent fewer *bits* of true information according to our definition in equation I.2. This is because information is related to uncertainty, and depends on the probability that a particular message occurs. We shall ignore that complication for now and refer only to potential information, assuming that all possible messages are equally likely to occur.

Example 1 A certain microcomputer control system has to distinguish between 12 possible conditions. The information represented is therefore $3 \cdot 6$ bits, and could be stored in a register with 4 stages, which has the potential to store 4 bits of information.

Example 2 A sequence of 6 decimal numbers has 10^6 possible combinations, and therefore corresponds to 20 bits of potential information.

Example 3 A certain digital voltmeter can resolve to within $\pm 0 \cdot 01$ V, and has a full-scale value of 10 V. A measurement using this instrument represents the choice of one value out of a total of 500, or a potential information content of 9 bits.

I.2.2 Sources of information

I.2.2.1 Human communication

Although information originates in the human mind, it is conveyed normally by speech, which can be measured and quantified. The great majority of communication today is still by voice, and progress is being made towards using the voice as a means of communicating between humans and machines. The true rate of information transfer using the human voice is modest, estimated at a few tens of bits/second. However, attempts to synthesise utterances at this rate lead to a very artificial kind of sound, devoid of most of the characteristics which we use to gauge a person's feelings and character. Telephone quality speech is generally produced at a rate of 64 kb/s, while high quality music digital recordings use a rate of 705·6 kb/s for the wider range of frequencies. Thus a telephone channel is capable of conveying a *potential* information rate of 64 kb/s, while the utterance being transmitted may possess only say 50 b/s of true information.

Pictures form a very efficient method of communicating information, since the eye can absorb large quantities of information very quickly. Picture communication ranges from simple cartoon sketches, through to detailed engineering drawings, and fast-moving video sequences. The ability of the eye to take information cannot be evaluated accurately, but some idea of this rate can be estimated by considering the potential information content of a conventional broadcast video picture frame.

A typical video picture frame contains a field of 576 active lines, each resolving 572 picture elements (or *pixels*), and making a total of 330 000. If each of these pixels in a monochrome picture is represented by an intensity chosen from a range of 128 possible values, the total number of different pictures which could be represented is difficult to calculate, but is greater than 10^{500}! However, although the number of different pictures increases exponentially with the resolution of the picture, the information content increases only as the logarithm of the total number, and in this case is only 2·6 Mb. Since picture frames occur at a rate of 25 per second, a monochrome video channel of this sort represents a potential information throughput of about 66 Mb/s. A colour picture requires a greater information throughput.

In practice, only a small fraction of the number of possible pictures is used, so that some compression of this rate is possible by signal processing. In order to appreciate this fact, consider a typical video scene, in which there are large areas of similar intensity, or colour. These areas are conveying little information, and can be represented, at a much lower transmission rate, by coding which takes account of the picture structure. Using this technique, an acceptable viewphone picture can be transmitted at a rate of 64 kb/s, although the system cannot accommodate quick movements of the image.

Textual representation of information is an essential part of human society. An A4-size page of typing holds typically about 2000 letters, or about 330 English words. In information terms, we could estimate the information content by considering the number of possible alphabetic characters, which would include 52 upper and lower case letters, 10 numerals and a number of miscellaneous punctuation and symbolic characters. In practice and for convenience, a single character as printed is represented by an 8-bit binary word, so an A4 typed page represents potential information of about 16 000 bits, or 2 kByte. Typing rates and reading rates vary widely.

I.2.2.2 Environmental information

While the above section dealt with communication between humans, this one relates to measurement situations where machines are used to assess and control the environment.

A simple example is that of a temperature control system for a building. The temperature sensor must produce information which is adequate for sensible control of the source of heat. On a time basis, the rate of generating information depends on the thermal time constant of the building. Significant readings cannot be taken at intervals of, say, a millisecond, since the temperature will not change by a measurable amount in this time interval. Then too, the temperature within a room is actually a three-dimensional spatial distribution, and careful thought must be given to how much information needs to be gathered.

At the other extreme of speed is a radar system. Radar operates by transmitting a short pulse of electromagnetic energy, and detecting the small amount of this energy which is reflected from the target, an aircraft for instance. Since propagation is at the speed of light c, then measuring the time interval between transmitted and received pulses gives an immediate estimate of distance to the target. Distance may be measured accurately in this way, but an estimate of target velocity is more difficult, due to the *uncertainty principle*. Broadly, one can measure the distance to a static target, or measure the velocity of a moving target, but the question "Where is this moving target?" has no meaning. A target at distance d from the transmitter, moving at velocity v away from the transmitter, moves a further distance $2vd/c$ during the measurement. Target position and velocity can therefore be estimated only within certain related limits—the information available is limited, and may be shared as required between estimates of position or velocity, but not independently.

These examples could be multiplied many times over, since automatic control is a vital part of society. It is important to realise that the quantity of information available from a process, object or event is limited intrinsically by physical constraints, and this restricts the kind of control which can be exercised.

I.2.3 Information and energy

Information is an abstract quantity, although it can be measured under carefully controlled circumstances. We can only assess quantity of information, and handle it, when it is represented by some physical quantity—which we call a *signal*. Most signals are electrical, but many are not. In the next chapter we shall begin to explore the relationships between information and signals.

Although information is abstract, it is not arbitrary, and is subject to certain rules imposed by the physical world. Information is like energy in the following ways:

     a) It is generated at a source
     b) It is always conserved
     c) It can be transferred from place to place, with some loss
     d) Its rate of transfer is limited.

Like energy, certain operations can be carried out on information:

     a) Storage
     b) Transmission
     c) Processing.

These operations can only be carried out on information when it is represented by signals, which have physical attributes. In some cases, the information content is compromised by the properties of the signals representing it. We now consider some aspects of these operations, with examples.

I.3   Operations with information

In order to appreciate some of the limitations imposed on operations with information, we now review the state of the art as defined by current technology. Orders of magnitude are important here, rather than actual numbers. Advances in microelectronics, and in optical storage and transmission, are occurring rapidly, so this discussion will give just a flavour of the quantities involved.

I.3.1   Storage

Information is stored generally in digital form. The elementary component of potential information is the *bit*, and 8 bits are termed a *byte*. A *word* contains 1, 2 or more bytes, and a *message* is composed of any number of words. It is convenient to handle information in *blocks* of a standard size, perhaps 512 bytes for instance, and a collection of such blocks forms a *file*.

Various media are available for storage, the most versatile being in magnetic form, tape or disc. Optical discs are rapidly coming in at the top of the storage hierarchy, and can hold 1000 MByte of data, any part of which can be accessed within say 40 ms. At present, optical storage is mostly *Write-Once-Read-Many-times (WORM)*, which is still very suitable for large databases. The audio *Compact Disc (CD)* is a good example of storage in this form, holding upwards of an hour's high-quality music on a 12 cm disc.

Figure I.1 gives a comparison between the storage capacities available and the types of message which might be stored. Storage is increasing rapidly in capacity and decreasing in cost, so that audio storage for pre-recorded announcements, for instance, is simple. Storage of video messages is also feasible, although the limitation there is the rate at which data can be recovered from the medium.

In order to quantify our review of information storage, and later its transmission and processing, we now define arbitrarily, two high-level items of information:

a *book* is 100k English words, represented by about 600 kByte
a *library* is 20k books

Thus, we notice that according to this arbitrary definition, a *book* could be

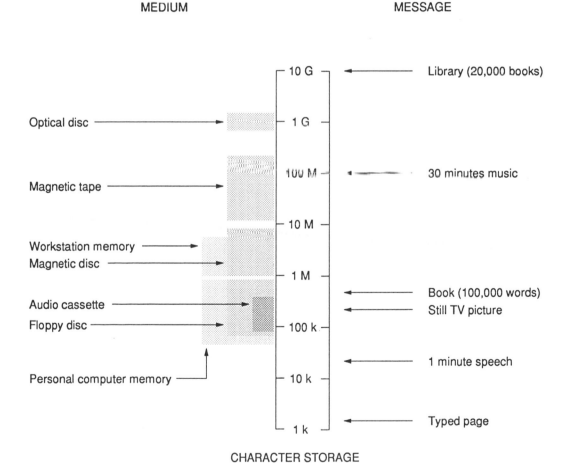

Figure I.1   A comparison of storage capacities available, and the use to which they can be put

7

stored easily on digital cassette tape or on floppy disc. A *library* can be stored on twelve optical discs.

The Oxford Dictionary lists 375 000 words, and occupies many volumes, but its content is estimated at only a few 100 MByte, so could be accommodated comfortably on a magnetic hard disc.

I.3.2   Transmission

In order to review the rate at which transmission of information can take place with current technology, we use the *book* as defined above, as a suitable unit of information.

Figure I.2 shows how the time of transmission varies with the transmission medium. Optimistically, letter-post takes a day, while using a trunk microwave or optical link, 2 ms is sufficient. In practice of course, a high-capacity transmission path would be shared between a very large number of users, so that each user sees a more usable rate. However, this comparison is interesting.

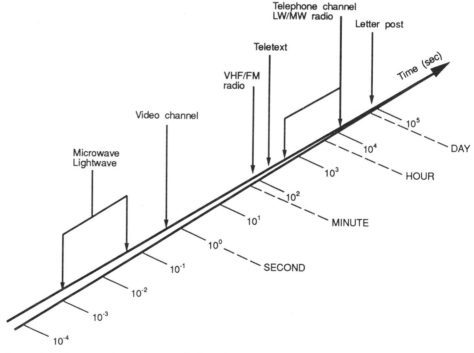

Figure I.2   A comparison of transmission times

I.3.3   Processing

Information processing is much more difficult to generalise upon than storage or transmission. Processing speed depends first on what is to be done, and then on the architecture of the processor which is to carry it out.

8

Table I.1 *Example of information compression*

Keyword	Token
CHAIN	D7
CLEAR	D8
COLOUR	FB
DELETE	C7
ENDPROC	E1
RENUMBER	CC

Hence we will content ourselves with a broad overview of the types of, and reasons for, various kinds of information processing.

In the first case, storage and transmission of raw information can often be made more efficient by *compressing* the information. A simple example is the strategy, used in most microcomputer systems, for storing keywords which are part of the operating language. For instance, the BBC micro uses the BASIC language, but the keywords are stored as *tokens* in hexadecimal number form, and occupy less space than the original keywords. Table I.1 shows some of these equivalents.

Compression is possible because the number of keywords is less than 255, which is the greatest number which can be expressed by two hexadecimal digits. Other forms of compression are used for general text processing, and for speech and video information, each of which contains *redundant* information. Thus, although raw telephone quality speech produces a rate of 64 kb/s, it is possible to transmit usable speech at 1 kb/s and even less. Video signals can, under certain restrictions, be transmitted by rates of 30 Mb/s for broadcast quality, and as low as 2 Mb/s or 64 kb/s for special conference services. This type of compression processing is known as *source coding*.

Information may also be encoded in order to hide it. This is the *encryption* or *encipherment* processing operation, which renders the information secure from unwanted intrusion or eavesdropping. Processing must be carried out carefully, so that the underlying structure of the information is hidden. Simple encryption can be achieved by the Ceasar Cipher, where a message is rendered so:

ARE YOU SURE THAT YOUR COMPUTER SYSTEM IS SAFE
FWJECTZEXZWJEYMFYECTZWEHTRUZYJWEXCXYJRENXEXFKJ

Although apparently garbled, this cipher which uses an alphabet shifted by 5 places, does not hide the essential characteristics of the language for a message of significant length. Encryption is a rather special form of information processing.

Information can be retrieved from a filing system by searching for keywords, which requires an accurate mapping between the original information and the keyword. Processing now amounts to a rapid search through the file, making comparisons as it proceeds. At 1 μs per comparison say, for information held in rapid-access memory, 1 MByte can be scanned in 1 sec, but the process could be made faster by a directed search.

Conversions of information from one form to another occur in language translation for instance, where language syntax is combined with possible meaning in order to make a valid translation. Conversions from text-to-speech and speech-to-text are also done, the former being easier than the latter. The first is employed for assisting blind readers, or for synthesising message announcements at railway stations for instance. The second form of conversion is being investigated with a view to making a word-processor with verbal input.

Information is frequently manipulated to improve perception, particularly in cases of robotic vision, where the image picked up by a video camera is processed so as to reduce its information content, and also to put it into a form which is more readily assessed by the robot processor.

In the radar application, signals probe the air space in order to determine what aircraft are present. A great deal of information is gathered, which must be reduced so that simple questions may be answered:

a) Is there an aircraft there?
b) Where is it?
c) How fast is it travelling?
d) Is a collision likely within the forseeable future?

This kind of assessment uses some complex signal processes, which we shall consider later.

I.4   Some example systems

We shall now outline a few systems in order to show how crucial is the concept of information, and in later chapters we shall examine them in greater detail in order to illustrate various aspects of signal processing. Clearly we are not going to study these examples exhaustively, but each one has been chosen to illustrate some particular point, and together they will give a broad overview of many signal properties and processes used in practice.

The objective of discussing these examples is to emphasise that, although signal processing uses precise mathematical tools, it can be used only as the basis upon which qualitative or value judgements are made, or by which system parameters are *estimated*.

I.4.1 Telecontrol

Telecontrol is the term used to describe the control of items of plant at a distance. Particular examples include electricity, gas and water supply networks, oil platforms, chemical manufacturing plants. In all these cases, the objects to be controlled are widely spread geographically, and control is to be exercised from some central location. Telecontrol requires first that the state of the remote plant is monitored and flow rates, etc. measured. Control operations are then carried out, based on the measured quantities and the demand pattern for the network or plant.

The prime requirement for such a communication network is reliability, since the consequences of a wrong control action can be immense. Some such networks have hundreds of controlled stations.

For an example system, consider one particular channel which controls an electricity supply circuit-breaker, or maybe a railway point machine. Control of the circuit breaker is ON/OFF, and is essentially 1 bit of information. Indication of the state of the circuit-breaker is three-state, ON/mid-way/OFF, and is essentially 1·585 bits. The third state is vital, in order to determine whether the action has been carried out correctly, or whether a fault has developed within the controlled switch.

These items of information must be transmitted reliably over a communication network. The information rate is very low, since the actions occur rarely, although the state indication will be examined frequently in order to make sure that the plant is in good health. Since reliability is crucial, the method of transmission will add *redundancy* to the information, so that errors in transmission are unlikely to result in a wrong action.

Consider also a typical *telemetered* quantity, say a current measurement. Suppose that this is over the range 0 to 100 A, and is to resolve an interval of 0·5 A. This measurement then generates 7·6 bits of information.

I.4.2 Room heating

Room heating is a familiar example of a control system: room temperature is estimated by several temperature sensors placed about the room, and heat input is controlled by an ON/OFF switch or valve.

Available information about conditions in the room is limited by the resolution, accuracy and placing of the sensors. However, the essential information about whether it is comfortable to sit within the room is only obtainable indirectly from the sensor readings, by some qualitative assessment. Deciding upon what target to set is difficult, particularly if several people are using the room, and they each have different perceptions of comfort.

Even when some control target has been decided, there remains the problem of using an ON/OFF style of heat input to achieve the temperature target; it is a very blunt instrument indeed! Some signal processing can help anticipate the movement of temperature and optimise the temperature profile.

I.4.3   Radar

Radar operates by transmitting a short pulse of electromagnetic energy, and then detecting the minute amount of this energy which is reflected from the target, an aircraft for instance. Since propagation is at the speed of light, then measuring the time interval between transmitted and received pulses gives an immediate estimate of distance to the target.

Distance to the target is *quantised* into discrete elements, called *range cells*, whose size is determined by the ability of the radar radio system to transmit information—its *information capacity*. A decision is made for each cell, as to whether there is a target (aircraft) present within that cell or not. This decision can involve some fairly complicated signal processing.

The transmission channel has limited the information which can be obtained from the target by a single measurement. However, by making the measurement repeatedly, the total amount of information available about the target has been increased, and signal processing can extract further implied information about the target.

Target distance is now known to within one cell out of N possible cells, and so represents information of $\log_2 N$ bits. This information can now be used to estimate the velocity of the target, and by extrapolation to estimate the likely position of the target in the future.

STUDY QUESTIONS

1   Explain why a quantitative measure of information is essential for engineering systems.

2   What is the information content of a current measurement, carried out over a full-scale range of 5 A, with a precision of ± 0.001 A?

3   Summarise the rates of information used by different methods of human communication.

4   Find some further examples of 'environmental information'.

5   Why is information compared to energy?

6   Estimate the storage capacity required for a railway station announcement system, using synthesised speech. Work out first, the number and lengths of announcements that might be necessary.

7   Summarise the examples of information processing which are given in the chapter. Produce one or two more examples.

1 Signals and Signal Processing

OBJECTIVES

To set the background for detailed examination of the signals which will follow in later chapters. In particular:

a) To gain a feel for the context in which various forms of signal are found, namely electrical, acoustic, thermal, mechanical, magnetic.
b) To identify some of the major considerations when signals are stored, transmitted and processed.
c) To examine some of the more obvious targets of signal processing.
d) To appreciate the significance of the block diagram or modular approach to describing signal processes, illustrated by simple filters.
e) To review the properties of some common types of signal.

COVERAGE

Largely descriptive, using practical examples which are either within the experience of the reader or which can easily be understood, to illustrate the key targets outlined in the objectives. The emphasis is on defining what is meant by Signal Processing, pointing out that the ultimate results are rarely stated numerically, and introducing some commonly used terms and concepts.

1.1 Some example systems

Information is abstract. It is

a) *either a concept which is to be communicated*
b) *or a property of the system being measured.*

Information is fundamental to all forms of communication, measurement and control.

Information can only be transmitted, stored or processed when it is represented by a *signal*, which has physical form. A signal may be formed from any physical quantity whose properties can be made to vary with time or distance; for example voltage, current, velocity, pressure, flux density, temperature, strain. Operations on signals are of course best carried out in electrical form, so a *transducer* is used to convert between one physical form of signal and another, frequently between an electrical form and some other.

Information is at the root of the operation, the reason *why*; while the corresponding signal is the mechanism of the operation, *how* it can be done.

Once information is represented by a signal, then the properties of that signal must be preserved through transmission, storage or processing. However, in some instances the signal can be distorted greatly without losing the essential information carried by it.

In the examples which follow, we endeavour to answer, by illustration, the question *what is a signal*?. Along the way we also comment on what criteria might be used to assess the quality of the signal, and how information content might be compromised.

1.1.1 Sound recording

Figure 1.1 shows the broad outline of the well-known system of recording a sound on magnetic tape. The signal starts off in acoustic form, a sound pressure wave, and is converted into electrical form by the microphone transducer. A further conversion takes place at the tape writing head, where the electrical signal is recorded as a magnetic flux pattern, fixed in the tape and stored. Some frequency compensation is required here, since on playback the magnitude of the emf generated from the magnetic flux pattern on the tape is proportional to frequency.

There are several causes of information loss at these transducers, in the conversion between one form of signal and another; for example:

a) Directional properties of the microphone, distorting the spatial sound pattern.

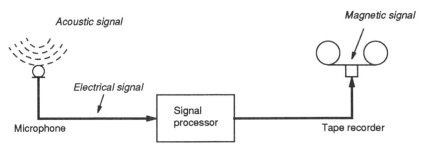

Figure 1.1 A sound recording system

b) Frequency response of the microphone, converting sound to voltage more efficiently at some frequencies than at others.

c) Distortion at the microphone, due to exceeding the allowable range of sound pressure for linear conversion; in a moving-coil microphone this might be due to the coil moving against its mechanical support.

d) Distortion at the recording head due to non-linearity between driving current and magnetic flux.

e) Reduction of the range of frequencies recorded, due to dust between the recording head and the tape.

These and other points are the focus for careful engineering design of this particular signal processing chain.

The signal impairments which can be tolerated depend on the application, upon the use made of the information. A telephone-answering machine places little emphasis on fidelity of reproduction; the meaning of a message is clear if the recording is to telephone quality. At the other extreme, a concert recording of a musician is subjected to critical appraisal with respect to frequency range, distortion, dynamic range and noise. Considerable subjective information is held in the minutest detail of the sound.

For extreme quality in recording and reproduction, digital signals are used. A *compact disc (CD)* stores the sound signal in the form of a sequence of numbers, which are read by optical means, and are not compromised by the reading process, unlike normal analog tape recording which suffers from noise and other distortion as mentioned above. Extremely high fidelity is retained by digital storage. However, it should be remembered that the sound is listened to through an amplifier and loudspeaker system, which can often compromise the quality of the recording.

1.1.2 Voice communication by radio

This system is an extension of the recording exercise, where the signal is transmitted directly through the system and is not stored. Figure 1.2 shows the outline system.

A further signal conversion is introduced, that of electrical to electromagnetic form, with the antenna acting as transducers. Information

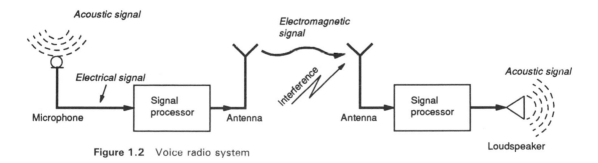

Figure 1.2 Voice radio system

transmission can be compromised and information lost, due to the following factors:

a) Bandwidth limitation of the channel. Compare the quality of AM broadcast radio with FM.
b) Interference from other stations.
c) Noise—hiss—within the receiver.
d) 'Atmospherics', natural interference.

Signal processing of various kinds can help reduce the effects of some of these problems.

Again, there are several criteria of quality by which to judge a radio transmission. Broadcast radio is required to be fair to good quality, allowing speaker identification and the transmission of music. Long-distance transmission by HF (in the range 3–30 MHz) propagates for several thousand miles over the horizon by multiple hops of the radio signal, and its effectiveness is judged only by the need to communicate; quality is secondary. Under these conditions interpretation of the meaning of the message is all important, and on some occasions this may be established only by repeated transmissions in response to requests from the person receiving it.

1.1.3 Communication with machines

Communication between humans is assisted by the associative processing abilities of the brain, and the large store of experience that the receiving person has had in interpreting similar messages in the past. Communication involving a machine has to be structured carefully so as to provide unambiguous interpretation of messages without this unique backup processing power.

Figure 1.3 shows such a communication path, involving an operator communicating with a local machine, and the local machine communicating with a remote machine. In terms of the signals used, the operator may employ a variety of techniques. The traditional input medium is the keyboard, where a coded key is pressed, which then passes a discrete codeword to the machine. The meaning of these codewords is agreed

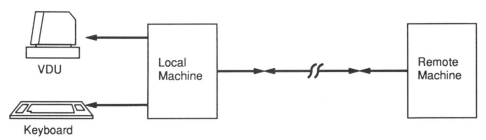

Figure 1.3 Communication with machines

beforehand between the operator and the machine structure, and the operator is forced to obey artificial and stilted rules for the benefit of the *dumb* machine.

The ASCII code (*American Standard Code for Information Interchange*) is generally used, and allows a total of 128 different codewords or *characters*, each of which is represented by an 8-bit binary word.

Variants of this kind of input include the touch-screen, mouse, and bit-pad, which allow a faster and more relaxing method of entering information, although still constrained by fairly clumsy rules. The process of entering information is helped by feedback from the machine in the form of a video screen, and the operator's intelligence permits recovery from the situation where a key is pressed in error, maybe asking the machine for diagnostic information.

If more processing power is available within the machine, then simpler methods of communication are possible. The *language* used can have a freer structure, so that the operator does not have to be so precise about instructions given, and commonly used sequences of instructions can be coded in compact ways. For use by an even wider group of people, voice input can be achieved, enabling computing machines to be accessed over an ordinary telephone. Automation of the telephone directory enquiries service is one such application, which is still in the research phase.

Communication between operator and machine is constrained therefore by two features:

a) The language structure that the machine can support
b) The communication channel between operator and machine.

The first lies in the province of *software engineering*, and is being helped by the provision of greater processing power, the development of better languages, and the advent of *Intelligent Knowledge Based Systems* (IKBS). The second point is within our interest here, and relates to the signal processing which is necessary in order to interpret the input signal, speech for instance.

Communication between machines is of a different order. Since no human is involved, there is no need to make the interface *user-friendly*, and the message interchanges are best made pedantic and regular. Messages such as *"are you ready?"*, *"here is information block X"*, *"your previous message is not understood"*, are typical. Signal processing is involved to the extent that the communication channel uses it.

1.1.4 Summary

A signal

a) *conveys information*
b) *represents an abstract concept*

c) *has many physical forms, but is normally electrical*

d) *may be stored, transmitted and processed.*

1.2 Signal processing targets

The general meaning of a *signal* has been explored above, and *signal processing* has been mentioned several times. We now examine the concept of signal processing, to see why signals need to be processed, and what targets are set.

Signals are processed so that

a) *the information they contain can be extracted*

b) or *the signal is rendered suitable for the medium over which it is to be transmitted or in which it is to be stored.*

We have already noted some of the processing factors which occur when recording signals on to magnetic tape (section 1.1.1). For radio transmission of signals, frequencies in the range of 1 MHz to several thousand MHz are used, so as to ensure propagation over large distances, and the information-bearing signal has to be *modulated* on to the radio frequency *carrier*. These are examples of the second category, of matching a signal to a medium.

Extracting information from a signal is a wide category, and we shall now examine several aspects of this operation, with simple examples. The overall

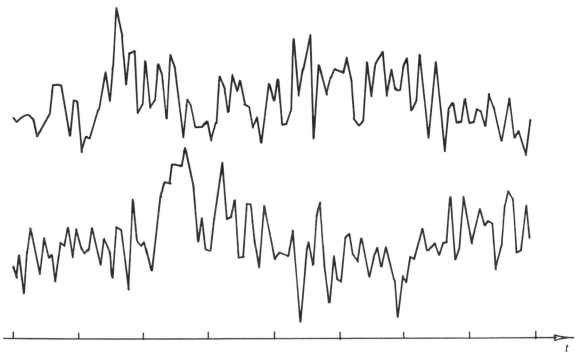

Figure 1.4 Signals from a wind tunnel model

18

problem is illustrated by the graphs in figure 1.4, which show recordings made during a wind tunnel test of a model building. The graphs represent pressure measurements at two points on the model. The information we wish to extract from these signals is whether there is any systematic trend to these signal records, are the two related in any way, and what does this test mean for the building model itself? Although these questions apparently are qualitative, the answers must be expressed in numerical form, and certain broad categories of signal processing can help to do this.

1.2.1 Analysis

Analysis is the first stage in signal processing. Behind analysis lies a signal *model*, which determines our view of the signal and the parameters that we wish to measure. The most fundamental decision is whether the signal is periodic in time, a single pulse waveform or a random function of time, and this will determine the method of analysis we adopt. Other assumptions will relate to whether the signal is almost sinusoidal (a *bandpass signal*) or contains a broad band of frequencies (a *baseband signal*). We shall later consider the basis of analysis for each of these signal types.

Having established the *model*, the signal is now examined and the model parameters determined. At this stage, we do not want to become too involved in detail, but consider the simple example of a periodic signal such as in figure 1.5, which might be part of a speech waveform or a distorted mains supply waveform. The parameters which we would wish to measure might be the signal period, positive and negative peak magnitudes and rms value.

The analysis of velocity signals measured at various points around the body of a motor car, for instance, can identify spurious resonances in the suspension system or the sources of unpleasant vibrations. Analysis of radar signals can determine the velocity of aircraft within the radio beam.

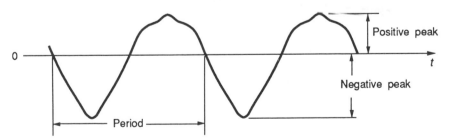

Figure 1.5   A simple periodic signal

1.1.2 Recognition

Analysis of a signal yields important parameter information, and enables it to be *recognised* as one particular signal out of a set of possible signals. The

19

set of possible signals must therefore be known in advance, and the measured parameters must be chosen so that members of the set can be located uniquely.

A common example is the binary transmission system, where one of two possible symbols is transmitted in order to represent binary information. The received signal, which may have extremely small energy and may be distorted, is known to be one of only two possible signals, and so can be recognised with good reliability.

Voice recognition is a further application of this principle, where an utterance is examined to discover what meaning it has. In order to operate at all, this system must have an agreed technique of analysis, and can only distinguish a limited number of different words. Since a variety of speakers will articulate the same word in different ways, such systems are also speaker-dependent. Maybe with greater processing power, it will become possible to make a voice recogniser with a far broader scope of application. Recognisers of say 50–100 words with a known speaker are in use in industry, and systems for services like directory enquiries are in an advanced laboratory stage of development.

1.2.3 Prediction

Having achieved an analysis of a certain signal, it is frequently necessary to predict how it will behave in the near future. Clearly, such prediction can never be totally correct, or else the signal would be wholly known and therefore not carry any information. Information is contained by the element of suprise in the signal.

A simple example is a temperature control system with ON/OFF control of the heat input. When the temperature limit is reached and the source of heat abruptly turned off, the air temperature in the room continues to rise due to heat stored within the fabric of the room, and therefore overshoots the desired temperature. By *predicting* what level the temperature is likely to achieve in the near future, the heat may be turned off earlier, so that the ultimate temperature reached is the desired value.

In the air-traffic control radar, prediction is vital in order to prevent collisions between aircraft. The present positions and velocities of aircraft within the controlled space are known, and a *predictor* estimates their future positions.

1.2.4 Modelling

We wrote earlier of the necessity of establishing a proper model of a signal so that it could be analysed (section 1.2.1). This same model can be used as the basis for studying a proposed hardware system in software model form. Mathematical operations are used instead of actual hardware, but the performance of the real system can be investigated in greater detail than for the real case. Parameters can be varied at will, and can even take on extreme values in order to see the effects of equipment faults.

In this way, novel radio receivers, signal processors or control systems may be studied without the expense of constructing complicated hardware, which then might not operate at all as required. In practice, signal processing is often carried out by digital machines, so the boundary between a mathematical model and the working system is often blurred.

1.2.5 Summary

Signal processing can be divided into the following broad areas of operation:

a) *Analysis*
b) *Recognition*
c) *Prediction*
d) *Modelling*.

Under this umbrella, there are subsidiary classes of signal processes which we shall discuss in more detail in later chapters.

1.3 Signal processes

A signal is most often a function of time; so consider a signal process which has an input signal $x(t)$ and an output signal $y(t)$. The function relating $y(t)$ to $x(t)$ could take many different forms. It might be a simple *averaging* or *smoothing* process, finding the mean or average value of the signal, or it might perhaps find the *peak* value of the input signal $x(t)$.

1.3.1 Filter process

An extremely common signal process is *linear filtering*, which restricts or shapes the frequency content of the input signal. Figure 1.6 shows the four classes of filter process. The input and output signals here are shown as the voltages V_1 and V_2 respectively. These actually represent signals of the form:

$$x(t) = V_1 \cdot \cos(2\pi f t) \tag{1.1}$$

The filter process therefore passes, without alteration, all signals of the form of equation 1.1 whose frequencies f lie within the *passband* of the filter. In the case of the *lowpass* filter, this means frequencies below the cutoff limit of f_c; while signals with frequencies above f_c are stopped or *attenuated*, and this is the *stopband*.

The transition between *passband* and *stopband* in real filters occupies a finite frequency range, and this is known as the *transition band*. We have indicated that in figure 1.6 in a very idealised way.

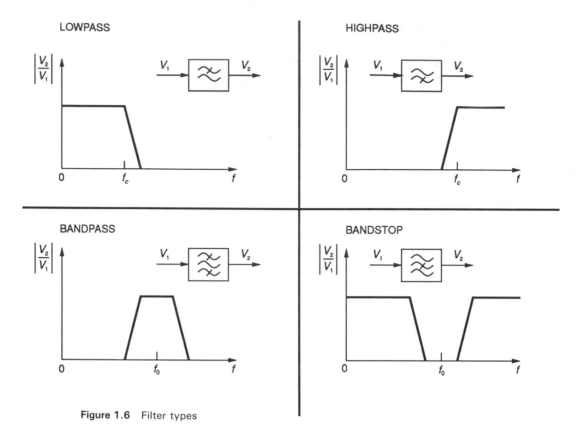

Figure 1.6 Filter types

These four filter types are used extensively in all forms of signal processing and often precede more complicated processes. A filter such as this imposes a restricted *bandwidth* on the input signal. Figure 1.7 shows how such filters affect a signal which contains many different frequency components of the style in equation 1.1.

The lowpass filtered signal changes much more slowly along the time axis than does the original signal, while the bandpass filtered signal shows a

(a) Original signal

(b) Lowpass filtered

(c) Bandpass filtered

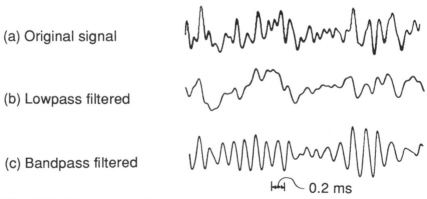

0.2 ms

Figure 1.7 Filtering a signal

characteristic frequency equal to the centre frequency of the passband. These statements are qualitative, which is adequate for now, but in later chapters we shall analyse the effects of filters like these in some detail.

1.3.2 Processes with two inputs

A signal process might have many inputs and indeed more than one output, but the case of two inputs is particularly fundamental. So let the inputs to the process be $x(t)$ and $w(t)$, and the output be $y(t)$.

A very important two-input process is *correlation* where the similarity of the two signals is assessed. The output is a single numerical measure of this similarity, rather than a time-varying function, and this is sometimes known as a *correlation coefficient*. We shall examine the content of the process itself at a later time. A direct application occurs when carrying out the *recognition* exercise for information signals. Suppose that there are N possible signals in the set which are used for representing information. When one signal is received, it is *correlated* with replicas of each of the possible signals, and the replica signal which gives the largest correlation coefficient is chosen as the most likely input signal.

A further important class of two-input processes is that where the input $w(t)$ has a single frequency and is of the form of equation 1.1, while the input $x(t)$ is an information-bearing signal. This form of combination is called *modulation* and occurs when a speech signal, for instance, is combined with a radio-frequency sinusoidal carrier signal of say 10 MHz, to form a signal which can be radiated from a radio antenna.

1.3.3 Block diagrams

Most signal processing tasks require the use of a number of signal processes combined in some way to achieve the main aim. It is usual to represent these in a block diagram form, as shown in figure 1.8.

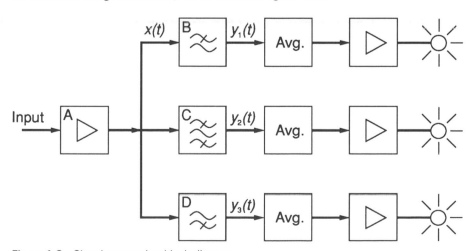

Figure 1.8 Signal processing block diagram

A single process is indicated by one block, providing a function which provides $y(t)$ given $x(t)$. Figure 1.8 describes a simple disco lights controller, which generates a pattern of flashing lights from a sound input. Block A is an amplifier, while blocks B, C, D are lowpass, bandpass and highpass filters respectively, splitting the sound signal up into three separate frequency bands. The output of each filter is then processed by an averaging device, which has yet to be defined precisely, and then drives a lamp through a power amplifier.

Such a description of a signal processing system allows the general flow of the signals to be seen, and also splits up the overall task into a series of individual non-interacting simpler processes.

1.4 Signals

1.4.1 Classical signals

Classical signals are used for testing signal processes, which are then described in terms of their responses to these signals. Testing in this context might refer to a physical test signal applied in a laboratory, or it may refer to a mathematical operation in a simulation of the real system.

The most commonly used test signal is the *sinusoid*:

$$x(t) = A \cdot \cos(\omega t) \tag{1.2}$$

It is easy to generate and analyse, and has the great benefit that, for linear systems, the output signal is also a sinusoid. Testing of this kind defines the *frequency response* of the system, which we shall deal with in great detail from chapter 3 onwards. Since we characterise sound signals for instance in terms of their frequency content (high, low, etc.), and use treble and bass controls on sound reproduction systems, the results of sinusoidal testing are immediately understandable.

The *impulse* is also widely used as a test signal, and generates an *impulse* response:

$$x(t) = A \cdot \delta(t) \tag{1.3}$$

where $\delta(t) = 0$ if $t \neq 0$ and $\int_{-\infty}^{\infty} \delta(t) \cdot \mathrm{d}t = 1$.

The output signal is not an impulse, but its shape gives important clues to the behaviour of the system in response to abrupt changes in input; something which is difficult to assess from the frequency response. For this reason, loudspeakers are usually characterised by their impulse response as well as their frequency response. In fact a *step* response is normally measured for greater convenience, but the step signal is only the integral of the impulse, and so these two responses are intimately related. Figure 1.9 illustrate these and other classical signals.

A rectangular pulse signal is also often used for system testing, and, when

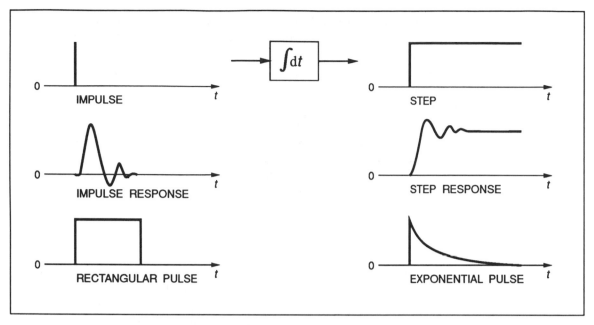

Figure 1.9 Some classical signals

the pulse width is extremely small, forms a practical approximation to the idealised impulse. When the pulse width is large, then it approximates to the idealised step signal.

A more subtle form of test signal is *white noise*, which is totally random in nature. However, the use of this signal depends on an understanding of the behaviour of random signals, so we shall not be discussing this method of testing until chapters 15 and 16.

1.4.2 Signal bandwidth

Information-bearing signals are intrinsically random, but all have a restricted bandwidth. All practical signals are *bandlimited*, that is they have

Table 1.1 *Bandwidths of some typical signals (cf. figure 1.2)*

Telephone	$0 \cdot 3$–$3 \cdot 4$ kHz
Music	15 kHz
AM radio receiver, audio bandwidth	$4 \cdot 5$ kHz
FM broadcast receiver: radio frequency bandwidth	180 kHz
audio frequency bandwidth	15 kHz
TV video signal (625 line)	$5 \cdot 5$ MHz
Radar radio frequency signal	≈ 10 MHz
Trunk microwave or lightwave circuit	≈ 1 GHz
Central heating temperature measurement	$\approx 0 \cdot 1$ Hz

been passed through a bandlimiting filter like one of those shown in figure 1.8. Notice that, apart from the sinusoid, the classical signals mentioned above are *not* bandlimited. Bandwidth is one of the most important distinguishing features of a signal, and Table 1.1 shows the bandwidths of certain commonly encountered signals. These bandwidths should be compared with the transmission rates over various media which are shown in figure I.2 of the Introduction. Notice that the systems which transmit a fixed message in shortest time are those with the greatest bandwidth.

Noise occurs naturally in most real systems and reduces the information content of the signal. This random effect can be minimised by signal processing, but not totally removed.

1.4.3 Dynamic range

Dynamic range is the ratio between the maximum and minimum values that a signal can take. We are not concerned here with the *peak* values of the signal so much as with the overall range or intensity of the signal. A large dynamic range is no problem for an idealised system, but it is often a limiting factor in the design of real systems. Most naturally ocurring signals have a large dynamic range.

For instance, at the input to a telephone mouthpiece, the mean sound level for different speakers in different moods varies over a range of about $30:1$. An orchestra at full power gives a sound intensity which is about 10^7 times that of the quiet note from a viola. A perfect sound recording system would encompass this large dynamic range of possible signals without hiding the small sound or distorting the large one. The human ear in good condition is capable of hearing sound intensities over a range of about $10^6:1$, between extremes of the threshold of pain and the threshold of hearing.

A further example is the radar system. Electromagnetic energy is radiated from an antenna and spreads out according to the *inverse square law*, so that at a distance d the energy density is proportional to d^{-2}. When the energy is reflected from a target, it spreads out in similar fashion. The signal received back at the sending point is therefore proportional to d^{-4}. So, if we comare the signal returns from two similar targets, one at a distance of say 10 km and the other at distance of say 100 km, then these two signal intensities will be in the ratio of $10\,000:1$. The same receiver is required to deal with both signals.

STUDY QUESTIONS

1   What is the relationship between information and signals, and how do they differ?

2   Devise your own example system after the style of section 1.1, to illustrate the summary points in section 1.1.4.

3   Explain the relationships between signal analysis, recognition, prediction and modelling. Give examples.

4   A certain signal has a frequency of 1·5 kHz. Through which of the following ideal filters will it pass?

    Lowpass filter: cutoff frequency 2·0 kHz
    Highpass filter: cutoff frequency 3·0 kHz
    Bandpass filter: bandwidth of 500 Hz, centre frequency 2·5 kHz
    Bandstop filter: stopband of 500 Hz, centre frequency 2·8 kHz

5   Find applications for lowpass, highpass, bandpass and bandstop filters.

6   Produce a signal processing block diagram for your example system of question 2 above.

7   Why are classical signals necessary, and how do they differ from information-bearing signals?

2 Signal Properties

OBJECTIVES

To investigate some of the general properties of signals, and to commence quantifying their parameters. In particular:

   a) To distinguish between long- and short-term properties.
   b) To point out the essential difference between periodic, pulse and random signals.
   c) To describe the corresponding properties of discrete-time or sampled signals.

COVERAGE

Starting with the broad appreciation of signals and signal processing established in the previous chapters, we now define certain parameters which can be used to analyse and describe various signals. In addition to continuous-time signals, we introduce discrete-time signals, and comment that they are much easier to deal with than continuous-time signals. There is also an introduction to random signals.

A signal represents information which generally is qualitative and maybe subjective, as we saw in chapter 1. It may be a voice signal for instance, whose meaning is to be determined by a *voice recogniser*. So it is vital that we have mathematically defined attributes and measures for each possible signal, so that one signal may be *analysed*, or compared with another via the *correlation* process.

2.1 Signal comparisons

Consider the three speech signals portrayed in figure 2.1. Although signal (a) is clearly of a different kind to signals (b) and (c), it is difficult to see

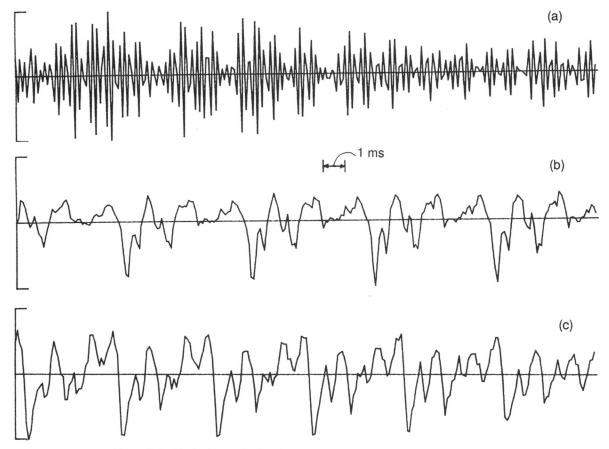

Figure 2.1 Typical speech signals

how this difference could be expressed quantitatively. To go further, we may well ask what sounds are being represented by these signals, and how could one go about the process of artificially generating those sounds?

The answers to these questions are complicated, and can only be obtained after a subjective study of the psychology of hearing and the mechanisms of speech production and understanding. However, three different types of numerical comparison can be made and are essential tools for signal processing of all kinds.

2.1.1 Long-term measurements

A long-term measurement on a signal waveform is an *average* or *mean*, and defines a single numerical parameter, normally based on the *amplitude* of the signal. The simple average of a signal waveform $x(t)$ over a period T is given by

$$\text{avg}[x(t)] = \frac{1}{T} \int_0^T x(t) \, dt \tag{2.1}$$

29

In the case of figure 2.1 for instance, each of the waveforms appears to have zero mean value, since the areas above and below the time axis appear to be similar. (Section 2.3.2 and chapter 10 will deal with this operation in more detail.)

There are other long-term measures apart from the simple mean or average; the *mean power* for instance, which is actually the *mean-square value* as we shall see in section 2.3.1.

Average measures may also be expressed in terms of a *probability*. It is often important to have a measure of the *peakiness* of a signal waveform, and this may be expressed as

the probability that the waveform exceeds a certain amplitude level

Such a function of amplitude is called a *distribution function*. Although it is an important measure, we shall not investigate its properties properly until much later, in chapter 14.

2.1.2 Short-term measurements

The three waveforms in figure 2.1 might conceivably have the same average value and the same mean power, but they are clearly very different from one another. In order to quantify these differences we must concentrate on the short-term fluctuations of the waveforms.

Signal (a) crosses the time axis much more frequently than either signals (b) or (c), so we might classify it as having a higher *frequency* than the other two. Our description of the short-term behaviour of a signal waveform is achieved by considering the *frequency content* of the signal, choosing a

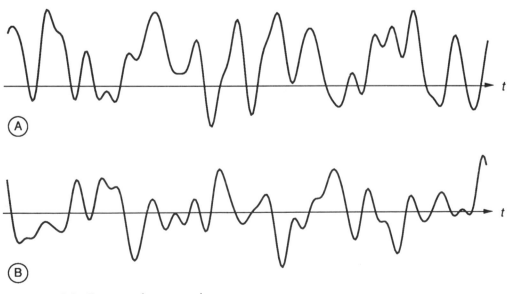

Figure 2.2 Two waveform examples

Table 2.1 *Sinusoidal components of waveforms in figure 2.2*

Signal A

Frequency	0	4	6	11	13	22	25	kHz
Magnitude	1·0	1·0	0·5	0·8	1·0	0·4	0·3	V
Angle	0	−90	90	−45	−40	150	0	°

Signal B

Frequency	4	6	9	13	20	25	kHz
Magnitude	1·0	0·5	0·7	1·0	0·6	0·3	V
Angle	90	0	45	50	60	90	°

model which builds up the waveform from a collection of sinusoidal components having different frequencies. This concept we shall explore in greater depth in chapter 4, but for now we might express the signal model as

$$x(t) = \sum_{r=1}^{N} a_r . \cos(\omega_r t + \varphi_r) \tag{2.2}$$

Figure 2.2 shows two examples of waveforms built up in this manner, and Table 2.1 gives the corresponding parameters of the sinusoidal components that have been used.

These two waveforms, although consisting of only 6 or 7 frequency components respectively, do begin to show the sort of complicated features which are characteristic of typical speech waveforms (figure 2.1). Notice too that relatively small changes in the composition of the components do change the waveform appreciably; this is a sensitive method of representing arbitrary waveforms.

2.1.3 Comparison measurement

When a limited set of signal waveforms is used for communication of information, or when a measurement is made to limited resolution, then the most powerful method of choosing which signal is present is by comparison with a set of expected signals.

Suppose for instance that signal (c) in figure 2.1 is one of a set of reference waveforms, and that one of signals (a), (b) or (c) is being received. Then if signal (b) is being received, it will compare more closely with reference signal (c) than would signal (a), but if signal (c) is received it will provide an exact match. In practical voice recognition, an exact match between a spoken waveform and a previously stored pattern waveform is almost impossible due to slight differences in pronunciation, so the *nearest* match is identified.

The operation of comparing signals is known as *correlation*, and the correlation coefficient r is defined for two signals $w(t)$ and $x(t)$ as

$$r \propto \text{avg}\,[w(t) . x(t)] \tag{2.3}$$

For the two waveforms in figure 2.2, this average happens to work out at $-1 \cdot 0$. Note that when $r = 0$, then the two signal waveforms are said to be *uncorrelated* or *orthogonal*, they are independent.

● **Example 2.1** The technique of correlation is used to analyse such a signal as that modelled in equation 2.2, to ascertain its frequency content.

Thus, analyse $x(t)$ for a frequency ω.

$$ r = k \cdot \text{avg}\left[\cos(\omega t) \cdot \sum_{r=1}^{N} a_r \cdot \cos(\omega_r t + \varphi_r) \right] $$

Then $\quad r = k \cdot a_r/2 \quad$ if $\omega = \omega_r$ and $\varphi_r = 0$
$\qquad\quad = 0 \quad$ if $\omega \neq \omega_r$

Varying ω continuously, then, enables discrete frequency components in $x(t)$ to be found. Since the operations involved are just multiplication and averaging, then these can easily be carried out, and the practical instrument is called a *Wave Analyser*. This style of analysis is known as the Fourier series (chapter 4 and appendix A2). ●

2.2 Signals and circuits

We note in passing that circuits and signals have a close affinity. In some cases they are interchangeable, since the need for certain signal processing circuits can be eliminated by proper signal design, particularly when using digitally generated signals. Circuits and signals are compared in Table 2.2 in terms of the quantities that they operate on, and the components that are used to model their behaviour.

Notice that energy and information have similar properties (section I.2.3). In particular, stored energy in a circuit cannot change instantaneously, and the mean rate of change (power) in a circuit must always be in balance. Information behaves similarly; stored information can be changed only at a finite rate, and mean flow rates of information in a system must be balanced.

Circuit components such as resistance, inductance, etc. are well known to the reader, but the *phasor*, which is the elementary building block of signal models, is a new concept to us and will be explored thoroughly in chapter 3.

Signal processing is carried out by designing a circuit which operates on

Table 2.2 *Comparison of circuits and signals*

	Circuits	*Signals*
Raw material	Energy	Information
Components	R, L, C, M, source	Phasor

the signal to obtain a desired effect, and for radio frequency systems this is the predominant technique. At lower frequencies, signal processing is mostly accomplished by digital means, and the design problem is how to design a suitable algorithm and then program a digital machine to execute it. Throughout this course, we shall develop the first principles of both techniques, although digital processing comes to the fore as a process which is easier to understand.

2.3 Signal classification

Signals may be grouped into classes which are defined by the type of mathematical analysis which is applied to them. Classical signals (section 1.4.1) fall naturally into these categories, since we select them in order to provide simple test signals for systems. Real signals are classified by the models that are used to represent them, since the signals themselves do not necessarily comply with the total requirements of any specific class.

2.3.1 Power and energy

The most fundamental classification procedure is based on ideas of *energy* and *power*.

A signal $x(t)$ has values which represent say voltage or current in an electrical circuit. It is reasonable therefore to regard $x^2(t)$ as a measure of signal *power*. Strictly of course, resistance should come into the picture when we speak of power, but generally we are comparing two signals which occur at the same point in a circuit or system, and hence the 'resistance' is a common scaling constant. (If this still seems somewhat arbitrary, then assume a reference resistance value of 1 Ω.) It also emerges that in signal theory $x^2(t)$ is an important quantity in its own right.

Since power is the rate of expending energy, it follows that we can also define the *energy E* of a signal:

$$E = \int_{-\infty}^{\infty} x^2(t) \, dt \tag{2.4}$$

Refer now to figure 2.3, where we see two distinct classes of signal. Those on the left have finite energy, while those on the right, which are intended to carry on for ever, have infinite energy. Infinite energy is not a helpful concept, but we can produce a finite measure by considering the rate of change of energy, the power, which is finite. Thus, the *mean power P* is

$$P = \frac{1}{T} \int_{-T/2}^{T/2} x^2(t) \, dt \tag{2.5}$$

For signals such as shown on the right of figure 2.3, the time limit T is extended to infinity, and the mean power P tends also to a limiting value.

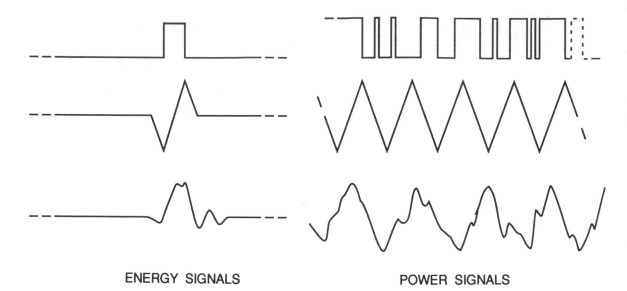

ENERGY SIGNALS   POWER SIGNALS

Figure 2.3 Signal classification–power and energy

These are called *power signals*. Signals on the left of the diagram, *energy signals*, exhibit zero mean power as T tends to infinity.

Energy signals are also called *aperiodic* or *pulse* signals, for obvious reasons. A radar signal is of this form since a short pulse of electromagnetic energy is transmitted and then reflected from the target. A 1 μs pulse having a transmitted peak power of, say, 100 kW represents real energy of 0·1 J. When considering the response of the receiver to just one pulse, we treat it as an energy signal.

The transmitted pulses are repeated, however, at a rate of say 1 kHz, and by considering the joint effect of these reflected pulses we are able to determine the velocity of the target as well as its position. In this case, the signal is regarded as a *pulse train* or *periodic* signal which has properties different from those of a single pulse. This pulse train has a mean transmitted power of only 100 W.

Mean power is a measure of the *mean-square-value (msv)* of the waveform, being the mean of the squared-waveform. Although power is often measured directly, it is frequently measured in voltage terms as the *root-mean-square* or *rms* value. Clearly the rms value is $\sqrt{P}$.

Table 2.3 *Signal classification*

	E	P
Energy signal	finite	0
Power signal	∞	finite

34

● **Example 2.2** A certain radar transmitter emits a pulse every $0\cdot4$ ms, of 1 MW for $0\cdot5$ μs. At the receiver antenna, the transmitted signal energy is reduced by 10^{14} times.

The transmitted pulse energy is $0\cdot5$ J.
The received pulse energy is therefore $0\cdot5\times10^{-14}$ J.

The mean received power is $(0\cdot5\times10^{-14})/(0\cdot4\times10^{-3})=12\cdot5$ pW
The peak received power is $(0\cdot5\times10^{-14})/(0\cdot5\times10^{-6})=10$ nW   ●

2.3.2 Periodic signals

Periodic signals are an important sub-class of *power* signals. Figure 2.4 shows a collection of several such signals.

A signal which is periodic in a time T is formally defined as follows:

$$x(t-nT)=x(t) \tag{2.6}$$

This statement means that the signal has the same value at intervals of nT sec along the time axis, where n is an arbitrary integer.

Calculation of the mean value or mean power is now simplified, since we have only to carry out the calculation over a single period of the waveform. The mean value is given by

$$\text{avg}\,[x(t)]=\frac{1}{T}\int_{0}^{T}x(t)\,\mathrm{d}t \tag{2.7}$$

For practical purposes this may be expressed in terms of areas. A_p is the area of the parts of the waveform above the time axis, the *positive area*, and A_n is the corresponding *negative area*. The mean or average value is simply

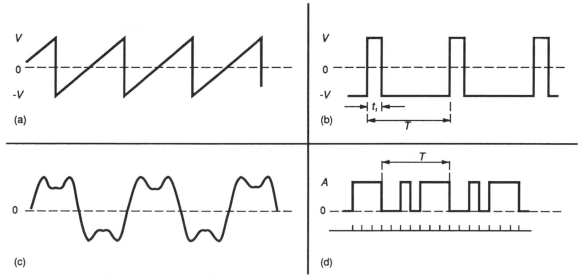

(a) (b) (c) (d)

Figure 2.4 Some periodic signals

35

expressed as

$$\text{avg}\,[x(t)] = (A_p - A_n)/T \tag{2.8}$$

Applying this procedure to the signals in figure 2.3, we see that by inspection they have the following mean values:

signal (a):   0
signal (b): $-V(1 - 2t_1/T)$
signal (c):   0
signal (d): $4A/7$

Similar arguments can be extended to the calculation of mean power, except that areas under the squared-waveform are taken, noticing of course that all areas are now positive. Calculation of these properties is straightforward when the waveforms are known, but measurement of them for an unknown waveform is quite a different matter (see chapter 10).

● **Example 2.3** Given a sinusoidal signal, calculate its mean value and power.

A certain current signal is $i(t) = I \cdot \cos(\omega t)$.

Then the instantaneous power is $i^2(t) = I^2[1 + \cos(2\omega t)]/2$.
These functions are illustrated in figure 2.5.

The mean value of the signal is plainly 0, due to its symmetry.

The mean power P is $I^2/2$, and the rms value is $I/\sqrt{2}$.

Notice that the peak power is I^2, which in some cases might be more important than the mean power.

These parameters are extremely important since they form the basis of much that we shall refer to later.   ●

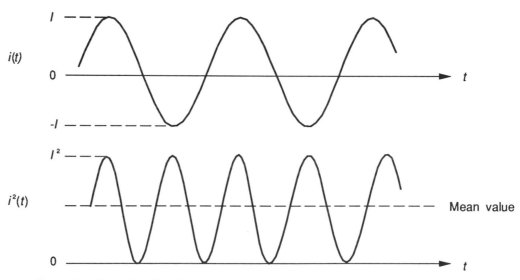

Figure 2.5 Sinusoidal signal power

Periodic signals have other desirable properties, the most important of which is that any bandlimited periodic waveform can be represented by a sum of sinusoidal components at different frequencies. This is the *Fourier series* which we will explore further in chapter 4 and appendix A2.

2.3.3   Discrete-time signals

The signal examples that we have illustrated so far (figure 2.1 for instance) are known as *continuous-time* signals, since the graph of the signal versus time is a continuous line. The signal has an amplitude value for all points in time. This is how most signals occur in the natural world; time is continuous.

Certain signals, however, occur only at discrete times, particularly when the measurement process takes some time to operate. Examples include observations from a digital voltmeter, measurements of rotational speed by counting revolutions of a wheel, and radar where the measurement only exists when a pulse is transmitted and received.

With the rapid increase in sophistication of digital machines such as microprocessors and signal processors, most signal processing is now done in digital form. A digital system by its very nature can only produce outputs at discrete points in time, and hence develops signals which exist only at discrete times. It is necessary therefore to develop a technique for describing and handling such signals.

Most signals of this type occur at regular intervals of time, say T_S. The value of the signal at the nth time interval is therefore described as: $x(nT_S)$. Since the sampling interval T_S is a constant for a given signal or system, we normally ignore it and refer to the nth signal value as $x(n)$. This is quite consistent with computer terminology, where successive values of $x(n)$ would be held as an array in a series of storage elements.

A discrete-time signal therefore consists of the set of values $\{x(n)\}$, for all values of n. This description is somewhat clumsy for pen and paper work, but it suits exactly the requirements of a digital processor and enables us to write a digital program to carry out our desired process.

Continuous-time signals may be converted to discrete-time signals by an *analog-to-digital converter (ADC)*, (see chapter 6 for further details). Using an ADC in fact changes the original *analog* signal in two ways:

a) The signal becomes discrete-time and only has values at $t = nT_S$.
b) The values are stored as digital words, and hence cannot represent all the possible values available in the original signal.

Figure 2.6 gives an example of a sampled analog waveform.

The continuous-time signal $x(t)$ is converted into a discrete-time signal $x(n)$, which in this case becomes as shown in Table 2.4. The set of values $\{x(n)\}$ is known as a *time-series* or *sequence*.

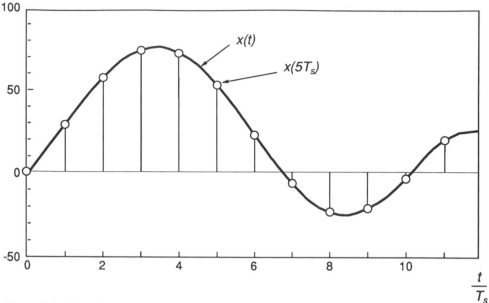

Figure 2.6 Sampled analog signal

Table 2.4 *Discrete-time signal*

n	0	1	2	3	4	5	6	7	8
$x(n)$	0	28·7	57·3	74·3	72·7	52·9	22·7	−6·2	−22·9

The sampling interval T_S may take on a variety of values, depending upon the type of signal being sampled:

For a speech signal, T_S may be 100 μs.
For a TV video signal, T_S may be 60 ns.
For a central-heating temperature signal, T_S may be 1 sec.

By sampling, any naturally occurring signal can be converted into discrete-time or digital form and then processed digitally. This operation has a profound influence on the signal processing business.

● **Example 2.4**   Consider a continuous-time or analog signal $x(t)$:

$$x(t) = a \cdot \cos(\omega t + \varphi)$$

Then the corresponding discrete-time signal is $x(n)$:

$$x(n) = a \cdot \cos(2\pi n\omega/\omega_S + \varphi) \qquad \text{where } \omega_S = 2\pi/T_S$$

For example, suppose that $\omega_S = 8\omega$, $\varphi = 45°$ and $a = 10$, then

$$\{x(n)\} = \{7\cdot07, 0, -7\cdot07, -10\cdot0, -7\cdot07, 0, ...\} \qquad ●$$

We may still calculate the mean value and mean power of a discrete-time signal, with a small change in the relationship of equation 2.1, etc. Thus

$$\text{avg}[x(n)] = \frac{1}{N} \sum_{n=0}^{N-1} x(n) \tag{2.9}$$

$$P = \frac{1}{N} \sum_{n=0}^{N-1} x^2(n) \tag{2.10}$$

Note that for *power signals*, the number of points N must tend to infinity, while for *periodic signals* N is the number of samples per period, and in other cases a finite value of N determines a *short-term* average. Compare with sections 2.3.1 and 2.3.2.

$$E = T_S \sum_n x^2(n) \tag{2.11}$$

Note that the T_S factor is necessary in order to make the 'energy' of the discrete-time signal $x(n)$ equivalent to the energy of the corresponding continuous-time signal $x(t)$. It is often ignored if all signals being considered are purely digital or discrete-time.

It turns out that not only is the discrete-time form of signal most convenient for signal processing, it is also very helpful for understanding the basic principles behind signal processing, and so we shall concentrate mostly on the discrete-time form of signal in later work.

For the signal fragment listed in Table 2.4, the following values can be calculated:

$\text{avg}[x(n)]$	$31 \cdot 1$
E/T_S	$18\ 789$
P	$2087 \cdot 7$
rms	$45 \cdot 7$

Notice that this calculation was carried out quite simply by using the statistical function on a hand-calculator.

● **Example 2.5** For the signal of example $2 \cdot 4$,

$$\{x(n)\} = \{7 \cdot 07, 0, -7 \cdot 07, -10 \cdot 0, -7 \cdot 07, 0, 7 \cdot 07, 10 \cdot 0, \ldots\}$$

Hence, using equations 2.9 to 2.11,

$$\text{avg}[x(n)] = 0 \text{ V}$$
$$P = 50 \cdot 0 \text{ V}^2$$

For one period of the waveform,

$$E/T_S = 400 \text{ V}^2$$

If $T_S = 100 \ \mu s$ and $f_S = 10$ kHz, then $E = 0 \cdot 04$ J.

Notice that these quantities do not necessarily correspond with those of the equivalent continuous-time signal ●

2.3.4 Random signals

Classical signals such as those mentioned in section 1.4.1 are fully described by an equation defining a function of time, such as

$$x(t) = a \, . \, \cos(\omega t + \varphi) \qquad (2.12)$$

Since the equation completely determines the behaviour of the signal for all time, these signals are known as *deterministic* signals. Whilst such signals are vital for testing systems, and for modelling real signals like those shown in figure 2.1, they are not information-bearing signals. In order to carry information, a signal must conceal some of its future values or attributes, since the element of surprise is essential to the representation of information. If a situation is completely known in advance, then it can provide no new information.

Information-bearing signals are therefore *random* in some sense, and the properties of such random signals will be examined in detail in chapter 14. For the time being we note that a random signal cannot be described by a deterministic function of time, but is instead represented by a *distribution* function, which tells us that the signal exceeds a certain amplitude level x for $y\%$ of the time. The distribution function enables us to calculate the mean value and mean power, etc. for a signal waveform, in just the same way as the equation for a deterministic signal does.

Figure 2.7 shows a fragment of such a discrete-time signal, with random sample values. Simple observation reveals that this signal has a well-defined mean value, which would be estimated using equation 2.9. The mean power or *mean-square-value (msv)* could also be estimated using equation 2.11. Since this is a discrete-time signal, no integration is involved in these calculations, hence the function of time which describes the signal is not important.

The mean value x_m, represents the *zero frequency* or *dc* value of the waveform and hence the quantity x_m^2 might be said to represent the 'dc power' in the waveform. Likewise, the *msv* gives the *total* power. Taking the difference of these two defines the 'ac power' or *fluctuation component*

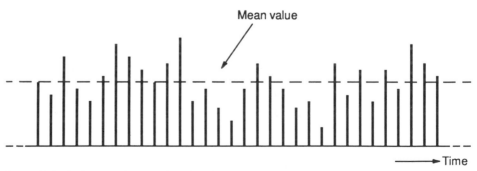

Figure 2.7 A discrete-time random signal

power in the signal, which we call the *variance*. Thus

$$\text{Variance} = (\text{Mean-square-value}) - (\text{Mean-value})^2 \qquad (2.13)$$

Variance has the same significance as in statistical analysis, and the *standard deviation* is given by

$$\text{Standard deviation} = \sqrt{(\text{Variance})} \qquad (2.14)$$

● **Example 2.6**   A certain continuous-time random signal is measured by standard laboratory instruments.

The mean value is measured as $3 \cdot 0$ V, using a moving-coil meter.

The rms value is measured as $4 \cdot 0$ V, using a true-rms meter.

The variance is then calculated as $(4 \cdot 0)^2 - (3 \cdot 0)^2 = 7 \cdot 0$ V^2 and the standard deviation is $2 \cdot 65$ V.   ●

Measurement of variance, then, is not difficult, but what significance does it have? While we shall not be examining this property until chapter 14, we can illustrate it in terms of the *Gaussian* or *Normal distribution*, which frequently occurs in practice. Standard tables of values turn a measurement of variance into the probability that a certain amplitude level is exceeded by the signal. In order to make the tables universal, they are given for the case where the mean value is zero and the standard deviation is unity. Table 2.5 shows some values for the Gaussian distribution.

Table 2.5   *Gaussian distribution*

Level	0·0	0·5	1·0	1·5	2·0	2·5	3·0	3·5
Probability	0·5000	0·3085	0·1587	0·0668	0·0228	0·0062	0·0013	0·0002

● **Example 2.7**   A certain Gaussian signal has a mean value of $3 \cdot 0$ V and a measured variance of $2 \cdot 25$ V^2. The standard deviation is, therefore, $1 \cdot 5$ V.

What is the probability that a level of twice the mean value is exceeded by this signal?

Twice the mean value is $6 \cdot 0$ V, which is $3 \cdot 0$ V more than the mean, or $2 \cdot 0$ standard deviations.

Referring to table 2.5, the probability of the fluctuation exceeding $2 \cdot 0$ standard deviations is $0 \cdot 0228$, or once in every $43 \cdot 8$ samples on average.   ●

● **Example 2.8**   A signal similar to that in example $2 \cdot 7$, has a variance of $1 \cdot 0$ V^2.

Since the standard deviation is now $1 \cdot 0$ V, the signal must move through three standard deviations in order to exceed the limit of twice the mean value, and this has a probability of $0 \cdot 0013$.   ●

These two examples illustrate how the Gaussian distribution table is used, and also make the point that this probability measure gives information as to how *peaky* the waveform is. The second example has a reduced variance, and hence is much less likely to exceed the 6·0 V level than the signal in the first example.

● **Example 2.9**   All electronic amplifiers are limited in the output voltage swing that can be developed across a load resistance, without 'clipping' taking place, i.e. chopping off the peak signal values.

A certain amplifier has a gain of 1000 and an output voltage range of ±10 V. The input signal has zero mean and is thought to be Gaussian.

What is the maximum variance of the input signal which will guarantee that clipping takes place for no more than 0·2% of the time?

From table 2.5, 3·0 standard deviations will satisfy the 0·2% requirement.

Since 3·0 standard deviations of the output signal must be $\leqslant 10$ V, the standard deviation of the output signal must be $\leqslant 3\cdot33$ V.

Now the input signal standard deviation is therefore $\leqslant 3\cdot33$ mV, so that its variance is $\leqslant 11\cdot1\ \mu V^2$.   ●

STUDY QUESTIONS

1   What three broad kinds of measure are used to analyse signals?

2   What similarities are there between circuits and signals?

3   List some signals which occur in communications or in measurement, and classify them into energy and power types.

4   What is the mean value and the mean power of the following signals?

> $50\,.\,\sin(2000t)$
> $5 + 10\,.\,\cos(500t)$

(Note that when a signal has more than one component, the total power is the sum of the individual powers.)

> $2 + 3\,.\,\cos(500t + 1) + 20\,.\,\sin(750t)$
> $\{x(n)\} = \{0, 0\cdot1, 2\cdot3, -0\cdot5, -0\cdot6, 0\cdot7, 0\cdot9, 0\}$

5   Calculate the first 8 terms of the discrete-time signal sequence for $x(t)$, when $T_S = 1$ ms and 4 ms:

> $x(t) = 3\,.\,\cos(2\pi\,.\,62\cdot5t)$

6 Determine the mean value, the mean-power and the energy per period of the signal in question 5.

7 A certain random signal has a mean value of $5 \cdot 0$ V, and a mean power of $25 \cdot 444$ V^2. Determine the variance, and the probability that a level of $6 \cdot 0$ V is exceeded by the signal.

3 Signal Models: Single-frequency Signals and Frequency Response

OBJECTIVES

To introduce the phasor as an essential element in signal modelling, and to discuss how it is applied to different forms of signal. In particular to apply it to

a) Sinusoidal signals
b) Discrete-time signals.

To give some processing applications of the phasor concept by calculating the frequency response

a) In analog circuits
b) In digital systems.

COVERAGE

Commencing with an understanding of complex quantities, the phasor is introduced as an idea and then applied to several practical cases of signal modelling, including that of discrete-time signals. The difference between a real signal and its complex model is emphasised. The phasor concept is then used to define the frequency response of a system.

In chapter 2 we noted that, in order to compare or analyse signals of the sort shown in figure 2.1, a *short-term* measure was required in order to describe the local fluctuations of such signals.

The universal model element which can be used to build up signals of any practical complexity is the *phasor*, which we will now describe and begin to use.

A signal can be classified according to the set of phasors which describe it adequately, and which will then allow quantitative comparisons to be made between this signal and other signals. The *phasor* is merely a tool for signal analysis, but it does enable complicated functions of time to be expressed in a mathematically compact form, and provides a physical model which can illustrate many of the properties of the real signal.

3.1 The phasor

3.1.1 Continuous-time

The phasor is a rotating vector in a complex plane. Figure 3.1 defines such a phasor. This particular phasor has a magnitude A and a rotational velocity of ω rad/sec. It may therefore be described by the following complex function of time, as explained in appendix A1:

$$x(t) = A \cdot \exp j(\omega t) \tag{3.1}$$

The phasor described in equation 3.1 has the value A at $t = 0$, but in the general case a phasor has an arbitrary starting angle φ at $t = 0$. Thus, a general phasor is of the form:

$$x(t) = A \cdot \exp j(\omega t + \varphi) \tag{3.2}$$

and

$$x(0) = A \cdot \exp j(\varphi) \tag{3.3}$$

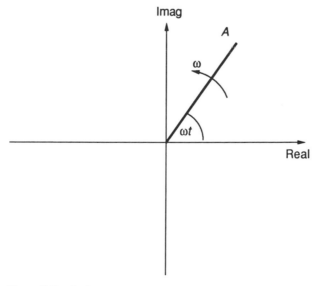

Figure 3.1 A phasor

45

A phasor then

a) is complex
b) has three attributes:

    (1) rotational velocity
    (2) magnitude
    (3) initial angle.

When using phasors, we frequently concentrate on their parameters of magnitude and angle, since frequency is often a common factor shared by a number of phasors. Thus, we might legitimately describe the phasor in equation 3.2 by a *complex magnitude* **A**, which is the value of the phasor at $t = 0$ (equation 3.3)

$$x(t) = \mathbf{A} \, . \exp \, \mathrm{j}(\omega t) \tag{3.4}$$

where $\mathbf{A} = A \, . \exp \, \mathrm{j}(\varphi)$ (a magnitude of A and an angle of φ).

We are saying by this definition that **A** is a complex constant, a vector having direction as well as magnitude.

The factor $\exp \, \mathrm{j}(\omega t)$ describes a *unit phasor* which is a function of both *frequency* ω and *time t*. In some applications we ignore one of the variables, and when the frequency is constant for instance, regard the phasor as being a function of time only. Alternatively when considering the response of a system to a varying frequency we may ignore the dependence upon time, since it is irrelevant for this case.

Whether we are considering the phasor in terms of its frequency or its time dependence, it is defined by a complex number **A**, magnitude A and angle φ.

● **Example 3.1** A certain phasor has a constant frequency of 100 rad/sec, a magnitude of 20, and a starting phase of 2·5 radians.
We might describe the phasor as

$$x(t) = 20 \, . \exp \, \mathrm{j}(100t + 2 \cdot 5)$$

The complex magnitude is $20 \, . \exp \, \mathrm{j}(2 \cdot 5)$ or $[-16 \cdot 02 + \mathrm{j}(11 \cdot 97)]$.
Since the frequency is constant, we are interested only in the variation with time.　　●

3.1.2　Discrete-time

Discrete-time, you will remember, is a regime where signals have non-zero values only at regular intervals of time with spacing T_S. In order to describe any signal in the discrete-time world, we simply express time t as a multiple of the sampling interval T_S.

A phasor, similar to that described in equation 3.1, except in discrete-time, is then

$$x(n) = A \, . \exp \, \mathrm{j}(n\omega T_S + \varphi) \tag{3.5}$$

Instead of a continuous variable t, we now have an integer or discrete variable n, so that the phasor advances in phase jumps of value ωT_S. Sometimes it is convenient to express this phase jump angle in other ways, call it θ:

$$\theta = \omega T_S \quad \text{or} \quad 2\pi \, . \, f/f_S \tag{3.6}$$

and

$$x(n) = \mathbf{A} \, . \exp \, \mathrm{j}(n\theta) \tag{3.7}$$

Notice that the *unit phasor* in the discrete-time case is $\exp \, \mathrm{j}(n\theta)$, but that the complex magnitude $\mathbf{A}$ is the same as for the corresponding continuous-time phasor.

● **Example 3.2** Taking the phasor in example 3.1, and a value for the sampling interval of $T_S = 3 \cdot 14$ ms, we obtain $\theta = 0 \cdot 314$ rad or $18°$. Thus

$$x(n) = 20 \, . \exp \, \mathrm{j}(0 \cdot 314n + 2 \cdot 5) \tag{3.8}$$

Figure 3.2 shows the first few positions taken up by this phasor. The complex magnitude is, again, $20 \, . \exp \, \mathrm{j}(2 \cdot 5)$. ●

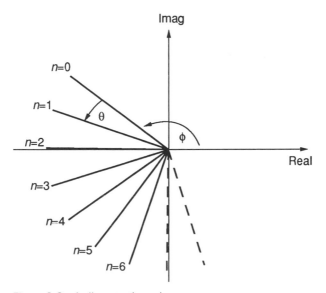

Figure 3.2 A discrete-time phasor

3.2 Phasor models of sinusoidal signals

Having defined the phasor in some detail, we are now in a position to apply it to the modelling of simple sinusoidal signals, which form the most important sub-class of classical signals (1.4.1). A general sinusoidal signal is

$$x(t) = a \cdot \cos(\omega t + \varphi) \tag{3.9}$$

In order to apply the phasor concept to signals such as this, we need to be aware of several complex number identities (see also appendix A1):

$$\cos \theta \quad = [\exp j(\theta) + \exp j(-\theta)]/2 \tag{3.10}$$
$$\sin \theta \quad = [\exp j(\theta) - \exp j(-\theta)]/2j \tag{3.11}$$
$$\exp j(\theta) = \cos \theta + j(\sin \theta) \tag{3.12}$$

So applying identity 3.10, we see that

$$x(t) = a \cdot [\exp j(\omega t + \varphi) + \exp j(-\omega t - \varphi)]/2 \tag{3.13}$$

A single sinusoidal signal is therefore represented by two phasors, which form a *conjugate* pair:

$$x(t) = \mathbf{A} \cdot \exp j(\omega t) + \mathbf{A}^* \cdot \exp j(-\omega t) \tag{3.14}$$

where $\mathbf{A}^*$ is the conjugate of $\mathbf{A}$.

The two phasors are described by a magnitude $a/2$, an initial phase of $\pm \varphi$ rad, and a frequency $\pm \omega$ rad/sec. Figure 3.3 shows these two phasors on the complex plane, for the case where the starting angle φ is zero. Notice that the resultant is real at all times, because the two phasors always have opposite angles and identical magnitudes, and that this resultant is the

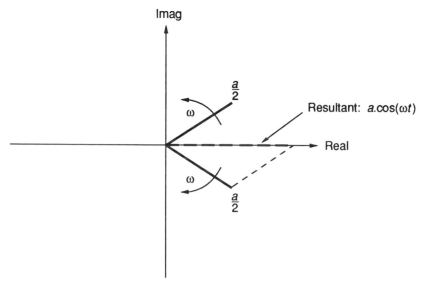

Figure 3.3 Phasor representation of a cosine signal

original signal (see equations 3.9 and 3.14). The differing signs of the two frequencies define *contra-rotating* phasors.

● **Example 3.3** Derive the phasors for the signal $y(t) = a \cdot \sin(\omega t + \varphi)$.
We note that $\sin \theta = \cos(\theta - \pi/2)$, and so

$$y(t) = (a/2) \cdot [\exp \, j(\omega t + \varphi - \pi/2) + \exp \, j(-\omega t - \varphi + \pi/2)]$$

or

$$y(t) = \mathbf{B} \cdot \exp \, j(\omega t) + \mathbf{B}^* \cdot \exp \, j(-\omega t)$$

where $\mathbf{B} = \mathbf{A} \cdot \exp \, j(-\pi/2)$.

Note that this amounts to the two phasors representing the cosine signal, but with each rotated $\pi/2$ rad against their direction of motion. Figure 3.4 shows this situation.   ●

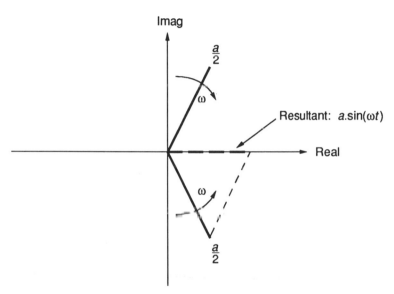

Figure 3.4 Phasor representation of a sine signal

Comparing these two examples, we can draw the following conclusions:

a) A real sinusoidal signal can always be represented by two contra-rotating complex phasors.
b) These two phasors are always conjugates, and so all the necessary information about the signal is contained within each one of them. For this reason, it is common to model simple signals by one phasor.
c) The phasor can be described by a constant complex number, by taking its time-varying and frequency-varying nature for granted. This is often the way in which phasors are used.

From this simple case, it is not clear that the phasor model of a signal will help our understanding of that signal nor help with analysing how systems will process it. In fact the phasor seems to complicate the description of a simple sinusoidal signal. However, we shall show shortly that this apparent complication can be put to good effect, and that this complex signal model is vital when describing and analysing all kinds of signal process.

3.2.1 Phasor addition

The majority of signal processes require addition of signals at some stage, and addition of sinusoids is cumbersome. Addition of phasors is conceptually clearer.

For example, consider the addition of two phasor signals $x(t)$ and $w(t)$ to form $y(t)$:

$$x(t) = \mathbf{A} \cdot \exp \mathrm{j}(\omega t) \tag{3.16}$$

$$w(t) = \mathbf{B} \cdot \exp \mathrm{j}(\omega t) \tag{3.17}$$

where $\mathbf{A} = A \cdot \exp \mathrm{j}(\varphi)$ and $\mathbf{B} = B \cdot \exp \mathrm{j}(\theta)$.

Forming the sum of the two components, $y(t)$:

$$y(t) = (\mathbf{A} + \mathbf{B}) \cdot \exp \mathrm{j}(\omega t) \tag{3.18}$$

Since the two phasors are of the same rotational velocity (or frequency), addition is just a matter of adding the *complex magnitudes* of the phasors.

Thus we write that

$$\mathbf{C} = \mathbf{A} + \mathbf{B}$$

and

$$y(t) = \mathbf{C} \cdot \exp \mathrm{j}(\omega t)$$

where $\mathbf{C} = [A \cdot \cos(\varphi) + B \cdot \cos(\theta)] + \mathrm{j}[A \cdot \sin(\varphi) + B \cdot \sin(\theta)]$

$$\text{or} \quad \mathbf{C} = A \cdot \exp \mathrm{j}(\varphi) + B \cdot \exp \mathrm{j}(\theta) \tag{3.19}$$

In order to represent this operation on a geometrical diagram, we note that all phasors have the same rotational velocity ω, which we then ignore. Figure 3.5 shows how to represent this addition.

Having turned the addition of phasors into a geometrical diagram, we can sometimes use the properties of geometry to obtain quick results, although this is not often the case!

● **Example 3.4** Two phasors are described by the complex magnitudes **A** and **B**, and are added to obtain **C**.

$$\mathbf{A} = 3 \cdot \exp \mathrm{j}(1) \qquad \mathbf{B} = 4 \cdot \exp \mathrm{j}(1 + \pi/2)$$

Then, resolving into real/imaginary form:

$$\mathbf{A} = 1 \cdot 621 + \mathrm{j}(2 \cdot 524) \qquad \mathbf{B} = -3 \cdot 366 + \mathrm{j}(2 \cdot 161)$$

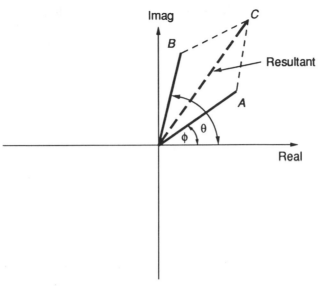

Figure 3.5 Phasor addition

Hence

$$\mathbf{C} = -1 \cdot 745 + j(4 \cdot 685) = 5 \cdot \exp j(1 \cdot 927) \quad \text{an angle of } 110 \cdot 4° \qquad \bullet$$

3.2.2 Addition of sinusoidal signals

Addition of phasors is only a first step; we actually want to add sinusoidal signals using this model. However, since each signal can now be represented by a pair of phasors, addition of two sinusoidal signals can be achieved by adding two pairs of conjugate phasors, which can be reduced to the addition of two phasors. Change the interpretation of our signals $x(t)$ and $w(t)$ to become

$$x(t) = a \cdot \cos(\omega t + \varphi) \tag{3.20}$$
$$w(t) = b \cdot \cos(\omega t + \theta) \tag{3.21}$$

Then in phasor form

$$x(t) = \mathbf{A} \cdot \exp j(\omega t) + \mathbf{A}^* \cdot \exp j(-\omega t)$$
$$w(t) = \mathbf{B} \cdot \exp j(\omega t) + \mathbf{B}^* \cdot \exp j(-\omega t)$$

where $\mathbf{A} = (a/2) \cdot \exp j(\varphi)$, etc.

The resultant is $y(t) = x(t) + w(t)$, and will be given in phasor form as

$$y(t) = \mathbf{C} \cdot \exp j(\omega t) + \mathbf{C}^* \cdot \exp j(-\omega t) \tag{3.22}$$

Hence

$$\mathbf{C} = \mathbf{A} + \mathbf{B} \quad \text{and} \quad \mathbf{C}^* = \mathbf{A}^* + \mathbf{B}^* \tag{3.23}$$

As mentioned above in section 3.2, one of the resultant phasors is sufficient to define a sinusoidal signal, so we finish up with a diagram identical to figure 3.5. Addition of single-frequency signals is no different from the addition of phasors, and is completely represented by phasor addition.

● **Example 3.5**   Two sinsuoidal signals of frequency 1000 rad/sec are to be added:

$x(t)$ has magnitude 20 and initial angle 30°
$w(t)$ has magnitude 16 and initial angle 0°

These two signals are represented by the following phasor complex magnitudes:

$$\mathbf{A} = 10 . \exp j(30°)$$
$$\mathbf{B} = 8 . \exp j(0°)$$

The resultant complex magnitude is $\mathbf{C}$, and is given by the vector addition of $\mathbf{A}$ and $\mathbf{B}$, which is best done in cartesian co-ordinates:

$$\text{real}[\mathbf{C}] = 10 . \cos(30°) + 8 . \cos(0°)$$
$$\text{imag}[\mathbf{C}] = 10 . \sin(30°) + 8 . \sin(0°)$$

Hence

$$\mathbf{C} = 16 \cdot 66 + j(5 \cdot 00) \quad \text{or} \quad 17 \cdot 39 . \exp j(16 \cdot 71°)$$

The resulting signal is therefore given by

$$y(t) = 34 \cdot 78 . \cos(1000t + 16 \cdot 71°) \qquad ●$$

3.3   Frequency response

3.3.1   A general system

When excited by a real input signal $x(t) = \cos(\omega t)$, a linear processing system generates an output signal which is at the same frequency, but changed in magnitude and phase:

$$y(t) = a . \cos(\omega t + \varphi) \tag{3.24}$$

The phase is advanced by φ rad and the signal amplitude is changed by a factor a. These parameters may be measured, and together define the *frequency response* of the system, which is expressed as a complex quantity $\mathbf{H}(\omega)$. Thus, in this case

$$\mathbf{H}(\omega) = a . \exp j(\varphi) \tag{3.25}$$

Now in calculations, the complex quantity $\mathbf{H}(\omega)$ sits uneasily alongside real expressions such as equation 3.24, so analysis is naturally done by employing a phasor model of the input signal. Thus

$$2 . x(t) = \exp \mathrm{j}(\omega t) + \exp \mathrm{j}(-\omega t) \tag{3.26}$$

$$2 . y(t) = a . \exp \mathrm{j}(\omega t + \varphi) + a . \exp \mathrm{j}(-\omega t - \varphi)$$
$$= \mathbf{H}(\omega) . \exp \mathrm{j}(\omega t) + \mathbf{H}^*(\omega) . \exp \mathrm{j}(-\omega t) \tag{3.27}$$

So the quantity $\mathbf{H}(\omega)$ is the *phasor response* of the system, but combines the two real parameters which we might measure in a practical situation, a and φ. Notice too that the response is dependent on frequency, and that includes negative frequencies of course.

In order to calculate the frequency response $\mathbf{H}(\omega)$, we apply a mathematical phasor to the system equations, and then calculate the corresponding output phasor. Dividing one by the other gives the frequency response. Frequency response is probably the most important description in use of how a signal process or system behaves.

3.3.2 Continuous-time circuit

While we are not going to investigate fully the application of phasors to electrical circuits, it will be useful to consider a simple example in order to show the benefits of phasor analysis and modelling.

Figure 3.6 shows a commonly used circuit, called a *first-order lowpass filter* or a *first-order lag*. We know that the current $i(t)$ in a capacitor of value τ F is related to the voltage across it, $v_2(t)$, by

$$i(t) = \tau \frac{\mathrm{d}v_2(t)}{\mathrm{d}t} \tag{3.28}$$

Adding volt drops around the loop, we form the Kirchhoff voltage equation:

$$v_2(t) + \tau \frac{\mathrm{d}v_2(t)}{\mathrm{d}t} = v_1(t) \tag{3.29}$$

(Notice that we have regarded the resistor as having resistance 1 Ω, and the capacitor as having a value of τ F. The circuit CR product, the *time constant*, is therefore τ sec.)

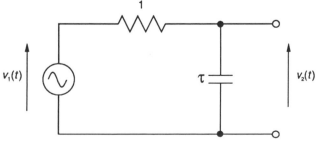

Figure 3.6 First-order circuit

Now we require a solution to the differential equation 3.29 for a sinusoidal input signal $v_1(t)$, which has a standard though cumbersome form. Alternatively we could represent the sinusoidal signal by its two phasors, and then solve the equation for each of them. Since the two phasors are conjugates we need to solve for only one member of the pair, so we use the positive frequency phasor:

$$v_1(t) = \exp \mathrm{j}(\omega t) \tag{3.30}$$

All voltages and currents in the circuit can also be expressed as phasors at the same frequency, but will have different magnitudes and phase angles. Consequently, we can write

$$v_2(t) = \mathbf{H} . \exp \mathrm{j}(\omega t) \tag{3.31}$$

where $\mathbf{H}$ is the complex amplitude of the output phasor.

Equations 3.30 and 3.31 can be combined with the circuit equation 3.29 to provide a solution for $\mathbf{H}$. Note that

$$\frac{\mathrm{d}}{\mathrm{d}t} [\exp \mathrm{j}(\omega t)] = \mathrm{j}(\omega) . \exp \mathrm{j}(\omega t) \tag{3.32}$$

Thus $\mathbf{H} . [1 + \mathrm{j}(\omega\tau)] = 1$

$$\mathbf{H} = \frac{1}{1 + \mathrm{j}(\omega\tau)} \tag{3.33}$$

$\mathbf{H}$ is the ratio of output phasor to input phasor, and represents the *response* or the *gain* of the circuit at this particular frequency. Since it is a function of frequency only, we normally write it as: $\mathbf{H}(\omega)$.

The frequency at which the low-frequency filter gain has fallen by 3 dB (or a fraction $1/\sqrt{2}$), is called the *bandwidth* of the filter, but this is not so abrupt as for the ideal filters of section 1.3.1. The 3 dB frequency in this case is then $\omega = 1/\tau$ or when $f = 1/2\pi\tau$.

● **Example 3.6**   A lowpass filter of the form in figure 3.6 has a time constant $\tau = 1$ msec. Calculate the output signal when the input signal is a sinusoid of frequency 2 krad/sec, and with a magnitude of 5 V.

At this frequency, $\omega\tau = 2$ (note that this factor is dimensionless). Therefore

$$\mathbf{H}(\omega) = 1/(1 + \mathrm{j}2)$$

The response at this frequency has magnitude $0 \cdot 45$ and phase angle $-63 \cdot 43°$.

The output signal will therefore have magnitude $2 \cdot 25$ and a phase relative to the input of $-63 \cdot 43°$. Thus

$$y(t) = 2 \cdot 25 . \cos(2t + \varphi - 63 \cdot 43°)$$

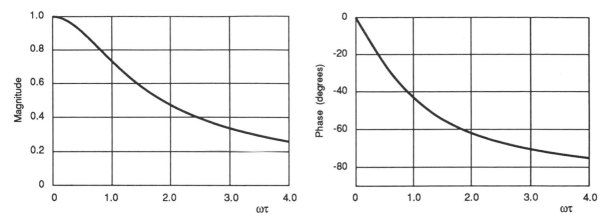

Figure 3.7 Response function for first-order lowpass filter

where t is measured in msecs, and φ is the angle of the input signal. ●

● **Example 3.7** Calculate the response function $\mathbf{H}(\omega)$ for the case where the time constant τ is unity.

From equation 3.33, the magnitude of $\mathbf{H}(\omega)$ is

$$1/\sqrt{[1 + (\omega\tau)^2]}$$

and the angle of $\mathbf{H}(\omega)$ is $\arctan[\omega\tau]$. These are plotted in figure 3.7. ●

These curves in figure 3.7 can be used as a template for a similar filter of any different time constant τ, merely by scaling the horizontal axis. Thus a figure of $\tau = 0\cdot01$ ms fixes the point $\omega\tau = 1$ at $\omega = 100$ krad/sec or $f = 15\cdot92$ kHz.

Although the response function has been calculated as a phasor response, yet it also applies to sinusoidal signals, as example 3.6 shows. In practical tests, we apply a sinusoidal signal.

3.3.3 Discrete-time system

To complete our discussion of the use of phasors to model single-frequency real signals, we now consider discrete-time signals.

Even when dealing with discrete-time, where the signals have non-zero values only at multiples of T_S, similar calculations can be carried out to those in the previous section. The practical difference is that these discrete-time signals will be processed by digital hardware, whereas the calculations in the previous section apply to circuit components in the real world of continuous-time signals.

Figure 3.8 shows the signal processing block diagram for the system we shall consider. For the moment it does not matter how it was arrived at, nor

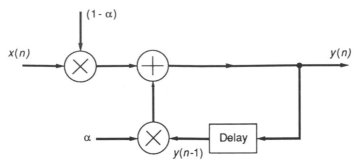

Figure 3.8 A discrete-time signal processing system

what its target process is, we shall use it merely as a vehicle to illustrate the use of discrete-time phasor signals.

In this block diagram we use circle symbols to denote addition and multiplication, and a square block to show signal delay. We shall define these operations now in terms of input signals $u(n)$, $v(n)$, and output signal $z(n)$:

Signal addition $\qquad\qquad$ $z(n) = u(n) + v(n)$ $\qquad\qquad\qquad$ (3.34)

Constant multiplication $\qquad$ $z(n) = \alpha \cdot u(n)$ $\qquad\qquad\qquad$ (3.35)

Signal delay $\qquad\qquad\quad$ $z(n) = u(n - 1)$ $\qquad\qquad\qquad$ (3.36)

So, applying these ideas to the system diagram of figure 3.8, we set up the *difference equation*, which is analogous to the *differential equation* in equation 3.29:

$$y(n) = (1 - \alpha) \cdot x(n) + \alpha \cdot y(n - 1) \qquad\qquad (3.37)$$

In order to derive its response function $\mathbf{H}(\omega)$, we first define the input signal as a discrete-time unit phasor:

$$x(n) = \exp \mathrm{j}(n\omega T_S) \qquad\qquad (3.38)$$

Although the phasor representing the output signal $y(n)$ is also discrete-time, stepping forward by angular increments ωT_S, its complex magnitude will be $\mathbf{H}(\omega)$ times that of the input phasor. Hence

$$y(n) = \mathbf{H}(\omega) \cdot \exp \mathrm{j}(n\omega T_S) \qquad\qquad (3.39)$$

Combining equations 3.38 and 3.39 with equation 3.37, we see that

$$\mathbf{H}(\omega) = \frac{1 - \alpha}{1 - \alpha \cdot \exp \mathrm{j}(-\omega T_S)} \qquad\qquad (3.40)$$

So the frequency response function is calculated in similar fashion to the continuous-time case of analog circuit components, but the resulting function is less tidy.

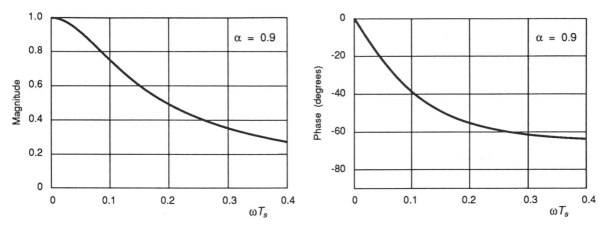

Figure 3.9 Response of discrete-time system

● **Example 3.8** Calculate the frequency response function for the system of figure 3.8, for the case where $\alpha = 0\cdot9$.

This result is plotted in figure 3.9.

Comparing it with figure 3.7, we see that it gives a similar type of lowpass response, although the frequency parameters are quite different. We shall not investigate this kind of system any more until chapter 7. ●

Again, the response of the system to a discrete-time sinusoidal signal is still defined by the phasor response, and is just an extension of the principle used for continuous-time signals in example 3.6.

Implementing such a signal process is very simple if a high-level computer language is used. The following example outlines a BASIC program which would *simulate* the filtering process on a microcomputer.

● **Example 3.9** N values of the input signal are already stored in an array $X(i)$, starting with $i = 0$, and the output signal is to be written into a corresponding array $Y(i)$. The difference equation of equation 3.37 is to be used.

For the purpose of this example we ignore certain organisational problems, so that the program below is not a fully operational signal process. However the filter algorithm is properly described. ●

```
100   REM: write the values of the input phasor
105   wt = omega * ts
110   FOR i = 0 to N − 1
120      X(i) = mag * COS(i *wt + phi)
130   NEXT
```

57

```
190  REM: now calculate the output phasor
200  const = 1 – alpha
210  FOR i = 1 to N – 1
220     temp = const * X(i)
230     Y(i) = alpha * Y(i – 1) + temp
240  NEXT
```

STUDY QUESTIONS

1 Define a phasor and list its three attributes.

2 Write down the complex magnitude of the phasor
$x(t) = 6 . \exp j(50t + 0 \cdot 4)$, in both cartesian and polar form.

3 Express the phasor of question 2 in discrete-time form, for the cases where T_S is $3 \cdot 14$ ms and $2 \cdot 0$ ms. Write down the complex magnitude of each of these phasors.

4 Using the identities 3.10 to 3.12, derive the two phasors which represent a signal

$$x(t) = a . \sin(\omega t - \theta)$$

5 Sketch on a complex plane, the phasors which adequately represent the following signals:

$$x(t) = 3 . \cos(500t + 45°)$$
$$w(t) = 4 . \sin(500t - 60°)$$
$$y(t) = 5 . \cos(500t + 120°)$$

6 Use phasor addition to obtain the sinusoidal resultant of adding together the three signals in question 5.

7 A lowpass circuit like that in figure 3.6 has an input signal $x(t)$. Use figure 3.7 to estimate the output signal $y(t)$ if the circuit time constant τ is $0 \cdot 1$ ms.

$$x(t) = 6 . \cos(\omega t + 30°) \quad \text{and } \omega \text{ is 20 krad/sec}$$

8 For the circuit shown in figure 3.10, form an integral equation relating input and output voltages, and then find and sketch the phasor response $H(\omega)$. What sort of filtering function does this circuit perform? (See figure 1.6.)

9 A discrete-time system has the following difference equation:

$$y(n) = (1 - \alpha) . x(n) - \alpha . y(n - 1)$$

Draw the block diagram of this process, then find and sketch the phasor response $H(\omega)$. Compare it with that found for question 8.

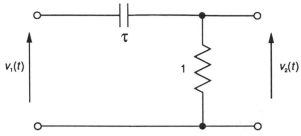

Figure 3.10 A circuit example

10 Magnetic recording tape stores a flux pattern which is proportional to signal current. It is then read by coupling the flux to a coil where the emf generated is proportional to the rate-of-change of flux linking with the coil. Write down an equation describing this process, and then determine the phasor response and the ratio of output voltages for frequencies of 300 Hz and 5 kHz, assuming that both input signals have the same magnitude.

4 Signal Models: Multi-frequency Signals and the Fourier Series

OBJECTIVES

To develop the phasor concept further, and show that it applies to signals which have several frequency components, allowing them to be modelled and their power to be calculated.

To show how the concept can be applied to processes of:

a) Linear filtering
b) Multiplication

COVERAGE

Using the phasor concept, a multi-frequency signal model is formed, and the total signal power calculated for cases where the mean is zero and also where the mean value is finite. The special case of the Fourier series is introduced.

The effect of filtering upon such a signal is examined for simple cases. Multiplication of signals is introduced, and the frequency-changing property demonstrated and applied to the spectrum analyser.

4.1 Multi-frequency signals

In chapter 3 we established the concept of the *phasor* and showed how it could be used to model real signals and to analyse the characteristics of signal processing systems by defining the frequency response. Armed with this background, we now extend the use of the phasor to include multi-frequency signals, of which there are various types.

4.1.1 Two-component signals

Before generalising to signals having many different frequency components, we will treat the case of only two components and discover the general principles which are involved. Such a signal is

$$x(t) = a \cdot \cos(\omega_0 t) + b \cdot \cos(\omega_1 t) \tag{4.1}$$

We are ignoring the starting phase of these two sinusoidal components, since that is less important for two components of different frequency than it is for two components of the same frequency. However, by using *complex magnitudes*, any relevant starting phase could be included.

We will represent these two frequency components by their phasors. Thus

$$x(t) = \mathbf{A} \cdot \exp \mathrm{j}(\omega_0 t) + \mathbf{B} \cdot \exp \mathrm{j}(\omega_1 t)$$
$$+ \mathbf{A}^* \cdot \exp \mathrm{j}(-\omega_0 t) + \mathbf{B}^* \cdot \exp \mathrm{j}(-\omega_1 t) \tag{4.2}$$

where $|\mathbf{A}| = a/2$ and $|\mathbf{B}| = b/2$.

Taking the positive-frequency phasors only, we see that the signal $x(t)$ is represented by a complex model $\mathbf{X}(t)$, where

$$\mathbf{X}(t) = \mathbf{A} \cdot \exp \mathrm{j}(\omega_0 t) + \mathbf{B} \cdot \exp \mathrm{j}(\omega_1 t) \tag{4.3}$$

This representation is shown in figure 4.1(a).

Notice that the simple nature of a phasor model is less clear in this case, since it is difficult to visualise the locus described by the resultant of these two phasors. However, the important feature is that we now have a compact mathematical model to represent signals of quite complicated waveform.

In order to simplify the model, we may refer one of the frequencies to the other, and express the phasor model in this way:

$$\mathbf{X}(t) = [\mathbf{A} + \mathbf{B} \cdot \exp \mathrm{j}(\omega_1 - \omega_0 t)] \cdot \exp \mathrm{j}(\omega_0 t)$$

or

$$\mathbf{X}(t) = \mathbf{E}(t) \cdot \exp \mathrm{j}(\omega_0 t) \tag{4.4}$$

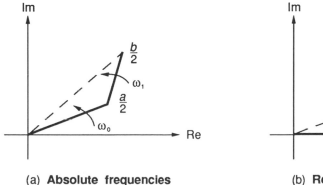

(a) **Absolute frequencies**

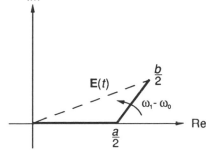

(b) **Relative frequencies**

Figure 4.1 Phasor model for two-frequency signal

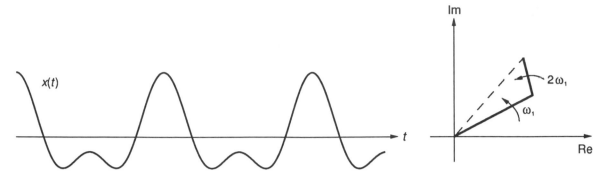

Figure 4.2 Example of two-frequency signal

$\mathbf{E}(t)$ is the complex magnitude of this resultant phasor, and is now a function of time. It is sometimes called the *complex envelope*, and this concept will be discussed in chapter 12. Although the idea is a valuable one in general, it is not much help here, so we shall not pursue it.

A simple example will serve to show the kind of signal we are talking about, and it is illustrated in figure 4.2:

$$x(t) = \cos(\omega_1 t) + 0 \cdot 6 \cdot \cos(2\omega_1 t) \tag{4.5}$$

4.1.2 Calculation of signal power

Calculation of signal power is straightforward. The total signal power is the sum of powers contained in each component, and this can readily be demonstrated as follows.

Consider a signal $x(t) = a \cdot \cos(\omega_0 t)$, whose mean power is then given by $(a^2/2)$. Note that each of the two representing phasors has magnitude $(a/2)$ and that the power contained in each one is $(a^2/4)$. Now let

$$x(t) = a \cdot \cos(\omega_0 t) + b \cdot \cos(\omega_1 t)$$

The mean power is given by

$$P = \text{avg}[x^2(t)] \tag{4.6}$$

where

$$x^2(t) = a^2 \cdot \cos^2(\omega_0 t) + b^2 \cdot \cos^2(\omega_1 t) + 2ab \cdot \cos(\omega_0 t) \cdot \cos(\omega_1 t) \tag{4.7}$$

The average or mean value of instantaneous power can be found by examining each of these terms individually:

1) $a^2 \cdot \cos^2(\omega_0 t) = (a^2/2)[1 + \cos(2\omega_0 t)]$, which has an average value of $(a^2/2)$.
2) $b^2 \cdot \cos^2(\omega_1 t)$ likewise has an average value of $(b^2/2)$.
3) $2ab \cdot \cos(\omega_0 t) \cdot \cos(\omega_1 t) = ab[\cos(\omega_0 + \omega_1)t + \cos(\omega_0 - \omega_1)t]$

Note that this expansion of the product of two cosines is of fundamental importance and is used widely (see appendix A1).

Observe that the expression has an average value of zero, since it consists of the sum of two cosines. Consequently

$$P = (a^2/2) + (b^2/2) \tag{4.8}$$

and is the *sum of powers at each individual frequency*.

● **Example 4.1** Calculate the power in the following signal:

$$x(t) = 2 \cdot \cos(100t + 0 \cdot 5) + 5 \cdot \sin(120t - 1 \cdot 3) \text{ V}$$

The mean power is $P = (2^2/2) + (5^2/2) = 14 \cdot 5 \text{ V}^2$.

Note that we have used *peak magnitudes* here, but rms values would do just as well, and indeed are more suitable since the rms value is usually measured rather than peak magnitude.

The rms values for this case are $1 \cdot 41$ V and $3 \cdot 54$ V respectively.   ●

A special case occurs when one of the two frequencies is zero. The signal now is of the form:

$$x(t) = a + b \cdot \cos(\omega_0 t) \tag{4.9}$$

Figure 4.3 illustrates this case. Several points distinguish it from the more general one where both frequencies are non-zero:

a) The mean value of the signal is now a, whereas it was zero before.
b) The phasor diagram must include both conjugate phasors at ω_0.
c) The zero-frequency phasor is magnitude a.
d) The power is still the sum of individual powers, but is now

$$p = a^2 + (b^2/2)$$

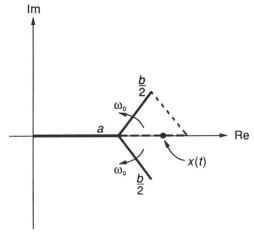

Figure 4.3 Two-frequency signal, including zero frequency

● **Example 4.2**   A certain signal is given by

$$x(t) = 10 + 6 \cdot \cos(200t + 0 \cdot 6) \text{ V}$$

Then the mean value is 10 V
the maximum value is 16 V
the minimum value is 4 V
the mean power is 118 V^2
and the rms value is 10·86 V.   ●

4.1.3   The Fourier series

Many intricate waveforms may be built up by adding together a number of sinusoidal components, and these in turn can be represented by their constituent phasors.

One particular form of signal model which contains many frequency components is the *Fourier series*. This uses a set of frequencies which are multiples of a *fundamental* frequency, and the nth frequency is said to be the nth *harmonic* of the fundamental frequency. Any periodic signal can be modelled this way, provided that sufficient harmonics are taken.

Representing a signal by the Fourier series means that it is represented by a sum of $2N$ harmonically-related phasors:

$$x(t) = \sum_{k=-N}^{N} \mathbf{C}_k \cdot \exp \mathrm{j}(k\omega_0 t) \tag{4.10}$$

ω_0 (or $2\pi f_0$) is the fundamental frequency, corresponding to the period of $x(t)$.

The waveform is now described entirely by the set of complex coefficients $\{\mathbf{C}_k\}$, and the fundamental frequency ω_0. For the time being it is not important to discuss how the coefficients are found for a given signal; it is enough to realise that this form of signal model is very flexible and easy to use.

We are using the signal model to represent or *synthesise* a moderately complicated waveform, which is quite a simple operation, but the complementary task of *analysis*, where we discover what set of coefficients will adequately represent a given waveform, is more difficult and requires further study. A brief treatment is given in appendix A2, where some further properties of the Fourier series are also presented.

Notice that this is a phasor model in the true sense, incorporating positive and negative frequency phasors. Since the resultant must always be real, we can assert that

$$\mathbf{C}_k = \mathbf{C}^*_{-k} \tag{4.11}$$

Consequently such a signal can be modelled adequately by considering only the positive-frequency phasors. Figure 4.4 shows such an example.

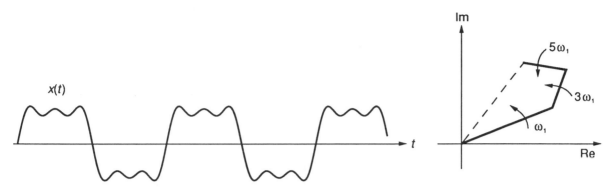

Figure 4.4 A Fourier series example

Although the resultant of these positive-frequency phasors is complex, the corresponding real signal is the projection of it on the real axis.

In order to justify further the use of this series, figure 4.5 gives a number of examples of Fourier series signals. Each example shows also the magnitude of harmonics that have been used, as a plot against ω/ω_0 (or k). This is known as the *frequency spectrum*, and describes the distribution of signal magnitudes along the frequency axis.

Signals modelled by the Fourier series are clearly periodic, but we can extend the model of equation 4.10 to include frequencies which are not harmonically related. Although the resulting signals must be periodic, the periodicity is not obvious. The describing equation becomes

$$x(t) = \sum_{k=-N}^{N} \mathbf{C}_k . \exp \text{j}(\omega_k t) \tag{4.12}$$

where $\mathbf{C}_{-k} = \mathbf{C}_k^*$ and $\omega_{-k} = \omega_k$ for $x(t)$ real.

The signal is now described fully by the set of coefficient couplets: $\{\mathbf{C}_k, \omega_k\}$.

In terms of this general model, whether or not the Fourier condition holds, we can summarise the signal properties:

a) The mean value is C_0, where $\omega_0 = 0$.
b) The mean power is P, and

$$P = \sum_{k=-N}^{N} |\mathbf{C}_k|^2 \tag{4.13}$$

c) No general statements may be made about the maximum or minimum values of the signal, except that the maximum positive peak value is $\leqslant \Sigma |\mathbf{C}_k|$ with $k \geqslant 0$.

Although the phasor representation is used extensively for theoretical studies, or for digital signal processing, yet real signals are usually analysed by a *spectrum analyser* which identifies only the magnitude of each real frequency component and the frequency at which it occurs. A

65

representation of the signal in these terms would be

$$x(t) = \sum_{k=1}^{N} A_k \cdot \cos(\omega_k t + \varphi_k) \qquad (4.14)$$

where φ_k represents the unknown phase.

This form of signal representation is used for example 4.3, and further examples of the technique in use are given in the next section.

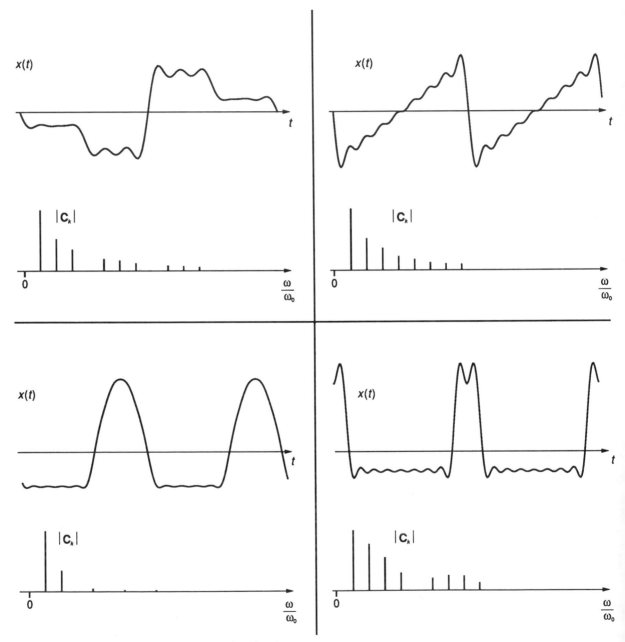

Figure 4.5 Fourier series signals

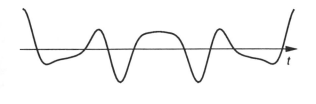 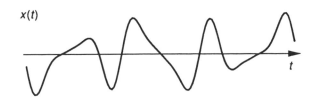

$x(t)$

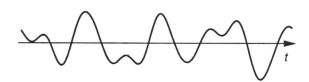 

Figure 4.6 Waveforms having the same power and frequency content

● **Example 4.3** Calculate the power in a signal which contains a number of different frequency components:

Frequency	0·5	1·0	1·75	2·5	kHz
RMS magnitude	0·7	2·0	1·2	0·3	V

Summing the powers at each frequency, $P = 6\cdot02 \text{ V}^2$. ●

Notice that example 4.3 does not specify the relative phase of each frequency component, so that this specification applies to an infinite range of signal waveforms having the same power and the same frequency distribution. A few of these waveforms are shown in figure 4.6. The first one is for zero phase in all components, and the others are for arbitrary phases.

4.1.4 Discrete-time signals

Our discussion has used continuous-time signals for convenience, but all the properties discussed above apply equally to discrete-time signals. The only

Figure 4.7 Frequency spectrum of a discrete-time signal

difference is that the phasors progress in phase jumps of ωT_S, instead of a regular angular velocity ω rad/sec. Their complex magnitudes have the same significance whether it is a continuous-time signal or a discrete-time signal.

In example 4.3 for instance, the rms magnitudes of the different components are used in the same way to calculate the signal power, whether the signal is expressed in continuous-time or discrete-time.

However, the frequency spectrum of a discrete-time signal is radically different from the spectrum of the corresponding continuous-time signal. We re-write the Fourier series form of equation 4.10 to take account of the discrete-time constraint:

$$x(n) = \sum_{k=-N}^{N} \mathbf{C}_k \cdot \exp \ j(k\omega_0 T_S \cdot n) \tag{4.15}$$

The phasor jump angle for the kth harmonic is $k\omega_0 T_S$, and evidently an ambiguity arises when

$$k\omega_0 T_S = 2\pi r \tag{4.16}$$

where r is an arbitrary integer.

Clearly the phasor frequency corresponding to this jump angle is indistinguishable from that when $k = 0$. A little further thought will show that the frequency spectrum of this signal must be periodic in frequency, with a period of $1/T_S$ Hz. This is illustrated in figure 4.7 for the case of a discrete-time signal having a time interval T_S of 100 μs, and a set of frequency components like those in example 4.3.

The periodic spectrum is extremely important for digital signal processing, and is subject to other constraints also, so it will be examined in great detail in chapter 6.

4.2 System examples

Having now established the phasor model for multi-frequency signals, we can now use it to analyse simple systems in order to demonstrate how valuable it is.

4.2.1 Linear filtering

Consider a linear filter having a phasor response $\mathbf{H}(\omega)$, and with input and output signals $x(t)$ and $y(t)$ respectively. Let

$$x(t) = \sum_{k=-N}^{N} \mathbf{C}_k \cdot \exp j(\omega_k t) \tag{4.17}$$

Now each of these phasors is operated on by the response $\mathbf{H}(\omega)$, and the output signal becomes

$$y(t) = \sum_{k=-N}^{N} \mathbf{H}(\omega_k) \cdot \mathbf{C}_k \cdot \exp j(\omega_k t) \tag{4.18}$$

The new set of coefficients for the output signal phasors becomes

$$\{(\mathbf{H}(\omega_k) \cdot \mathbf{C}_k), \omega_k\}$$

and the output signal characteristics are found easily by combining the phasor response of the system with the input phasors.

● **Example 4.4** The input signal to a network consists of two phasors having the following parameters:

Frequency	100	200	rad/sec
Magnitude	1·0	2·5	V
Angle	0°	30°	

The network has the phasor response

$$\mathbf{H}(\omega) = \frac{10}{1 + j(\omega/100)}$$

Calculate the output phasors.

Calculate first the response at the two phasor frequencies:

$$H(100) = 7 \cdot 07 \cdot \exp j(-45°)$$
$$H(200) = 4 \cdot 47 \cdot \exp j(-63 \cdot 43°)$$

Combine this information with the input phasor parameters to obtain the output phasors:

Frequency	100	200	rad/sec
Magnitude	7·07	11·18	V
Angle	−45°	−33·43°	

●

● **Example 4.5** A certain filter network has a response $\mathbf{H}(\omega)$ and an input signal $x(t)$; determine the output signal.

$$x(t) = 3 + 2 \cdot \cos(1000t)$$

$$\mathbf{H}(\omega) = \frac{2}{1 + j(\omega/1000)}$$

The input signal has two frequency components: $\omega = 0$ and 1000 rad/sec. Then

$$H(0) = 2$$
$$H(1000) = 1 \cdot 41 \cdot \exp \mathrm{j}(-45°)$$

Operating on each of the two input signals with this response function, we see that

$$y(t) = 6 + 2 \cdot 82 \cdot \cos(1000t - 45°) \qquad \bullet$$

● **Example 4.6**   A discrete-time system with a sampling interval T_S, has a phasor response $\mathbf{H}(\omega)$, and an input signal $x(n)$.

$$\mathbf{H}(\omega) = \frac{1}{1 - 0 \cdot 7 \cdot \exp \mathrm{j}(-\omega T_S)}$$

$$x(n) = 3 + 2 \cdot \cos(0 \cdot 5n)$$

Consider first the frequency component at zero frequency, $H(0) = 3 \cdot 33$, so the corresponding output term is $10 \cdot 0$.

Consider now the cosine term. The phasor advances by $0 \cdot 5$ rad at each sampling time, which corresponds to (ωT_S) or a phasor frequency of $\omega_1 = 0 \cdot 5 / T_S$ rad/sec. Hence

$$H(\omega_1) = \frac{1}{1 - 0 \cdot 7 \cdot \exp \mathrm{j}(-0 \cdot 5)} = 1 \cdot 96 \cdot \exp \mathrm{j}(-41 \cdot 0°)$$

The corresponding output signal is then

$$y(t) = 10 \cdot 0 + 3 \cdot 92 \cdot \cos(0 \cdot 5n - 41 \cdot 0°) \qquad \bullet$$

4.2.2   Frequency band splitting

In addition to the three numerical examples above, we will also invoke an illustrative system example from section 1.3.3, where we examine that part of the overall system which splits the incoming signal into three discrete frequency bands. The block diagram is reproduced in figure 4.8, together with example signals.

The filter parameters are for example, as follows:

Filter B	lowpass	0–500	Hz
Filter C	bandpass	500–1500	Hz
Filter D	highpass	1500–∞	Hz

Ideal filter characteristics are assumed, and signal phase is ignored. Notice that the highpass filter (D) does not have any upper limit. In practice of course the input signal is bandlimited, and so this does not provide any difficulty.

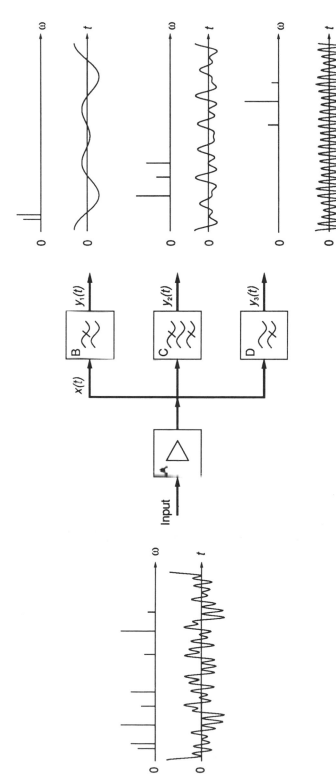

Figure 4.8 An example system, showing frequency band splitting

71

Table 4.1 *Distribution of frequency components*

Frequency	0·2	0·3	0·7	1·1	1·4	2·2	2·7	3·1	kHz
RMS magnitude	0·5	0·7	1·0	0·4	0·7	0·3	1·2	0·2	V
Lowpass output	0·5	0·7	—	—	—	—	—	—	V
Bandpass output	—	—	1·0	0·4	0·7	—	—	—	V
Highpass output	—	—	—	—	—	0·3	1·2	0·2	V

Table 4.1 shows the frequency components in an assumed input signal, and also shows how they are distributed by the bank of filters.

From these signal descriptions, we can calculate the power content immediately:

Input signal power	$3 \cdot 96 \; V^2$
Lowpass filter output power	$0 \cdot 74 \; V^2$
Bandpass filter output power	$1 \cdot 65 \; V^2$
Highpass filter output power	$1 \cdot 57 \; V^2$

As well as calculating the distribution of signal power, we can also plot the actual waveforms developed, and figure 4.8 includes this information, assuming zero initial phase for all frequency components. The output signals are calculated as in section 4.2.1.

Notice that although this discussion has been conducted in terms of the continuous-time variable t, exactly similar considerations apply in discrete-time. In fact, for this example, the waveforms have been calculated using a discrete-time process to model the continuous-time system.

4.3 Systems with multipliers

Many signal processes employ linear filtering as discussed above, but a great number also use the multiplication operation, which possesses important properties. In order to illustrate some of these properties, we take first the case where the two inputs are both single-frequency phasors, and then expand the ideas to include sinusoidal signals and multi-frequency signals. However, we attempt only to give a brief introduction to the extremely powerful properties of this process, which have far-reaching implications in many fields.

4.3.1 Basic principle

Figure 4.9(a) shows the first case we have in mind. Thus, the process multiplies the input phasor $x(t)$ by a *local phasor* $w(t)$.

$$x(t) = \mathbf{A} \cdot \exp j(\omega t) \tag{4.19}$$

$$w(t) = \mathbf{B} \cdot \exp j(\omega_0 t) \tag{4.20}$$

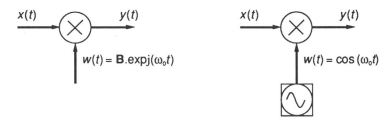

(a) **Phasor inputs**   (b) **Sinusoidal local input**

Figure 4.9   The simple multiplier process

Then $y(t) = x(t) \cdot w(t)$
and

$$y(t) = (\mathbf{A} \cdot \mathbf{B}) \cdot \exp j[(\omega + \omega_0)t] \tag{4.21}$$

The multiplication process has therefore increased the frequency of the input phasor by an amount ω_0. The process has a *frequency-translating* or *frequency-changing* property.

Consider now the case where the local input is sinusoidal as shown in figure 4.9(b), hence:

$$w(t) = \cos(\omega_0 t) \tag{4.22}$$

The output signal is now

$$y(t) = \mathbf{C} \cdot \{\exp j[(\omega + \omega_0)t] + \exp j[(\omega - \omega_0)t] \} \tag{4.23}$$

where $\mathbf{C} - \mathbf{A}/2$.

So two output phasors or frequencies are produced for each input phasor. The complex magnitude of the input phasor is carried through unchanged apart from the scaling factor of $\frac{1}{2}$, but the frequencies of the output phasors are the sum and difference respectively of the input frequency and the local frequency.

4.3.2   A frequency-changing system

This process is so important that we shall now investigate it further, using the simple system block diagrams of figure 4.10. The system as shown contains a multiplier and a lowpass filter, which for now we shall assume to be ideal. Note that the convention $(\omega \pm \omega_0)$ in figure 4.10 denotes the frequencies $(\omega + \omega_0)$ *and* $(\omega - \omega_0)$.

Consider first the case where the local signal is in phasor form. Although this might be thought unrealistic, many applications of the principle do employ this kind of local signal, as we shall explain in chapter 12.

In order to make our point clearly, we shall make the local phasor have negative frequency, and so we write

$$w(t) = \exp j(-\omega_0 t) \tag{4.24}$$

73

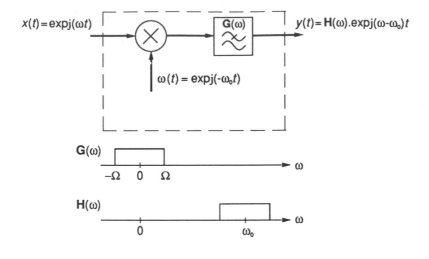

(a) **Phasor local signal**

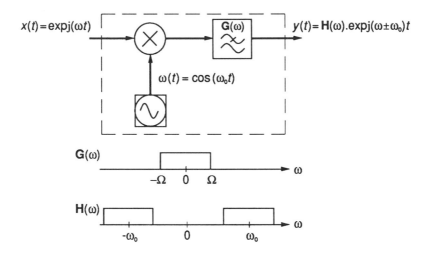

(b) **Sinusoidal local signal**

Figure 4.10 Simple frequency-changing system

The lowpass filter has a bandwidth $\pm \Omega$ rad/sec, and a phasor response $\mathbf{G}(\omega)$.

$$\mathbf{G}(\omega) = 1 \quad \text{if } -\Omega < \omega < \Omega$$
$$\qquad = 0 \quad \text{elsewhere} \tag{4.25}$$

Since the filter is ideal, with zero response outside its passband, an output signal can appear only when the input signal satisfies the condition:

$$-\Omega < (\omega - \omega_0) < \Omega$$

or $\qquad (\omega_0 - \Omega) < \omega < (\omega_0 + \Omega)$ $\qquad\qquad$ (4.26)

Clearly, this is a *bandpass* characteristic, with a bandwidth $\pm\Omega$ about ω_0 rad/sec. The effect of the frequency-changing system is to convert a lowpass filter into a bandpass filter at the input, with the important property that the centre frequency of the bandpass filter can be changed by varying the local phasor frequency.

The overall frequency response is effectively $\mathbf{H}(\omega)$, as shown in figure 4.10. Strictly we cannot use an overall frequency response to describe such a system, because the output signal is at a different frequency to the input signal. In practice though, such a system is used to determine the distribution of signal power at different frequencies (the spectrum analyser of section 4.3.3), or to frequency-change a modulated signal such as described in chapter 5, and so it does behave like a frequency-selective system as far as the input signal is concerned.

In practice, a sinusoidal local signal is often used, and since this has two phasors, the system generates two passbands to an input phasor. They are

$$(\omega_0 - \Omega) < \omega < (\omega_0 + \Omega) \quad \text{and} \quad (-\omega_0 - \Omega) < \omega < (-\omega_0 + \Omega) \qquad (4.27)$$

If the input signal is also sinusoidal, having conjugate phasors, then four different output phasors are generated, which combine to form two conjugate pairs. Although it is quite straightforward to evaluate these, we shall not pursue this development now.

● **Example 4.7** A frequency-changer after the style of figure 4.10(a) has a local phasor at -300 rad/sec and a bandwidth of ± 20 rad/sec.

The input signal consists of phasors at the following frequencies:

Frequency	290	310	350	500	rad/sec
Magnitude	3	5	2	7	V

After frequency-shifting, the phasor frequencies become

$$-10 \qquad 10 \qquad 50 \qquad 200 \qquad \text{rad/sec}$$

Only the first two are within the bandwidth of the lowpass filter, so the output components are at -10 and $+10$ rad/sec, with magnitudes 3 and 5 respectively. $\qquad$ ●

● **Example 4.8** Frequencies are normally expressed in Hz rather than rad/sec.

Consider a frequency-changer of the style of figure 4.10(b), having a sinusoidal local signal of 10 kHz and a bandwidth of ± 2 kHz.

Let the input signal be real, with sinusoidal components at 10, 11, 30 kHz.

The input signal then has phasors at

$$-10 \qquad +10 \qquad -11 \qquad +11 \qquad -30 \qquad +30 \qquad \text{kHz}$$

The local signal has phasors at ± 10 kHz.

Multiplying these two signals together yields frequency components at

$$
\begin{array}{cccccc}
-20 & 0 & -21 & +1 & -40 & +20 \qquad \text{kHz} \\
0 & +20 & -1 & +21 & -20 & +40 \qquad \text{kHz}
\end{array}
$$

The only frequencies which are passed through the lowpass filter are therefore 0 and ± 1 kHz. ●

4.3.3 A spectrum analyser

One major application of this principle is the *spectrum analyser*, an instrument which evaluates the frequency components contained in a signal. A multi-frequency signal is applied to a system such as figure 4.10(b), and the local frequency ω_0 is varied over the desired frequency range. When ω_0 is within $\pm \Omega$ rad/sec of one of the input signal frequencies, then an output appears from the lowpass filter. When ω_0 is exactly equal to the input frequency, then the output frequency is zero. A measurement of the magnitude or power of this output gives the corresponding magnitude or power of the input component. Figure 4.11 shows the complete principle.

Notice the similarity between this structure and the *correlation* of section

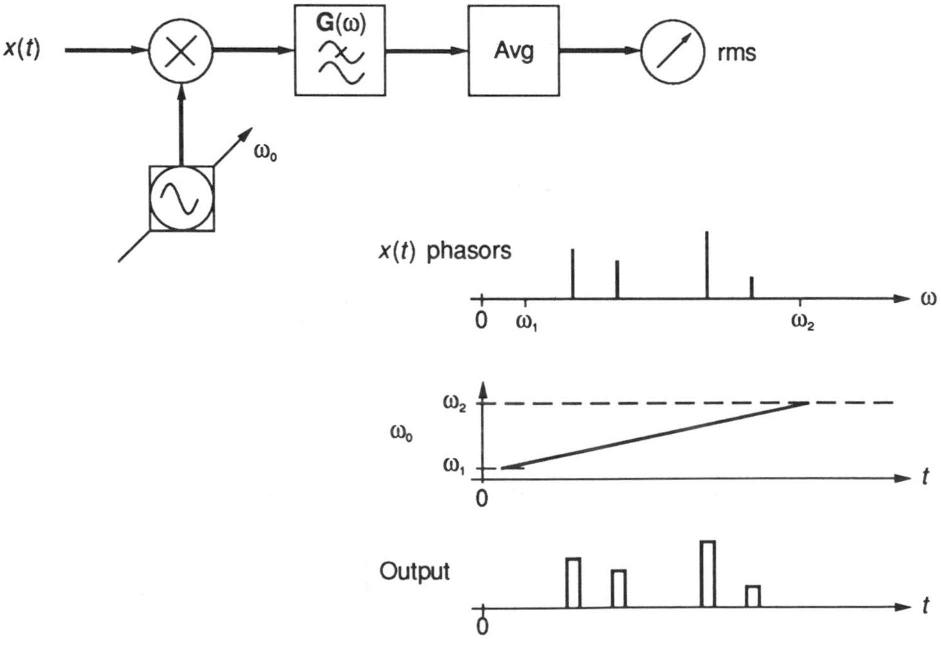

Figure 4.11 Principle of spectrum analyser

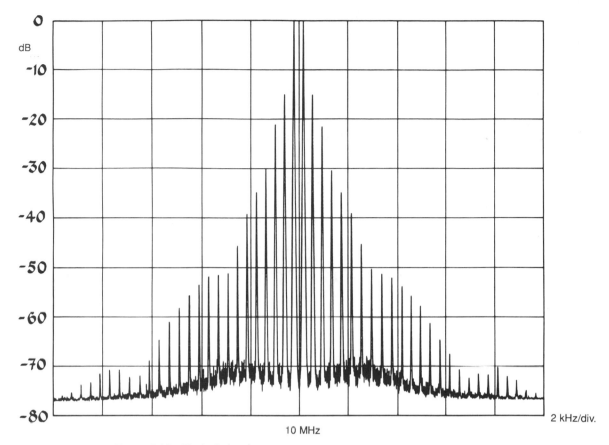

Figure 4.12 Typical signal spectrum

2.1.3. Here, we are analysing the input signal to discover what components it has which are *similar* to a sinusoidal term of frequency ω_0.

In use, the local frequency ω_0 is swept smoothly over a range ω_1 to ω_2, and as each input frequency component ω_k is approached, so the lowpass filter characteristic is traced out. Clearly the bandwidth $\pm\Omega$ must be small, in order to allow the resolution of two frequency components which are close together.

The averaging operation, which evaluates the magnitude of the input frequency component, is crucial to the performance of the analyser. This will be discussed in chapter 10.

This system may be implemented in either continuous-time or discrete-time form. Discrete-time or digital implementation is favoured for frequencies below, say, 1 MHz, but for higher frequencies analog circuits must be used.

Such an instrument is vital for the study of many signals, particularly of speech waveforms, radar signals and radio spectra. Figure 4.12 shows a typical spectrum obtained from a commercial instrument.

STUDY QUESTIONS

1 Draw the phasor model of the following signals $x(t)$, using both absolute and relative frequencies.

   a) $x(t) = 5 \cdot \cos(100t) + 3 \cdot \cos(150t)$
   b) $x(t) = 5 + 3 \cdot \cos(200t)$

2 What is the mean value, the power and the rms value of each of these signals?

3 A certain signal is composed of the following components:

Frequency	0·6	1·2	1·8	3·3	5·1	5·5	kHz
RMS magnitude	0·2	0·3	0·5	1·3	0·5	0·4	V

Calculate the signal mean power and rms value.

4 The signal of question 3 is applied to a bank of filters, having the following details:

Filter A	lowpass	0–1·0	kHz
Filter B	bandpass	1·0–2·0	kHz
Filter C	bandpass	2·0–4·0	kHz
Filter D	highpass	4·0–∞	kHz

Write expressions for the output signal from each filter, and calculate their powers.

5 A certain filter has the phasor response $\mathbf{H}(\omega)$. Find the output signals when the two signals of question 1 are applied to the filter.

$$\mathbf{H}(\omega) = \frac{6}{1 + j\omega/100}$$

6 The signals of question 1 are sampled with an interval $T_S = 2$ ms. Write expressions for the discrete-time form of these signals.

7 A discrete-time system has a phasor response $\mathbf{G}(\omega)$, and has the signals of question 6 applied to it one at a time. Write expressions for the corresponding output signals, and calculate their power.

$$\mathbf{G}(\omega) = \frac{1}{1 - 0 \cdot 7 \cdot \exp j(-\omega T_S)}$$

8 A frequency-changing system has a local phasor of angular frequency 150 rad/sec, and an ideal lowpass filter of bandwidth ± 10 rad/sec. What outputs appear if the two signals of question 1 are applied one at a time?

9 A multiplier has a local signal of $w(t) = \cos(1000t)$, and an input signal which has sinusoidal components at 0, 1000, 1800 rad/sec. What frequencies are present in the output?

5 Signal Models: Modulated Signals

OBJECTIVES

To introduce the idea of modulated signals, and to show how the phasor model is used to represent these signals. In particular, to demonstrate

a) Amplitude modulation
b) Angle modulation.

To show how Amplitude-Modulated (AM) signals can be generated and demodulated using systems with multipliers.

COVERAGE

Starting with the concept of a phasor with time-varying parameters, amplitude and angle modulation are considered. Explicit expressions are formed for cases where the modulating signal is sinusoidal. Then the multi-frequency phasor tool is used to demonstrate further properties of modulated signals.

Some applications are mentioned, and the corresponding signal processing systems for modulation and for demodulation are briefly discussed.

The phasor concept was introduced in chapter 3, and applied to simple sinusoidal signals. The idea has been developed in chapter 4 to include signals of arbitrary complexity, modelled by a set of phasors having different frequencies, and simple applications of these concepts have been covered.

We now extend the use of the phasor still further, to include sinusoidal signals whose parameters are time-varying or *modulated*. Two major parameters can be varied: *amplitude* and *angle*. Such signals find extensive use in most kinds of communication systems and in many measurement systems.

5.1 Amplitude-modulated signals

5.1.1 Amplitude-modulated phasor

A real amplitude-modulated signal is simply described as

$$x(t) = a(t) \cdot \cos(\omega_0 t + \varphi) \tag{5.1}$$

The starting phase of the sinusoid is φ, and the frequency is constant at ω_0 rad/sec. The quantity $a(t)$ is described as the *envelope* of the signal, and is the information-carrying component, while $\cos(\omega_0 t)$ is known as the *carrier* because it carries the envelope function.

Splitting this sinusoidal signal function into two contra-rotating phasors, we obtain

$$x(t) = \mathbf{A}(t) \cdot \exp \mathrm{j}(\omega_0 t) + \mathbf{A}^*(t) \cdot \exp \mathrm{j}(-\omega_0 t) \tag{5.2}$$

where $\mathbf{A} = [a(t)/2] \cdot \exp \mathrm{j}(\varphi)$.

Note that $\mathbf{A}(t)$ corresponds to the complex magnitude of section 3.1.1, etc. The phase φ will be ignored for these initial discussions, but it would become important when such a signal is being demodulated.

Let us choose a simple specific form for the envelope function $a(t)$ by way of an example:

$$a(t) = 1 + b \cdot \cos(\omega_m t) \tag{5.3}$$

Notice that

a) The envelope $a(t)$ is always positive if $b \leqslant 1$, which is often the case.
b) ω_m is the *modulation frequency*, and represents the information signal being carried by the amplitude-modulated (AM) signal.
c) $\omega_m \ll \omega_0$.

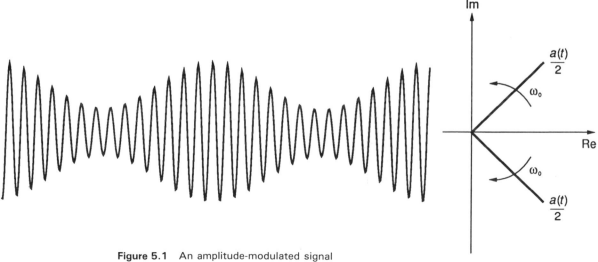

Figure 5.1 An amplitude-modulated signal

Figure 5.1 illustrates this case and shows the two phasors as well as the time waveform of the composite signal. Observe that this signal is represented accurately by the two phasors, whose time-varying magnitudes are given by $a(t)/2$. We also note that the maximum and minimum envelope values are $(1 + b)$ and $(1 - b)$ respectively. The coefficient b is sometimes known as the *depth of modulation*.

The phasor with time-varying magnitude is a simple but useful model, which can improve our intuitive understanding of AM signals. It is, however, too complicated to be helpful as a visual aid, so we examine the frequency spectrum of this signal, in terms of several phasors of different frequencies.

5.1.2 Multi-frequency representation

The form of envelope $a(t)$ in equation 5.3 is a signal that we have modelled before in section 4.1.2, and consists of a real component at zero frequency, and a pair of conjugate phasors at ω_m rad/sec. Inserting this information into equation 5.2, we represent the modulated signal in terms of its constituent positive-frequency phasors:

$$\mathbf{x}(t) = \mathbf{A}(t) \, . \, \exp \, j(\omega_0 t) \tag{5.4}$$

and

$$\mathbf{A}(t) = [1 + (b/2) \, . \, \exp \, j(\omega_m t) + (b/2) \, . \, \exp \, j(-\omega_m t)] / 2 \tag{5.5}$$

Note that in this chapter we are using $\mathbf{x}(t)$ to represent the complex phasor model of the real signal $x(t)$. The complete set of positive-frequency phasors (since $\omega_m \lll \omega_0$), is then

$$(1/2) \, . \, \exp \, j(\omega_0 t)$$
$$(b/4) \, . \, \exp \, j(\omega_0 + \omega_m)t$$
$$(b/4) \, . \, \exp \, j(\omega_0 - \omega_m)t \tag{5.6}$$

These three phasors define three frequencies, one of which is the carrier frequency. The other two are known as *sidebands*, and $(\omega_0 + \omega_m)$ is the *upper sideband* while $(\omega_0 - \omega_m)$ is the *lower sideband*. This is a most important concept since it defines the *bandwidth* of the AM signal, which is $2\omega_m$, or twice the modulating-signal bandwidth.

Having expressed the signal in terms of a set of phasors, various other properties can now be derived. For instance, we may easily calculate its power by adding together the power carried by each phasor (see section 4.1.2).

Although complete, this set of phasors is somewhat difficult to visualise, as indicated in figure 5.2(a)! However, a simpler picture emerges when we plot just the *envelope* function $\mathbf{A}(t)$ in figure 5.2(b), which is real. Each phasor is now given relative to the carrier phasor $\exp \, j(\omega_0 t)$, which is a common factor in each phasor.

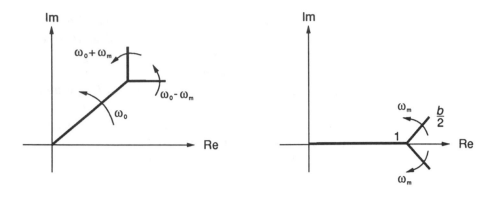

(a) **Positive-frequency phasors**   (a) **Envelope function phasors**

Figure 5.2   Amplitude-modulated signal phasor diagrams

Although visually complicated, the phasor model is mathematically accurate and complete, so is used extensively in analysis. In many practical cases however, particularly where digital signal processing or signal simulation is involved, we use the envelope as a complete model of the information-bearing part of the signal—the important part. The carrier frequency ω_0 is of course important in the technology used to transmit and demodulate the AM signal, but adds nothing to our understanding of the behaviour of the signal, since it is constant. We frequently ignore it for analysis purposes.

● **Example 5.1**   Examine the structure of an amplitude-modulated (AM) signal, with a carrier frequency of 10 kHz, a modulation (signal) frequency of 200 Hz, and a depth of modulation 0·6.
   The AM signal is then

$$x(t) = [1 + 0 \cdot 6 \cos(2\pi \cdot 0 \cdot 2t)] \cdot \cos(2\pi \cdot 10t)$$

where t is in msec.
   The envelope is $a(t) = [1 + 0 \cdot 6 \cos(2\pi \cdot 0 \cdot 2t)]$

with maximum and minimum values of 1·6 and 0·4 respectively. The signal power is determined from the set of phasors which make up the total signal. Working from equation 5·6, we see that the total power is

$$P = 2[(1/2)^2 + 2 \cdot (0 \cdot 6/4)^2] = 0 \cdot 59 \text{ V}^2$$

Notice that the total power could also be calculated from the integral of $x^2(t)$, but that the phasor route is simpler.
   The upper and lower sidebands are 10·2 and 9·8 kHz respectively.   ●

5.1.3 Generation and demodulation

Generation of the AM signal is just a matter of multiplying together the envelope and the carrier signal as indicated by equation 5.1, and which is called the *modulation* process. We have already covered the theory behind this operation in section 4.3.1.

Recovery of the envelope component, or the *demodulation* process, is carried out similarly by using the frequency-changing principle of section 4.3.2, but with slight modification. Suppose first, that the input signal is the phasor:

$$\mathbf{x}(t) = \mathbf{A}(t) . \exp j(\omega_0 t) \tag{5.7}$$

After multiplication by $w(t) = \cos(\omega_0 t)$, the result is

$$\mathbf{y}(t) = [\mathbf{A}(t)/2] [1 + \exp j(2\omega_0 t)] \tag{5.8}$$

This signal therefore contains two *carriers*, at frequencies of 0 and $2\omega_0$ rad/sec, both of which are modulated by the envelope $\mathbf{A}(t)/2$. The lowpass filter of figure 4.9 eliminates the twice-carrier-frequency term, leaving a result equal to the original envelope signal. If the input signal is a real sinusoidal signal, instead of the corresponding complex phasor model, then

$$x(t) = a(t) . \cos(\omega_0 t) \tag{5.9}$$

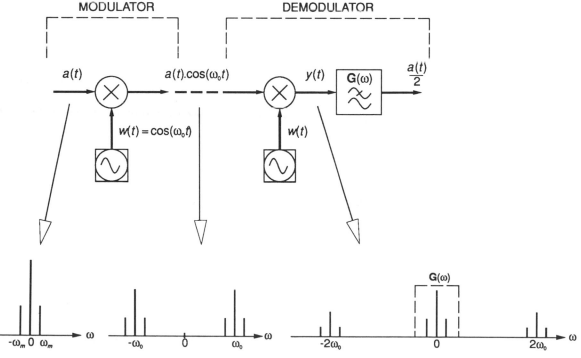

Figure 5.3 AM modulation and demodulation

83

and
$$y(t) = [a(t)/2] [1 + \cos(2\omega_0 t)] \tag{5.10}$$

Carriers now exist at frequencies of 0 and $\pm 2\omega_0$ rad/sec. The end result is the same as for the phasor input case, equation 5.8 and the operation of the overall system is summarised in figure 5.3.

In principle therefore, generation and demodulation of AM signals is carried out using the multiplication signal process, with some additional lowpass filtering in the case of demodulation. We shall not discuss here the technology used to implement these processes in practice, since we are concerned only with the basic principles of signal modelling and processing. However, we can assert that such processes are readily implemented in a variety of ways, and over the frequency range of, say, 10 Hz up to 100 GHz. A relatively new field is the implementation of AM principles at optical frequencies, in the order of 10^{14} Hz.

5.1.4 Some applications

The most direct application of this principle is for broadcast radio transmission, where each station broadcasts an AM signal at its own carrier frequency, and the user has a receiver which can be tuned to any one of these station frequencies. Tuning the receiver amounts only to changing the frequency of the local signal, known as the *local oscillator* frequency.

In practice, the technology of such receivers is considerably more complicated than defined by this simple principle, and some envelope detectors operate in a different way than this, but the overall principle is always similar.

As well as the medium of radio, AM is used widely to squeeze many different transmission channels over a cable system. When several distinct signals are to be conveyed between two fixed points, it is possible to transmit several carrier frequencies over one cable, each modulated with a different signal, and so avoid the necessity of laying extra cables or using additional conductor-pairs.

Section I.4.1 of the Introduction outlined the principle of a certain telecontrol system, and we shall use this example to illustrate the technique. The example relates to the control of an electrical circuit-breaker, at an *Outstation* situated some distance from a *Control Centre*. Three information channels are required, one from Control Centre to Outstation controlling the circuit-breaker, and two from Outstation to Control Centre to indicate the position of the breaker and to measure the value of current at the circuit-breaker.

We are not yet going to discuss the form of information-bearing signals which are to be used, but will just consider how these three channels of information can be combined on one cable. Clearly the AM principle could be used, and we allocate somewhat arbitrarily, the carrier frequencies given in Table 5.1.

Table 5.1 *Telecontrol system channel allocation*

Direction	Purpose	Carrier freq.
CC to Outstation	ON/OFF control	1·5 kHz
Outstation to CC	ON/midway/OFF indication	3·0 kHz
Outstation to CC	Current measurement	10 kHz

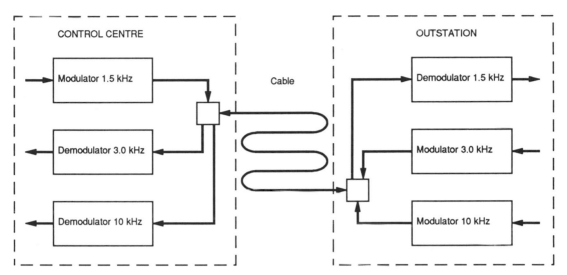

Figure 5.4 Telecontrol transmission system

Figure 5.4 shows the overall layout of the transmission system, using these carrier frequencies. Thus, using a different carrier frequency for each information channel, a single pair of cable conductors is able to carry three independent signal channels. Many more channels are possible in practice, and 12 or 60 voice channels on a single pair of wires has been common.

5.1.5 Discrete-time

Amplitude-modulated signals can be processed by digital systems, when expressed in discrete-time form. However there is a particular problem when using a digital computer to simulate or process an AM signal.

The original signal is

$$x(t) = a(t) \cdot \cos(\omega_0 t) \tag{5.11}$$

where $a(t) = 1 + b \cdot \cos(\omega_m t)$.

In discrete-time form, with an interval between samples of T_S,

$$x(n) = a(n) \cdot \cos(\omega_0 T_S n) \tag{5.12}$$

where $a(n) = 1 + b \cdot \cos(\omega_m T_S n)$.

So in principle there is no reason why we should not calculate these values for our simulation. However, examine the two signal frequencies ω_0 and ω_m. The number of sample values which are needed for each period of the sinusoid is given by

$$N_0 = 2\pi / \omega_0 T_S \qquad\qquad\qquad (5.13)$$
$$N_m = 2\pi / \omega_m T_S \qquad\qquad\qquad (5.14)$$

and $\quad N_0 / N_m = \omega_0 / \omega_m \qquad\qquad\qquad (5.15)$

In the real world, there is usually an enormous difference between these two frequencies. The carrier frequency ω_0 corresponds to, say, >1 MHz, whereas for normal speech applications, the modulating frequency ω_m is less than 3 kHz, giving the ratio N_0 / N_m of $>330 : 1$. Suppose that we wish to calculate, say, 10 sample values for each period of the modulating frequency; then in order to define the modulated-carrier waveform properly, for each period of the modulating signal, we would have to calculate 3300 samples in total.

Remember our comment in section 5.1.2 that the carrier frequency is constant and therefore conveys no useful information, which means that we can safely ignore it for analytical purposes. We do the same here, and deal only with the *envelope* function $a(t)$, which reduces our computational load by a factor N_0 / N_m.

● **Example 5.2** Consider the discrete-time form of the signal in example 5.1. Take first, an interval between samples T_S of 20 μs, giving 5 sample values for each period of the carrier frequency.

The signal then becomes

$$x(n) = [1 + 0 \cdot 6 \,.\, \cos(0 \cdot 025 n)] \,.\, \cos(1 \cdot 26 n)$$

In order to calculate the values for a time span of 100 ms, or 20 periods of the modulating signal, 5000 calculations must be done, which cover 1000 periods of the carrier frequency.

However, if the envelope alone is simulated, the effective sampling interval T_S can be increased to say 1 ms, allowing 5 sample intervals for each period of the modulating signal. The signal can now be evaluated over a time span of 100 ms, with only 100 calculations, according to the expression for the envelope of

$$a(n) = 1 + 0 \cdot 6 \,.\, \cos(1 \cdot 26 n)$$

Note that the figure of 5 sample values per sinusoidal period has been chosen arbitrarily—we shall discuss the limit of this parameter in chapter 6. ●

5.2 Angle-modulated signals

5.2.1 Angle-modulated phasors

Although we shall not investigate angle modulation in as much detail as amplitude modulation, it is still important to show that the phasor concept can be used here too with great effect. Magnitude and angle are the two attributes of the phasor which can be modulated or varied. Angle modulation of a sinusoidal signal may be described in this form:

$$x(t) = \cos[\omega_0 t + \varphi(t)] \tag{5.16}$$

The corresponding phasor is

$$\mathbf{x}(t) = \exp j[\omega_0 t + \varphi(t)] \tag{5.17}$$

Such a signal plainly has constant magnitude, unity in this case, but a varying angle given by the arbitrary function $\varphi(t)$. The phasor model is therefore of a constant magnitude phasor, rotating at a rate which is varying with time.

Now the rotational velocity of the phasor at any instant of time is given by the rate-of-change of the phasor angle. Thus

$$\omega(t) = \omega_0 + \frac{d\varphi(t)}{dt} \tag{5.18}$$

This general relationship is most important, since it defines the *instantaneous frequency* of the signal. If the phase term $\varphi(t)$ is constant, then the instantaneous frequency is simply ω_0; but if not, then the frequency fluctuates as a function of $\varphi(t)$. The effect is modelled clearly by the variation in angular velocity of the phasor, and so the phasor is a vital link in understanding this process, which can be quite difficult to visualise.

Let us take a specific example, where the angle $\varphi(t)$ changes according to a sinusoidal function:

$$\varphi(t) = b \cdot \cos(\omega_m t) \tag{5.19}$$

Applying this condition to equation 5.18, we discover that the instantaneous frequency is

$$\omega(t) = \omega_0 - b \cdot \omega_m \cdot \sin(\omega_m t) \tag{5.20}$$

Thus, the instantaneous frequency in the case has a sinusoidal variation about the average value of ω_0 rad/sec, with *maximum deviation* of $\pm b \cdot \omega_m$ rad/sec. Figure 5.5 illustrates the signal waveform for this example.

In popular jargon, such a modulation scheme is often referred to as *frequency modulation (FM)*, and there is some support for this name since equation 5.20 shows that the frequency is indeed being modulated. In practical systems, for reasons which we need not go into now, the carrier frequency is often modulated directly, which is true Fm, but *phase*

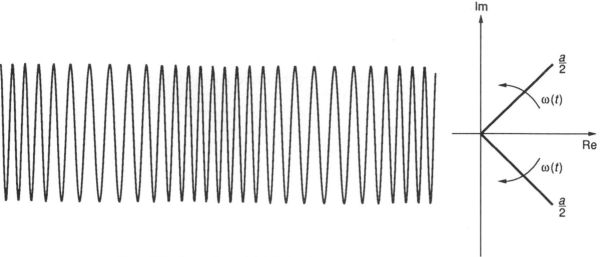

Figure 5.5 An angle-modulated signal

modulation (PM) is also used, and sometimes in conjunction with FM. So, because the precise situation is more complicated than we admit at present, it is best to stick to the generic term *angle modulation* to describe both or either of these schemes.

One further point can be brought out, concerning the phasor model. Equation 5.17 can be re-written as follows:

$$\mathbf{x}(t) = \mathbf{A}(t) \cdot \exp \, \mathrm{j}(\omega_0 t) \qquad (5.21)$$

where $\mathbf{A}(t) = \exp \, \mathrm{j}[\varphi(t)]$.

In this form, we have separated the modulated-signal phasor into a unit-phasor with angular velocity ω_0 rad/sec, and a *complex envelope* $\mathbf{A}(t)$ which is $\exp \, \mathrm{j}[\varphi(t)]$. The complex envelope is a generalisation of the complex magnitude which we have used before, and imparts angle modulation to the unit-phasor rather than magnitude modulation. In practice we can use the complex envelope to model modulated signals of arbitrary complexity and possessing both amplitude and angle modulation simultaneously; but we shall not pursue that line now. It is enough to show that the complex envelope has meaning, and to sow the seeds for further understanding at a later date in chapter 12.

Modulated-signal power can be calculated easily, since the model is of two contra-rotating conjugate phasors, each of magnitude $(1/2)$. The total power is therefore $2 \cdot (1/2)^2$ or $0 \cdot 5$ V^2 for unity magnitude. This is of course the same as for an unmodulated sinusoid.

● **Example 5.3** Examine an angle-modulated signal, possessing similar parameters to that in example 5.1. Thus, the carrier frequency is 10 kHz, and the modulating frequency is 200 Hz. The input signal has a peak magnitude of 2 V, and the constant b corresponds to $0 \cdot 3$ kHz/V.

Referring to equation 5.20, we calculate that the maximum deviation is $2 \cdot 0 \times 0 \cdot 3$ kHz, or $0 \cdot 6$ kHz. The instantaneous frequency therefore fluctuates between $9 \cdot 4$ and $10 \cdot 6$ kHz.

These latter parameters may also be expressed in terms of angular velocity, and become $59 \cdot 1$ krad/sec and $66 \cdot 6$ krad/sec respectively.

The signal power is $0 \cdot 5$ V^2, if the carrier is unit peak magnitude. ●

5.2.2 Generation and demodulation

Angle-modulated signals are not generated with simple operations like multiplication, nor are they demodulated with simple frequency-changing techniques as for AM. In fact the whole process of angle modulation is altogether more messy and elaborate than for AM, because it is a non-linear operation, so we shall not investigate this in any detail now.

In principle however, generation is simple if a *voltage-controlled oscillator* is available. This is a sinusoidal generator whose frequency is proportional to some control input voltage. Such devices are quite easy to construct using conventional electronic techniques, and discrete-time versions of these can also be programmed.

Demodulation requires a circuit or system which gives an output voltage proportional to frequency. Again, there are several techniques available, but it is not our intention to investigate them here. Figure 5.6 shows the schematic of such a transmission system in block diagram form.

Having generated an angle-modulated signal at a specific carrier frequency, then this signal can be shifted to a different carrier frequency by an adaptation of the frequency-changing techniques introduced in section 4.3.2. Consequently, multi-channel transmission systems after the style of figure 5.4 can also be made with angle-modulated signals, with certain intrinsic advantages over AM.

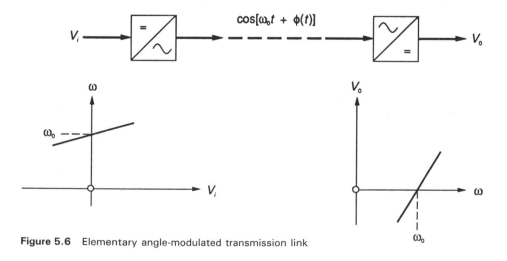

Figure 5.6 Elementary angle-modulated transmission link

5.2.3 Some applications

As an example of this technique, consider the telecontrol example introduced in section I.4.1 and referred to again in section 5.1.4. One of the quantities to be transmitted is a signal conveying the measurement of an electric current, and when talking about AM we allocated a notional frequency of 10 kHz to this signal channel.

A simple method of transmitting the desired information, is to frequency-modulate the carrier frequency of 10 kHz. For instance, a range of 10·0 to 10·4 kHz might be used, so that the deviation is 4 Hz/A. This makes a particularly straightforward system, more so if the signal variable is changing slowly.

Before leaving this topic, figure 5.7 shows an approximate method of angle modulation which is often used to generate such signals for a broadcast transmitter. We are not so concerned to learn about this particular method as to use it to illustrate the extremely valuable role played by complex phasors in a system of this kind.

In figure 5.7, the signal source at ω_m rad/sec represents the information-bearing signal, while the source at ω_0 rad/sec is the carrier-frequency. The output signal is

$$y(t) = -\sin(\omega_0 t) + b \cdot \cos(\omega_m t) \cdot \cos(\omega_0 t) \tag{5.22}$$

While this equation may be evaluated as it stands, it can be examined much more easily through its phasor representation. Thus, express $y(t)$ in terms of its positive-frequency phasor model:

$$\mathbf{y}(t) = (1/2)\exp \, \mathrm{j}(\omega_0 t + \pi/2)$$
$$+ (b/4)\{\exp \, \mathrm{j}[(\omega_0 + \omega_m)t] + \exp \, \mathrm{j}[(\omega_0 - \omega_m)t]\,\} \tag{5.23}$$

This too may be simplified by using just the complex envelope of this signal, since $\exp \, \mathrm{j}(\omega_0 t)$ is a common factor:

$$\mathbf{y}(t) = \mathbf{A}(t) \cdot \exp \, \mathrm{j}(\omega_0 t) \tag{5.24}$$

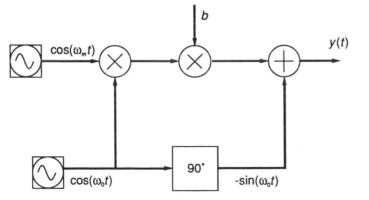
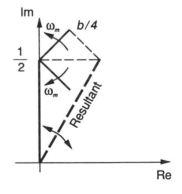

Figure 5.7 An angle-modulated signal generator

and

$$\mathbf{A}(t) = j/2 + (b/4) . \exp j(\omega_m t) + (b/4) . \exp j(-\omega_m t)$$
$$= (b/2) . \cos(\omega_m t) + j/2 \qquad (5.25)$$

This state of affairs is illustrated by the phasor diagram of figure 5.7, which shows that the resulting signal is modulated in both angle and amplitude. Reference to the diagram allows the maximum deviation of both angle and amplitude to be determined:

$$|\mathbf{A}(t)| = (1/2)\sqrt{[1 + b^2 . \cos^2(\omega_m t)]}$$
$$\arg[\mathbf{A}(t)] = \arctan\{1/[b . \cos(\omega_m t)]\} \qquad (5.26)$$

If $b \ll 1$ then the amplitude fluctuations will be small. The maximum phase deviation is $\pm \arctan[1/b]$.

The major point being made is that this fairly complicated situation has been analysed in a straightforward manner by the phasor concept. QEF!

5.2.4 Discrete-time

There is little further to say under this heading, except that the discrete time form of angle-modulated signals may be analysed as simply as for the AM case. Again, the carrier frequency is an unnecessary constant as far as the analysis goes, so we would normally deal with the complex envelope alone.

Although such operations are not intrinsically difficult, we shall not pursue this line of thought until it is necessary in order to analyse a particular application. Chapter 12 deals more generally with the idea of processing complex signals

STUDY QUESTIONS

1   Sketch the phasor diagrams and the corresponding waveforms for the following AM signals:

    a) $x(t) = [1 + 0 \cdot 4 . \cos(4t)] . \cos(24t)$
    b) $x(t) = [1 + 0 \cdot 5 . \cos(2t)] . \cos(24t)$
    c) $x(t) = \cos(2t) . \cos(24t)$

2   Calculate the mean signal power in each of the cases in question 1.

3   Calculate all the relevant parameters for an amplitude-modulated signal with a carrier-frequency of $1 \cdot 2$ MHz, a modulation (signal) frequency of 3 kHz, and a depth of modulation of $0 \cdot 3$.

4   The signal in question 3 is to be simulated on a digital computer. Deduce suitable sampling intervals T_S assuming a minimum of 5

sample values per period, for the following cases:

    a) Complete AM signal simulated

    b) Envelope function only simulated.

How many sample values must be calculated in each case, in order to simulate a signal record length of 10 ms?

5 Define 'instantaneous frequency' for an angle-modulated signal. Derive the instantaneous frequency for the following signal:

$$x(t) = 3 \cdot 5 \,.\, \cos[2000t - 2 \cdot 5 \,.\, \sin(10t)]$$

6 A certain angle-modulated signal has a carrier frequency of 10 MHz and a modulated frequency of 3 kHz. The form of modulated signal is given by

$$x(t) = 3 \,.\, \cos[\omega_0 t + 5 \,.\, \cos(\omega_m t)]$$

Determine the maximum deviation of frequency from 10 MHz and the signal power.

7 Write an expression for the 'complex envelope' of the signal in question 6.

6 Sampling and Analog-to-Digital Conversion

OBJECTIVES

To introduce the signal process of converting between analog (continuous-time) and digital (discrete-time) domains, and vice-versa, with a view to providing information which can immediately be used in practice.

To examine some of the constraints imposed by this process, in particular

    a) Sampling frequency
    b) Sample-and-hold frequency distortion
    c) Quantisation.

COVERAGE

The conversion process is first introduced in broad outline, distinguishing clearly between the different signal domains and their characteristics. Discussion ranges over some applications, and reasons why this process is so important.

The inherent constraint imposed by the sampling rate is derived, and the sampling theorem developed, which leads on to discussion of its significance for signals internal and external to the digital process. The sample-and-hold function is explained, and its effect on the frequency response of the system is presented.

Digital quantisation is introduced, quantisation noise defined, and the effect of this on signal-to-noise ratio is calculated.

We have introduced the distinction between continuous-time signals that occur in the physical world, bounded by the normal laws of physics, and discrete-time signals which occur with certain methods of measurement or processing. In this chapter we examine the interface between these two domains.

Most naturally occurring signals are continuous functions of time; for example speech, video, room temperature. Signal processing, storage and transmission are increasingly carried out in discrete-time, since digital hardware is used for these operations, and is much more flexible, more stable and becoming cheaper than many analog processing circuits. Conversions to and from these different domains are frequently required, and it is this interface that we investigate now.

Note first that we are regarding the Analog-to-Digital (A-D) converter as a signal process, trying to discover how it affects the signals that it handles; we are not concerned here with the technology that accomplishes this aim. That is a different discussion altogether.

6.1   The mixed analog and digital system

6.1.1   Overview

Figure 6.1 shows an example of a continuous-time signal $x(t)$ which, when sampled at regular intervals, forms the sample values $\{x(nT_S)\}$. An analog-to-digital (A-D) converter must therefore carry out this sampling process, and generate a succession of sample values $x(n)$ that can be held in digital storage. Digital representation of these signal samples is limited by the number of bits in the digital words, so a b-bit word can store any one out of a set of only 2^b possible word values. This imposes a restriction on the process, as we shall see in section 6.4.

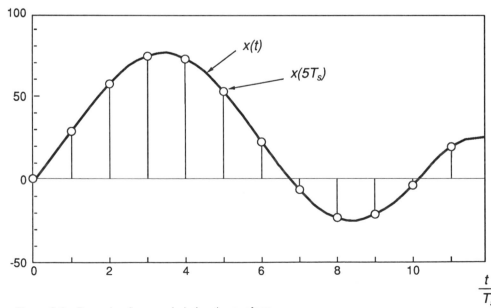

Figure 6.1   Example of a sampled signal waveform

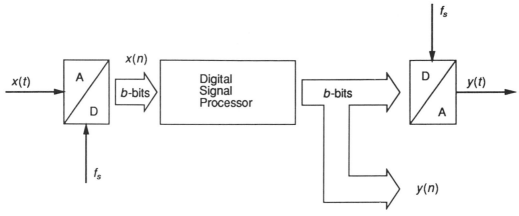

Figure 6.2 Analog interfaces to a digital signal processor

The overall picture is then as in figure 6.2. Notice the following points:

a) The input signal is a function of continuous-time, and is usually a voltage signal possessing the qualities of true energy and power.
b) The digital signal is a collection of numbers, and strictly has no true energy or power, although we interpret the quantity $x^2(n)$ as 'power' in the sense that it is a useful signal property.
c) The output signal may be either digital or analog, depending upon the outcome desired from the process.

The digital hardware shown between the A-D and D-A converters, may be a microprocessor, a mainframe computer, or a Digital Signal Processor (DSP) which is purpose built for real-time handling of signals. General-purpose digital processors do not usually have adequate speed to carry out signal processing tasks in real time.

Alternatively, the digital hardware may be a radio or line transmission system or a storage system like magnetic tape.

Processes which utilise a digital output, include

a) Analysis of a signal into its constituent frequency components.
b) Correlation or comparison of an incoming signal with a number of stored signals, to see which one it is most like.
c) Reception of a transmitted and coded signal, and interpretation of what data has been sent.

In these cases, the output $y(n)$ will be in logical form, and will be added to the system database storage or channelled to one of the system output devices like a printer or VDU screen. The output is in quite a different form to the input analog signal.

Processes which require an analog output include

a) Linear filtering, where the output signal is a filtered version of the

95

input signal, and will then be passed forward for further processing in the analog or continuous-time world.

b) Modulation and demodulation for on-line applications.

In these cases, the digital process is part of a larger system for transmitting information, for storing information, or for applications like audio or video reproduction where the result is finally 'processed' by a human observer.

The prime reasons for contemplating digital hardware for signal processing are

a) Programmability of the process algorithm, which can be changed to develop a better process, or may be changed on-line to *adapt* to a new situation.

b) Repeatability, since the performance of the hardware is not affected by temperature or ageing effects.

c) Cost, since VLSI* digital processing devices are becoming extremely cost-effective, and even general-purpose DSP devices offer the same hardware for a wide variety of applications. Production problems are generally eased by adopting digital solutions.

d) Complexity, since some digital processes could not be conceived in analog terms, for example picture processing or adaptive filtering.

6.1.2 An example — the digital oscilloscope

The digital oscilloscope is becoming commonplace and has several advantages over the conventional analog oscilloscope. Broadly, the diagram of figure 6.2 applies to this instrument, with the inner digital process being storage. However, the sampling rates at input and output are usually quite different, offering the opportunity to *time-scale* the signal.

a) **Transient operation** Consider, for example, the signal due to shock excitation of a structure like a car suspension system, which gives a transient signal, dying away in about 1 sec. Sampling and storing this signal at a rate of, say, 1 kHz will store 1000 sample values over the interval of 1 sec.

The stored signal is now permanent, and so can be replayed as many times as we please, and at what speed we choose. Replaying it repetitively at a sampling rate of, say, 100 kHz enables the transient to be observed on a conventional oscilloscope with a time-base of 10 ms. Replaying it once at a sampling rate of, say, 10 Hz enables the transient to be committed to hard-copy via an X-Y plotter.

A further asset is that the stored signal may be analysed directly, and at leisure, to discover the maximum and minimum values, the frequency content and so on. The digital oscilloscope can also be an analysing instrument.

* VLSI stands for Very Large Scale Integration, and means a semiconductor integrated circuit which contains several hundred thousand components.

b) **Repetitive sampling**   Sampling a periodic signal enables us to reduce the effective period of the signal, and so display it on an oscilloscope with a lower bandwidth. This is the *stroboscopic* effect, well-known as a technique of 'freezing' moving machinery.

The diagram in figure 6.3 shows the principle. The original signal $x(t)$ has a bandwidth in excess of that which the oscilloscope can display, but the A-D converter has a sampler which can take a virtually instantaneous sample value. Taking this sample once per period of the waveform, at a delay time which is incremented through the period, the oscilloscope screen displays the entire waveform, but in a relaxed time scale.

For instance, let the signal period be T_P, and let the sample interval T_S be $(T_P + t_i)$. The nth sample taken is therefore

$$x(nT_S) = x(n(T_P + t_i)) \tag{6.1}$$

However, the signal is periodic, so

$$x(T_P + t) = x(t) \tag{6.2}$$

and hence

$$x(nT_S) \equiv x(nt_i) \tag{6.3}$$

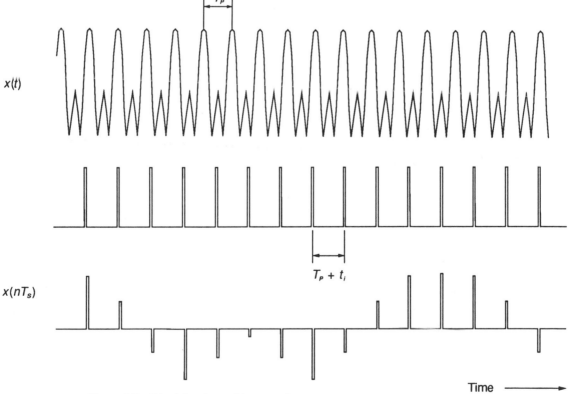

Figure 6.3   Principle of repetitive sampling

97

The time scale of the signal has therefore been changed and the oscilloscope displays the original signal but on a relaxed time scale. The oscilloscope receives samples at intervals of approximately T_P, while the effective time between them is t_i. Provided that the signal is truly periodic, then samples may be taken at intervals which are multiples of T_P in order to increase the time between samples and make them more easily handled. If T_P is such a value that the oscilloscope can display it, then continuous real-time sampling can take place.

This arrangement is very flexible, and enables very wideband signals to be displayed with a modest oscilloscope. For instance, oscilloscopes with real-time bandwidths of 20–50 MHz are common, but bandwidths of 100–300 MHz are much more expensive, and 350 MHz represents the upper limit without considerable expense. Real-time bandwidths of 500–1000 MHz are possible, but at great expense. However, sampling oscilloscopes, such as we have described here, can offer sampling bandwidths in the range 1–20 GHz, although these instruments are certainly not cheap!

6.1.3 Constraints on the A-D conversion process

Intuitively, the most accurate way of representing an analog signal in digital form is to use the shortest possible sampling interval T_S, and a digital word with the maximum number of bits b. In practice however, these parameters are restricted.

The sampling interval T_S, and hence the sampling frequency f_S ($= 1/T_S$), needs to be limited for two reasons:

a) Higher-speed sampling systems are more expensive, since all electronic circuits increase in difficulty as the frequency is increased.

b) The number of samples generated in a given time increases in proportion to the sampling frequency f_S, so there is a danger of swamping the signal processing system with unnecessary data.

So, what is the *minimum* sampling frequency that can be used? This we explore in section 6.2, where we derive the Sampling Theorem.

Likewise, the number of bits in the digital word is subject to some limitation, since the cost and complexity of the digital hardware increase roughly in proportion to the length of the words. Again, we wish to choose the minimum acceptable word size, b bits. Criteria for assessing this factor are given in section 6.4, where we examine the issue of *quantisation*.

Of course, in any particular case we may well use non-optimum parameters for convenience, but the assessments which follow will enable us to make a good judgement as to what limiting values are essential.

One further topic which we shall investigate is the effective frequency response of the A-D and D-A processes. Although at first it looks as though there will be no significant distortion of the signal if we obey the constraints mentioned above, in fact there is some frequency-shaping imposed by the *sample-and-hold* operation at the output of the D-A converter. This we comment on briefly in section 6.3.2.

6.2 Sampling frequency limit

6.2.1 An absolute limit—the Nyquist frequency

Simple considerations will enable us to determine the minimum sampling frequency f_s ($\omega_s = 2\pi f_s$), or alternatively the maximum allowable input frequency for a given sampling frequency. Let the input signal which is to be sampled be the unit phasor, which allows complete generality since this component is the basis of all sinusoidal signals (chapters 3, 4, 5):

$$x(t) = \exp j(\omega t) \tag{6.4}$$

When sampled at f_s Hz, with an interval of T_s sec ($f_s T_s = 1$), the corresponding discrete-time signal becomes

$$x(n) - \exp j(\omega T_s n) \tag{6.5}$$

This discrete-time phasor steps forward by an angle ωT_s each time a sample

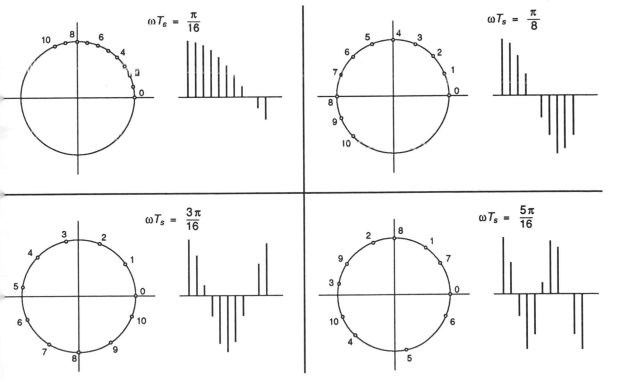

Figure 6.4 Discrete-time phasor at several sampling frequencies

99

occurs. If the sampling frequency is increased, the angle gets smaller and the waveform is described more adequately, as indicated in figure 6.4 which shows the locus of the stepped phasor and its real component. (Remember that a real cosine signal is modelled by two conjugate phasors of this kind, and hence its magnitude is equal to twice the real part of each phasor.)

However, if the sampling frequency is reduced so as to produce a more acceptable sampling system, what lower limit is there on f_S? Alternatively what upper limit is there on the phasor jump angle ωT_S?

There are two distinct limits, dictated by the nature of circular functions.

Limit 1 $\omega T_S < 2\pi$

This is a common-sense limit, since sine, cosine and tangent functions are periodic in 2π. Figure 6.5 shows the situation where $\omega T_S = 2\pi + \delta\varphi$. Clearly

$$\exp j[(2\pi + \delta\varphi)n] = \exp j(\delta\varphi n) \tag{6.6}$$

As far as the discrete-time signal is concerned then, a phasor jump angle of $\delta\varphi$ can represent any input signal having a frequency of the form $(r\omega_S + \delta\varphi/T_S)$, where r is an arbitrary integer. Unless the input signal frequency is prevented from being above ω_S, there is ambiguity in the discrete-time signal.

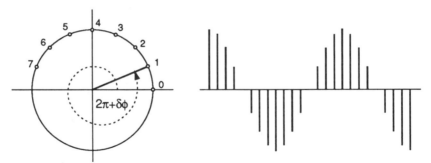

Figure 6.5   Sampling near f_S

For real input signals which are modelled by the sum of two conjugate phasors, we can assert that an input frequency of $(\omega_S \pm \delta\omega)$ cannot be distinguished from an input frequency of $\delta\omega$ rad/sec. This region of input frequency must therefore be avoided by ensuring that $\omega < \omega_S$.

$$\cos[(\omega_S \pm \delta\omega)T_S n] = \cos(\delta\omega T_S n) \tag{6.7}$$

since $\omega_S = 2\pi/T_S$.

(This is the principle of repetitive sampling which is used to good effect in the digital oscilloscope; section 6.1.2.)

Limit 2 $\omega T_S < \pi$

Figure 6.6 illustrates the case where $\omega T_S = \pi + \delta\varphi$, and shows the peculiar real waveform which results from this case. We note that

$$\exp j[(\pi + \delta\varphi)n] = \exp j[-(\pi - \delta\varphi)n] \tag{6.8}$$

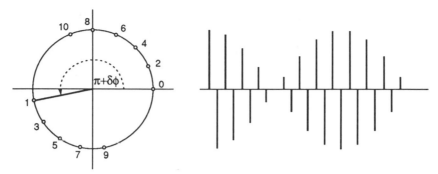

Figure 6.6 Sampling near $f_S/2$

The angle may be measured in either of two directions, so that a jump angle of $(\pi + \delta\varphi)$ may represent an input frequency of $(\omega_S/2 + \delta\varphi/T_S)$ or one of $-(\omega_S/2 - \delta\varphi/T_S)$. This ambiguity can only be resolved if the input frequency is limited so that $|\omega| < \omega_S/2$ or $|\omega T_S| < \pi$.

For real signals, we combine two conjugate phasors and obtain for this case:

$$\cos[(\omega_S/2 \pm \delta\omega)T_S n] = (-1)^n . \cos(\delta\omega T_S n) \tag{6.9}$$

So if we have a discrete-time signal like that on the right-hand side of equation 6.9, we have no idea to whether the input signal frequency was $(\omega_S/2 + \delta\omega)$ or $(\omega_S/2 - \delta\omega)$. The ambiguity can only be resolved by restricting $|\omega|$ to $\omega_S/2$.

Figure 6.7 shows the complete set of phasors corresponding to the case of equation 6.9, and shows that these two input signals do generate the same discrete time signal.

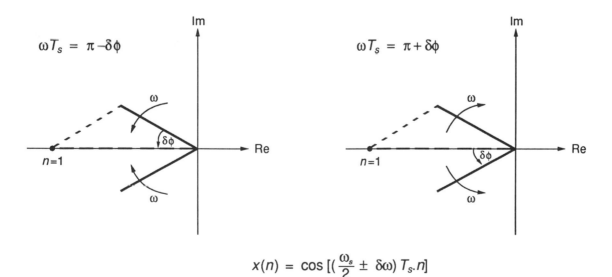

$$x(n) = \cos[(\frac{\omega_s}{2} \pm \delta\omega) T_s . n]$$

Figure 6.7 Phasor diagrams for sampling near $f_S/2$

$f_S/2$ (or $\omega_S/2$) is known as the *Nyquist frequency*, or the *folding frequency* because of the symmetrical properties of equation 6.9. It represents the upper limit of input frequency, necessary to avoid ambiguity in the corresponding discrete-time or sampled signal, and gives rise to the Sampling Theorem in its simplest form:

A signal having a maximum frequency of W Hz, or a bandwidth of W Hz, must be sampled at a rate of at least $2W$ samples/sec in order for it to be represented accurately in sampled form.

6.2.2 The Nyquist frequency and aliasing

The Nyquist frequency has been derived as that upper limit on input signal frequency which avoids ambiguity in the corresponding discrete-time signal. For the purposes of discussion, we assumed that it was possible to sample a *complex* input signal. Although we shall show in Chapter 12 that signals are often processed in complex form, an individual analog-to-digital converter must operate on real signals, and if an input signal violates the Nyquist frequency limit, then curious effects occur.

If the input signal is allowed to have frequency components which are above the Nyquist frequency, then these components are *folded back* or *aliased* to below the Nyquist frequency, and give a form of distortion which is most unpleasant for audio applications and quite destructive for video applications.

In order to see the full effect of aliasing, consider what happens when the input signal is $\cos(\omega t)$. Provided that $|\omega| \leqslant \omega_S/2$, then no problems occur, and the input signal frequency is faithfully reproduced in the discrete-time signal. However, if $\omega = (\omega_S/2 + \delta\omega)$, then the input signal and the discrete-time signal may be written as

$$2 . x(t) = \exp j\,[(\omega_S/2 + \delta\omega)t] + \exp j\,[-(\omega_S/2 + \delta\omega)t] \qquad (6.10)$$

$$2 . x(n) = \exp j\,[-(\pi - \delta\omega . T_S)n] + \exp j\,[(\pi - \delta\omega . T_S)n] \qquad (6.11)$$

The jump angles in the two conjugate phasors correspond to *apparent* input frequencies of $\pm(\omega_S/2 - \delta\omega)$, instead of the actual input phasors at $\pm(\omega_S/2 + \delta\omega)$. The apparent frequency is now $(\omega_S - \omega)$.

Table 6.1 summarises the rules that apply in various cases, based on equations 6.10, 6.11 and 6.7, and figure 6.8 illustrates them in graphical form.

The *apparent* frequency is the one which will be observed when the discrete-time signal is converted back into continuous-time through a digital-to-analog converter, and will not be the same as the input signal frequency if the Nyquist limit is violated.

In order to prevent this confusion, an *anti-alias* filter is usually placed at the input of an A-D converter. This filter is lowpass with a bandwidth of $\omega_S/2$ rad/sec or $f_S/2$ Hz. Frequency components which otherwise would

Table 6.1 *Effect of sampling*

Region	Range	Apparent frequency
Normal	$0 < \omega < \omega_S/2$	ω
Alias	$\omega_S/2 < \omega < \omega_S$	$\omega_S - \omega$
Repeat	$\omega_S < \omega < 3\omega_S/2$	$\omega - \omega_S$
	$3\omega_S/2 < \omega < 2\omega_S$	$2\omega_S - \omega$

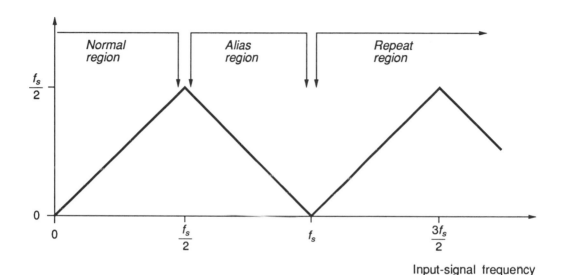

Figure 6.8 Frequency aliasing effect for a sampled system

cause aliasing are therefore excluded from the input to the converter. Operation is limited to the *normal region* in the graph, and there is no ambiguity about the frequency of the signal.

● **Example 6.1** A digital system has a sampling frequency of 10 kHz. Determine the output frequency for the following input frequencies:

a) 3 kHz, *b*) 6 kHz, *c*) 7 kHz, *d*) 13 kHz

The Nyquist frequency is equivalent to $f_S/2$, and is 5 kHz. Therefore

a) 3 kHz is below this limit, hence the output is 3 kHz.
b) 6 kHz is above this limit, hence the output is $10 - 6 = 4$ kHz.
c) 7 kHz is above this limit, hence the output is $10 - 7 = 3$ kHz.
d) 13 kHz is above the sampling frequency, hence the output is $13 - 10 = 3$ kHz.   ●

● **Example 6.2** Consider a speech-like signal, sampled at 8 kHz and consisting of the following frequency components:

Frequency	1·5	3·5	5·0	6·0	kHz
Magnitude	0·5	1·0	0·3	0·2	V rms

Determine the pattern of the output signal frequency components, and the proportion of the signal power which is aliased.

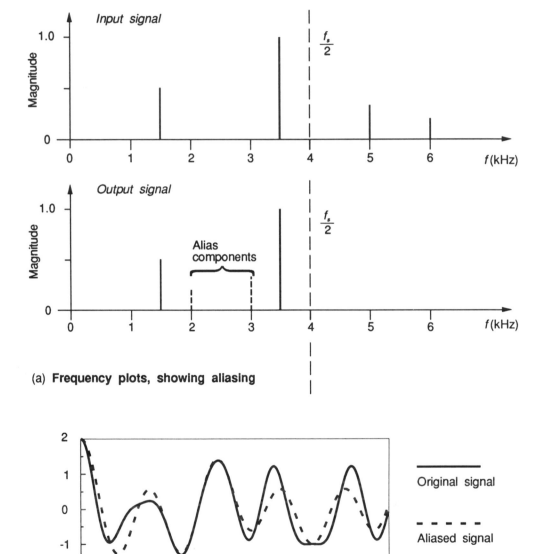

(a) **Frequency plots, showing aliasing**

(b) **Signal waveforms**

Figure 6.9 Simple example of aliasing

The Nyquist frequency is 4 kHz, so the following pattern emerges:

Input frequency	1·5	3·5	5·0	6·0	kHz
Output frequency	1·5	3·5	3·0	2·0	kHz

Input components at 5·0 and 6·0 kHz are therefore aliased, giving rise to an aliased-signal power of 0·13 V^2. The total signal power is 1·38 V^2, and so the proportion which is aliased is 0·094 or 9·4%.

Aliasing could have been avoided if

a) The input was restricted by a lowpass anti-alias filter; but that would result in signal distortion by removing two frequency components.
b) The sampling frequency was increased to greater than twice the highest input frequency, or 12 kHz.

Figure 6.9 illustrates the example graphically.   ●

6.3   Converting back into analog form

Once the digital processing of a signal has been completed, it is often necessary to reproduce the processed signal in the continuous-time or analog world outside the processsor. There are two distinct aspects to this conversion, the first relating to the Nyquist frequency limit, and the second relating to the nature of the digital-to-analog converter.

6.3.1   Output frequencies

We have concentrated so far on the principal frequency components in the discrete-time signal, produced by sampling an input signal. However, there is an infinite number of frequency components produced by the sampling process. Figure 6.10 shows a diagrammatic representation of a simple input signal and the corresponding sampled or discrete-time output signal produced from an A-D converter and a D-A converter in cascade.

Although we have shown the input signal in a $\cos(\omega t)$ form, for simplicity assume for the moment that it is possible to sample and recover an input signal of the form:

$$x(t) = \exp j(\omega_1 t) \tag{6.12}$$

The sampled values are

$$\{x(n)\} = \{\exp j(\omega_1 T_S n)\} \tag{6.13}$$

Now the nth sample value scales the magnitude of the timing pulse at time (nT_S) in the signal $s(t)$ to provide the output signal $y(t)$. Effectively then

$$y(t) = x(t) . s(t) \tag{6.14}$$

105

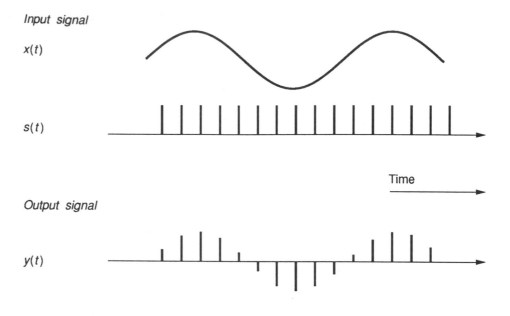

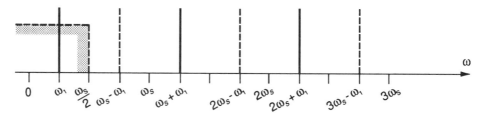

Figure 6.10 Input and output of the sampling process

Now the timing signal $s(t)$ is periodic and so may be represented by a Fourier series (see appendix A2). Since this signal has abrupt changes in magnitude when the pulse occurs, we would expect a large number of frequency components to be present. We have no need to calculate the coefficient values, so we represent them by $\{S_k\}$:

$$s(t) = \sum_{k=-\infty}^{\infty} S_k \cdot \exp j(k\omega_S t) \tag{6.15}$$

and

$$y(t) = \sum_{k=-\infty}^{\infty} S_k \cdot \exp j(k\omega_S + \omega_1)t \tag{6.16}$$

The output signal $y(t)$, after emerging from the D-A converter, is therefore represented by an infinite series of complex phasors, of frequency $(k\omega_S + \omega_1)$. This is also shown in figure 6.10.

If the input signal were $\cos(\omega_1 t)$, then an additional set of phasors would be added to the output signal, at frequencies $(k\omega_S - \omega_1)$. These are also shown in figure 6.10, as dashed lines.

A lowpass filter is therefore required at the output of a D-A converter in order to convert the sampled or discrete-time waveform into a smooth analog waveform, and this filter is known as an *interpolating* or *reconstruction* filter. It is lowpass, with a bandwidth of $\omega_S/2$ rad/sec or $f_S/2$ Hz. The complete system diagram therefore is usually as shown in figure 6.11.

The *interpolating* filter must distinguish between frequencies below the Nyquist limit, which are required, and those above the Nyquist limit which must be removed. In the case of telephone-quality speech, the sampling rate is 8 kHz and the bandwidth of the anti-alias and interpolation filters is about $3\cdot4$ kHz. These figures allow a margin of $0\cdot6$ kHz between the edge of the filter passband and the Nyquist frequency, which is necessary because realisable filters require a finite *transition band* between passband and stopband.

In the case of the audio compact-disc digital recording format, the sampling frequency is $44\cdot1$ kHz, and the required useful bandwidth is about 20 kHz. Now it turns out that a filter of any sort whose response changes rapidly from passband to stopband has a poor phase characteristic and poor transient response. Since the audio compact-disc format is used for exceptionally high-quality sound, these distortions must be avoided. The strategy is to increase the sampling rate while the signal is still in digital form, retaining the original signal bandwidth. Suppose that four times oversampling is carried out, to a sampling frequency of $176\cdot4$ kHz. The interpolation filter has now to distinguish between 20 kHz which is at the upper end of the useful audio range, and 156 kHz [i.e. (176 − 20) kHz] which is the first major interfering frequency component (see figure 6.12) Design of such a filter can be carried out to cause very little phase distortion in the frequency range 0 to 20 kHz. The technique of oversampling is somewhat more complicated than we have admitted here, but is by no means difficult.

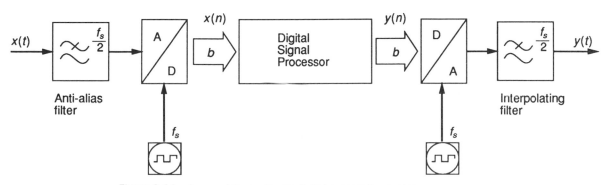

Figure 6.11 A complete analog-to-digital processing system

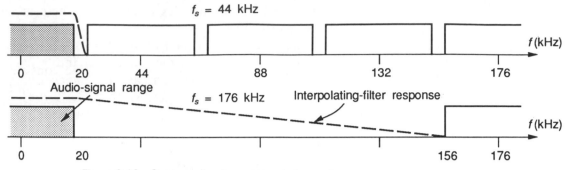

Figure 6.12 Oversampling in a compact-disc system

One further point before we leave this section. Examine the plot of frequency components in figure 6.10, and visualise what happens when the input frequency ω approaches and then exceeds $\omega_S/2$. The full-line components and the dashed-line components cross over, leading to alias distortion. This explanation of aliasing is the one most often used, and can be generalised for all forms of signal.

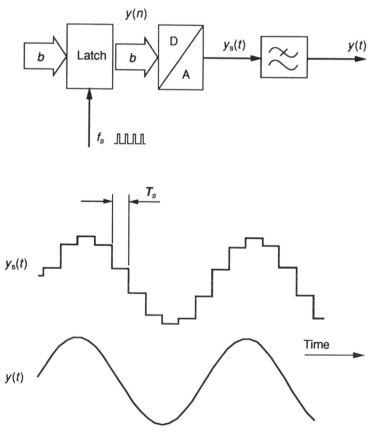

Figure 6.13 Output sample-and-hold

6.3.2 Output sample-and-hold

An important practical feature of D-A conversion is the frequency response of the *Sample-and-Hold (S-H)* operation. Although we have consistently explained that the discrete-time system represents the signal by a sequence of point values, and hence discrete samples, yet the output as communicated to the analog world is *stepped*.

Figure 6.13 shows the scenario where each output sample $y(n)$ is held in a digital register for the whole of the sampling interval T_S before being changed to the next value. The output signal before the final filter therefore maintains a constant value between sampling instants.

It can be shown (section 7.3.3) that the phasor response of the sample-and-hold process is the function $\mathbf{G}(\omega)$:

$$|\mathbf{G}(\omega)| = \frac{2}{\omega T_S} \sin\left(\frac{\omega T_S}{2}\right) \tag{6.17}$$

This phasor response is sketched in figure 6.14 and shows an attenuation of about 4 dB at the Nyquist frequency, $\omega = \omega_S/2$. When the complete signal process, including A-D and D-A conversion, is used for filtering operations, this distortion must often be compensated for.

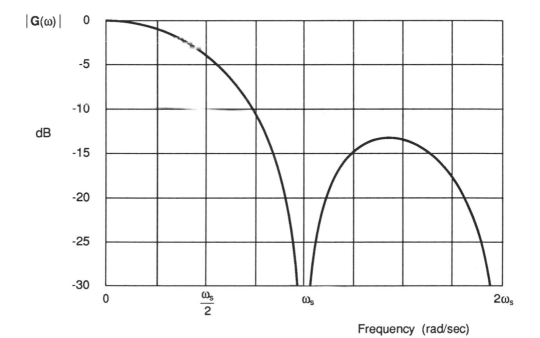

Figure 6.14 Frequency response of sample-and-hold

6.4 Quantisation

6.4.1 Discrete amplitudes

The system diagram of figure 6.11 reminds us that the sample values are represented by digital words of b bits. Although the input signal is analog and can take up any value within its range, the corresponding samples are restricted to a scale having only 2^b discrete values, and cannot represent the input signal exactly.

Put another way, the input signal before sampling is represented by a real number with the precision of several decimal places, while the signal after sampling is represented by an integer number. The digital signal is said to be quantised, and the difference between the original value and the quantised value is called a *quantisation error*.

Let us now represent this state of affairs by some simple algebra, and then illustrate more exactly how it appears in practice.

Let the input signal be $x(t)$, then the exact sample value at time nT_S is $x(n)$. For the moment we assume that the signal is restricted to positive values, so that $0 \leqslant x(n) \leqslant x_m$. For simplicity we drop the index, since we are considering only a single sample value, and re-write the condition just mentioned as

$$0 \leqslant x \leqslant x_m \tag{6.18}$$

The corresponding discrete or digital amplitude scale is defined by a b-bit binary word, and so the digital version of the sample value is an integer m, and

$$0 \leqslant m \leqslant N - 1 \tag{6.19}$$

where $N = 2^b$.

Since the signal range $0 \Rightarrow x_m$ is divided into $(N-1)$ equal intervals or *quanta*, each interval has a value Q:

$$Q = x_m / (N - 1) \tag{6.20}$$

$$m = \text{int}\,[x/Q] \tag{6.21}$$

where int [ ] denotes 'the integer value of'.

Now the integer m is the digital value held in the digital processor, but when the equivalent output signal is generated, it is scaled by the value of each quantum, Q. So the quantised output signal x_q becomes

$$x_q = mQ \tag{6.22}$$

● **Example 6.3**   Figure 6.1 shows a typical signal, where $x_m = 255 \cdot 0$.

The table below gives the actual values for the first few samples, and the corresponding integer values for 8-bit and 7-bit words

n	0	1	2	3	4	5	6
$x(n)$	0	28·7	57·3	74·3	72·7	52·9	22·7
m ($N = 256$)	0	29	57	74	73	53	23
m ($N = 128$)	0	14	29	37	36	26	11

●

● **Example 6.4** Now calculate the quantised output signals, assuming straight-through digitisation and then conversion back into analog form with the same full-scale value.

First we calculate the equivalent quantising interval, Q.

For $N = 256$, $Q = 1·0$, while for $N = 128$, $Q = 2·01$.

n	0	1	2	3	4	5	6
x_q ($N = 256$)	0	29·0	57·0	74·0	73·0	53·0	23·0
x_q ($N = 128$)	0	28·1	58·2	74·3	72·3	52·2	22·1

●

This second example illustrates the overall effect of quantising: that even when a signal is digitised and then converted directly back into analog form, the result is not identical to the input. Quantisation errors have occurred. We investigate the gross effect of these errors in the next section.

Before leaving the present discussion, we do need to make one further point. Comparison of these tables of values with the graph shown in figure 6.1 shows that we truncated the series at an opportune position! Subsequent values are *negative*, and our initial treatment of the quantised scale has assumed wholly positive sample values.

There is of course a variety of different digital codes which will represent negative as well as positive values, and the *2's complement* code is widely used for arithmetic processes. Indeed, many A-D converters operate according to this code structure, but an alternative is frequently used, that of the *offset binary* scale where $N/2$ is added to each value to ensure that all the digital numbers are positive. The maximum range is now approximately $\pm N/2$, and the scale zero corresponds to an integer of $N/2$. This offset can easily be removed by the digital machine before commencing a calculation based on the set of input samples.

So for the binary offset code:

$$(-x_m/2) \leqslant x < (x_m/2) \tag{6.23}$$

$$m = \text{int}\,[x/Q] + N/2 \tag{6.24}$$

$$x_q = (m - N/2)\,.\,Q \tag{6.25}$$

For the case discussed in example 6.4, we get the set of integer codes, and corresponding analog output values, shown in Table 6.2. The max-min range of the analog output is assumed to be 256·0.

111

Table 6.2 *Offset binary code*

n	0	1	2	3	4	5	6	7	8
$x(n)$	0	28·7	57·3	74·3	72·7	52·9	22·7	$-6\cdot2$	$-22\cdot9$
m, $(N=256)$	128	157	185	202	201	181	151	122	105
x_q	0	29·0	57·0	74·0	73·0	53·0	23·0	$-6\cdot0$	$-23\cdot0$

6.4.2 Defining quantisation noise

We have established in the previous section that sample values of an analog waveform cannot be represented exactly by digital words. Our discussion has been limited to single samples, but now we look at the effect of quantisation on the whole signal. Figure 6.15 shows a broader view of how quantisation affects a signal.

The upper graph shows an original analog signal, and superimposed is a set of quantised sample values. The quantum interval Q is shown for comparison, and is made quite large so as to emphasise the effect. The lower graph shows the quantising error $\varepsilon(n)$, on an expanded vertical scale.

Although the original signal is a smoothly varying waveform, the quantising error varies unpredictably. It is almost impossible to make a

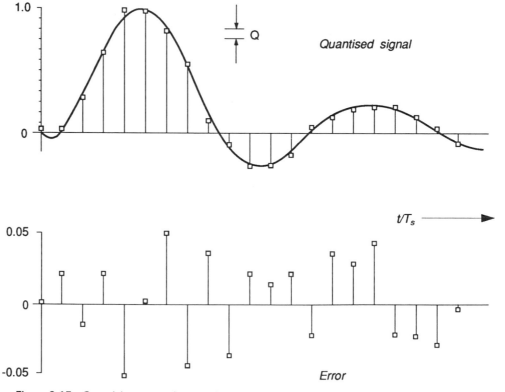

Figure 6.15 Quantising errors for a typical signal

statement about the value of error at any particular sample time, but we ought to be able to make meaningful statements regarding the long-term averages of the error. Since the error is equally likely to be positive as negative:

$$\text{avg}[\varepsilon(n)] = 0 \qquad (6.26)$$

The maximum and minimum values of the error are $\pm Q/2$ respectively.

The mean-square value is clearly non-zero, and will be a useful measure of the long-term effects of quantising errors. When viewed over a long time, we call the error *quantisation noise*, since it adds a noise-like random signal to the original signal.

In order to estimate the mean-square value of the error $\varepsilon(n)$, we consider a signal record consisting of M sample values, where M is a very large number. Now the range of the error $\pm Q/2$ can be divided into a number of elementary intervals $\delta\varepsilon$ and there will be $(Q/\delta\varepsilon)$ of these intervals.

Take one particular interval, which corresponds to an error value ε. Out of M samples of the signal $(M.\delta\varepsilon/Q)$ signal samples will have an error which falls within the band of $\delta\varepsilon$ around a value ε, assuming that all values of error are equally likely to occur.

The contribution of each one of these errors to the total mean-square value will be ε^2, and there are $(M.\delta\varepsilon/Q)$ of them. Their mean-square value contribution will therefore be

$$(\varepsilon^2 . \delta\varepsilon . M/Q)/M = \varepsilon^2 . \delta\varepsilon/Q \qquad (6.27)$$

The net effect of all such elementary intervals on the error scale is obtained by letting $\delta\varepsilon \to 0$ and integrating:

$$\text{avg}[\varepsilon^2] = \frac{1}{Q} \int_{-Q/2}^{Q/2} \varepsilon^2 . d\varepsilon \qquad (6.28)$$

$$= \frac{Q^2}{12} \qquad (6.29)$$

Thus the rms value of the error, over a long period of time and many samples, is given by $Q/\sqrt{12}$. Alternatively the mean-square value of error is $Q^2/12$. This significant result we examine in the next section.

6.4.3 The effect of quantisation noise

The most convenient way of expressing the effect of quantising noise is to calculate the ratio between signal level and rms quantising noise. Thus a large value for the *signal-to-noise ratio* means that the signal is relatively unaffected by the quantising noise. A small value on the other hand implies that the quantising noise will be significant.

Observe first that the quantising noise depends only on the size of the quantum step, and is independent of the signal level. Given a certain A-D converter, then the signal-to-noise ratio improves as the signal is increased, and worsens as the signal level is reduced.

● **Example 6.5** A certain A-D converter can accommodate a range of input signals of $0 \to 1 \cdot 0$ V. Calculate the rms value of quantisation noise for digital words having various numbers of bits.

Number of bits b	8	10	12	16	
Quantum Q	3910	977	244	15·3	μV
Quantisation noise $Q/\sqrt{12}$	1100	280	70	4·4	μV rms

●

Quantisation noise therefore can be made to be very small by choice of a modest number of bits in the digital word. However, the signal-to-noise ratio is influenced also by the signal level, and if the signal is small, as is sometimes the case where the input signal has a large *dynamic range*, then the ratio can be quite adverse.

● **Example 6.6** Calculate the rms signal-to-noise ratio for the cases in example 6.5, if the signal is sinusoidal with an rms value of 20 mV.

Number of bits b	8	10	12	16	
RMS signal-to-noise ratio	18·2	71·4	285	4545	
	25·2	37·1	49·1	73·2	dB

(see appendix A7 for an explanation of the *decibel* or dB.)   ●

These relationships can now be extended to form a general statement of the effect of quantisation noise. Consider an A-D converter which generates a b-bit binary word and has an input voltage range of $\pm V_m$. Our introductory discussion has been with unipolar signals, but we now move to the more usual case of *bipolar* signals. The quantum interval Q is therefore

$$Q = 2V_m/(2^b - 1) \tag{6.30}$$

Note that the total input range is $2V_m$ and that $2^b \gg 1$.

A sinusoidal input signal having an rms value V_r generates a signal-to-noise ratio of

$$\text{SNR} = (\sqrt{12}) \cdot V_r/Q \tag{6.31}$$

or

$$\text{SNR} \approx 2^{(b-1)}(\sqrt{12}) \cdot \frac{V_r}{V_m} \tag{6.32}$$

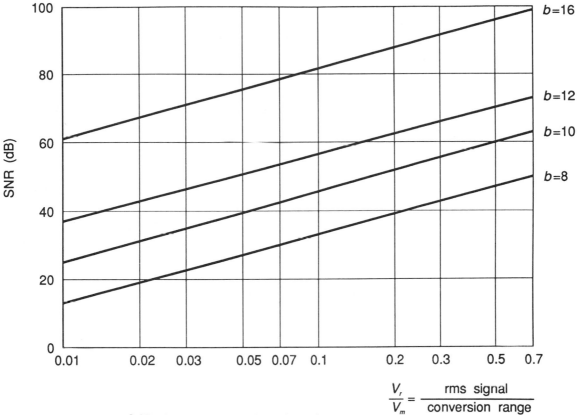

Figure 6.16 Signal-to-quantisation-noise ratio

Note that the maximum possible value of V_r is $V_m/\sqrt{2}$, which makes the peak value of the waveform just reach the extremes of the A-D converter range.

Expressing equation 6.32 in decibel form, which is often preferred:

$$SNR - 4 \cdot 8 + 6b + 20 \log_{10}\left(\frac{V_r}{V_m}\right) \ dB \tag{6.33}$$

For the maximum-range signal, this expression reduces to

$$SNR = (1 \cdot 8 + 6b) \ dB \tag{6.34}$$

Figure 6.16 shows a plot of this function for certain popular values of word length b. This graph may be used to estimate the requirements for a given application.

● **Example 6.7** Suppose that rms quantising noise is to be kept below 1/100th of the rms signal, that is an SNR of better than 40 dB.

If all relevant signals are to have an rms value greater than $0 \cdot 2$ of the A-D range, then an 8-bit conversion satisfies the criterion.

If the input signal is to be allowed to drop to $0 \cdot 01$ of full range, then a 12-bit conversion would be necessary. ●

● **Example 6.8** High-quality audio channels, used by the BBC to relay music-quality sound between studios and transmitter sites, operate at a sampling rate of 32 kHz and use 13-bit words. Calculate the SNR for a signal which has a level of −30 dB relative to the maximum level possible.

First calculate the SNR for the maximum signal, which from equation 6.34, is 80 dB. A signal whose level is −30 dB relative to the maximum signal reduces the SNR by 30 dB. Hence the required figure is 50 dB.   ●

In many signals such as speech, the sample values only rarely approach the upper reaches of the A-D range, so that the converter is operating at a relatively low SNR for most of the time, and not making best use of the number of bits available. (See section 1.4.3 for a brief discussion of the dynamic range of speech signals.)

The problem, then, is that the quantising noise is constant, whether the signal is large or small. Consequently for speech signals, the quantising scale is often made to vary with signal level in order to maintain an approximately constant SNR. The exact derivation of an optimum non-linear quantising scale is beyond us at the moment, but we note from equation 6.31 that making $Q \propto V_r$ seems to achieve this aim. The output/input characteristic of the quantiser in this case is therefore approximately logarithmic. This type of *compression* law is used for converting voice signals for telephone use, where the dynamic range of signal is typically 72 dB and the SNR is required to be at least 40 dB. Reference to figure 6.16 shows that 12-bit words would be required for a linear quantising scale, but 8-bit words are adequate in practice, when used with logarithmic compression.

STUDY QUESTIONS

1 Distinguish between analog and digital signal processes. What are the major reasons for using digital signal processes?

2 State the *sampling theorem*, and explain its significance for practical signal sampling.

3 A certain digital system has a sampling interval T_S of 20 μs. What is the Nyquist frequency? Determine the output frequencies when components of 10 kHz and 30 kHz are sampled.

4 Explain why an anti-alias filter is normally needed when converting analog signals into digital form. When is such a filter not necessary?

5 A certain signal is modelled by the following set of components:

Frequency	1·0	2·0	3·8	5·5	7·5	kHz
RMS magnitude	0·6	1·4	1·8	0·6	0·4	V

Choose the lowest sampling frequency which allows the alias power to be not more than 10% of the total signal power.
   Make a sketch of the frequency plots for both analog and sampled signals, for the sampling frequency you have chosen.

6 Calculate the integer values generated by the A-D process, from the following input signal samples:

n	0	1	2	3	4	5	6	7
$x(n)$	0·1	0·4	0·9	0·6	0·0	−0·2	−0·3	−0·1

The A-D converter has a maximum range of $\pm 1 \cdot 0$ V, and generates a b-bit offset binary code. Carry out the calculation for (a) $b = 4$, (b) $b = 6$.

7 Calculate the quantising errors for the two cases in question 6, assuming that the output D-A converter has similar parameters to the A-D converter.

8 Define quantising noise, and quote its mean-square value.

9 Starting from question 8, deduce the expression for signal-to-quantisation-noise ratio.

10 Background noise affects the quality of signal recorded by a microphone. Consider a system where the signal-to-acoustic-noise ratio is 40 dB. How many bits are required in the A-D converter, if the quantising noise is to be no greater than the acoustic noise, and the minimum rms signal level is to be a fraction δ of the converter positive range?

   (a) $\delta = 0 \cdot 2$ (b) $\delta = 0 \cdot 01$

7 Linear Filtering: Time and Frequency

OBJECTIVES

To investigate further the very important signal process of linear filtering.

To describe the close relationships that exist between signals and processes, in terms of time and frequency. In particular to

a) Describe signals in terms of elementary functions of time—the unit-pulse and the impulse.
b) Investigate the significance of the unit-pulse response and impulse response of linear filtering processes.
c) Develop a relationship between this response and the frequency response.

COVERAGE

Starting with the meaning of linearity, the linear equations which describe discrete-time and continuous-time systems are discussed. The signal-flow graph is introduced as a tool for describing such equations.

The main principles are presented in terms of discrete-time systems, since digital processing is widespread, and the concepts are a little easier to see than for analog or continuous-time systems. The unit-pulse is an elementary 'time' function which can be used to model discrete-time signals, and the unit-pulse response describes a linear filtering system. Unit-pulse response and input signal are combined in the convolution operation to calculate the output signal. Carrying out this calculation with a phasor input defines the frequency response of the system, and relates it to the unit-pulse response, the relationship being known as the DFT.

We then study some of the general properties of time and frequency descriptions of a system, and emphasise the important concept of *Reciprocal Spreading* between time and frequency domains.

Continuous-time system descriptions are introduced by similarity with the discrete-time system properties. The impulse is the equivalent elementary time function, and convolution and frequency response are developed from it in similar fashion to the discrete-time case.

In previous chapters we have used the concepts of linear filtering to show how bands of frequencies can be selected from a given signal (sections 1.3.1, 4.2.1, 4.2.2). In order to understand these operations we have defined the *phasor response* or *frequency response* of the process. We look now at linear filtering from a different point of view, from the time-function description of signals, and then show how this time response is related to the frequency response.

For instance, when processing speech signals, the waveform is the important thing, it defines the utterance. We may choose to analyse it in terms of its frequency description, but the time-function description is all important.

Signals and processes may be described using functions either of time or of frequency, and the relationship between these two descriptions is central to all consideration of signal processing. So in this chapter we are taking a wide view of the meaning of time and frequency, presenting classical bookwork but in a non-rigorous fashion, and laying foundations for crucial fundamental ideas. We shall try to show the practical implications of each concept as we go along, but a concrete application will have to wait until Chapter 9.

These ideas are central to the design of a linear signal process which will generate a specified output signal from a given input signal; although in this chapter we shall merely introduce the ideas for application in later chapters.

Both discrete-time and continuous-time regimes are governed by similar concepts, even though the precise details differ because of the way in which signals are described in the two regimes. Discrete-time leads to a better understanding of the underlying principles, so we shall carry out that analysis first. Continuous-time analysis will then follow by similarity.

Note that we are trying to lay foundations, to present ideas and techniques which are useful, but not to prove rigorously all the relationships. There are many textbooks which address the formal mathematical aspects of this subject, and the reader is directed to these for background reading.

7.1 Introduction

First, we outline the meaning of the term *linear*, and show how this occurs in signal processes of both the discrete-time and the continuous-time sort.

7.1.1 Linearity

The principle is that the response of a system or process to the sum of two or more signals is the same as the sum of the responses to the individual signals. Expressed algebraically, consider two input signals x_1 and x_2, which produce output signals y_1 and y_2 respectively. The process is represented by some function $F[\ ]$.

So, when the two signals are connected to the input of the process individually,

$$y_1 = F[x_1] \quad \text{and} \quad y_2 = F[x_2] \tag{7.1}$$

Applying them simultaneously, the system is linear if the following relationship is satisfied:

$$y_1 + y_2 = F[x_1 + x_2] \tag{7.2}$$

We have already assumed this condition in our previous discussions of filters (e.g. section 4.2.1), and in fact it does underpin a great deal of signal processing.

Operations which preserve linearity include

- addition and subtraction
- multiplication by a constant
- delay
- differentiation and integration.

● **Example 7.1**   Determine whether or not the following signal processes are linear.

a) $y = a \cdot x$

Then $\quad y_1 + y_2 = a(x_1 + x_2)$

This process is therefore linear.

b) $y(t) = K \int x(t) \cdot \mathrm{d}t$

Then $\quad y_1(t) + y_2(t) = K \int [x_1(t) + x_2(t)] \cdot \mathrm{d}t$

And this process is linear.

c) $y = x^2$

Then $\quad F[x_1 + x_2] = (x_1 + x_2)^2$
$$= x_1^2 + x_2^2 + 2 \cdot x_1 \cdot x_2$$

This process is therefore NOT linear.   ●

7.1.2 Linear equations

Linear signal processes are ones which use the set of operations listed above to generate an output signal from a given input signal.

Discrete-time signals are described by a *difference equation*, relating the output sample values $\{y(n)\}$ to the set of input sample values $\{x(n)\}$. For instance,

$$y(n) = a \cdot x(n) + b \cdot x(n-1) + c \cdot x(n-2) \tag{7.3}$$

The operations used here are multiplication by a constant, addition and delay $[x(n-1)]$. Previous output values may be used also in the calculation of the present output sample, and the process is then known as *recursive* because it feeds on its own output. An example of a recursive difference equation is

$$y(n) = a \cdot x(n) + b \cdot y(n-1) \tag{7.4}$$

Such equations are graphically portrayed by means of a *Signal-flow graph*. In such a diagram, nodes represent signal values, while branches between nodes represent multiplication by a constant or a simple signal process like delay, differentiation, etc. Where two or more branches meet at a node, it is implied that the signal value is the sum of all the *flows* into that node. Figure 7.1 illustrates these general properties, and figure 7.2 shows the signal-flow graphs that correspond with the two difference equations 7.3 and 7.4.

Linear equations are constructed for continuous-time systems, using differential equations. Thus, an example is given in equation 7.5, and the corresponding signal-flow in figure 7.3.

$$y(t) + k \cdot \frac{\mathrm{d}y(t)}{\mathrm{d}t} = x(t) \tag{7.5}$$

The signal-flow graph is not a magic solution to the equation, it merely allows us to appreciate the structure of it, and in complicated cases may

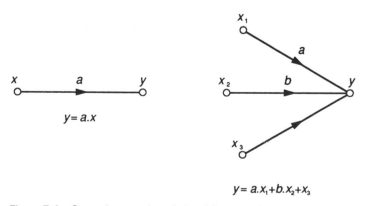

Figure 7.1 General properties of signal-flow graph

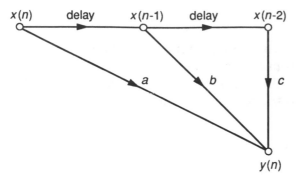

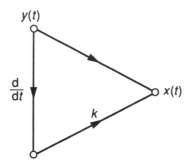

Figure 7.2 Two signal-flow graph examples

Figure 7.3 Signal-flow graph for continuous-time system

allow ambiguities to be resolved. In digital processing it is a tool which helps turn the process definition into a computer program or some digital hardware. We shall use this technique from time to time to describe signal processes.

7.2 Discrete-time systems

Discrete-time systems use digital hardware to implement signal processing, and so are extremely important. The signal is described by a set of sample values $\{x(n)\}$, and relationships are described by difference equations.

7.2.1 Unit-pulse response

The *unit-pulse* is an elementary function which can be used as a building block to describe discrete-time signals. It is a function of the index number n, and hence broadly of 'time'. We define the unit-pulse $p(n)$ in the following manner:

$$p(n) = 1 \quad n = 0$$
$$= 0 \quad n \neq 0 \tag{7.6}$$

This then is a discrete-time signal which has only one non-zero value, at $n = 0$.

Such a signal can be used as input to a signal process, yielding a discrete-time output signal which we call the *unit-pulse response*, $g(n)$. This response contains all the information that we need to know about the process, provided that it is linear.

Consider the difference equation 7.3:

$$y(n) = a \cdot x(n) + b \cdot x(n-1) + c \cdot x(n-2)$$

Apply the unit-pulse $p(n)$, so that $x(n) = p(n)$, and calculate the output sequence $y(n) = g(n)$ in response to the unit-pulse: see Table 7.1.

Notice that the coefficients a, b, c appear at the output in succession, and can therefore be measured in an experiment, when a unit-pulse is applied to a discrete-time system. Notice also that the number of sample values in the unit-pulse response is finite, so that this is known as a *Finite-Impulse-Response (FIR)* process or filter.

In the case of the simple example above, the unit-pulse response would be written as

$$\{g(n)\} = \{a, b, c, 0, 0, 0, ...\} \tag{7.7}$$

In the general case where such a process uses the past N values in the input sequence to calculate each output value, we may write the difference equation as

$$y(n) = \sum_{r=0}^{N} a_r \cdot x(n-r) \tag{7.8}$$

The unit pulse response would therefore be described as

$$g(n) = a_n \qquad 0 \leqslant n \leqslant N \tag{7.9}$$

Notice that we are prohibiting the index n from being negative, thus making sure that the unit-pulse response starts only *after* the input pulse has arrived. This seemingly obvious condition is known as *Causality*, since the output is the *effect* of the input *cause*. There are occasions in signal processing where a process can be *non-causal*, and the unit-pulse response may then be written as

$$g(n) = a_n \qquad -N \leqslant n \leqslant N \tag{7.10}$$

A different type of unit-pulse response occurs when we use a recursive

Table 7.1 *Unit-pulse response, FIR process*

n	0	1	2	3	4	5
$x(n) = p(n)$	1	0	0	0	0	0
$y(n) = g(n)$	a	b	c	0	0	0

Table 7.2   *Unit-pulse response, recursive process*

n	0	1	2	3	4	5	6	
$x(n) = p(n)$	1	0	0	0	0	0	0	
$y(n) = g(n)$	a	$a \cdot b$	$a \cdot b^2$	$a \cdot b^3$	$a \cdot b^4$	$a \cdot b^5$	$a \cdot b^6$	...

difference equation like that in 7.4, which yields the table of values given in Table 7.2. In this case there is no end to the unit-pulse response since it continues to feed upon previous values. A recursive process therefore gives rise to an *Infinite-Impulse-Response (IIR)*.

Notice that two conclusions can be drawn from this example:

a) The coefficient b in equation 7.4 must be less than unity, or else the unit-pulse response will grow without limit.

b) The unit-pulse response for this case follows an exponential law

$$g(n) = a \cdot \exp(-\beta n) \quad \text{where } \beta = -\ln(b)$$

● **Example 7.2**   Calculate the unit-pulse response for the following difference equations:

a) $y(n) = 4 \cdot x(n) + 6 \cdot x(n-2) - 7 \cdot x(n-3)$

The response is the sequence:

$$\{g(n)\} = \{4, 0, 6, -7, 0, 0, 0, 0, ...\}$$

b) $y(n) = 2 \cdot x(n) - 4 \cdot x(n-1) - 0 \cdot 9 \cdot y(n-1)$

$$
\begin{aligned}
\text{Then} \quad y(0) &= +2 \cdot 0 \\
y(1) &= -4 \cdot 0 - 0 \cdot 9(2 \cdot 0) = -5 \cdot 80 \\
y(2) &= -0 \cdot 9(-5 \cdot 80) = +5 \cdot 22 \\
y(3) &= -0 \cdot 9(5 \cdot 22) = -4 \cdot 70 \\
y(4) &= -0 \cdot 9(-4 \cdot 70) = +4 \cdot 23 \\
y(5) &= -0 \cdot 9(4 \cdot 23) = -3 \cdot 81
\end{aligned}
$$

Hence $\{g(n)\} = \{2 \cdot 0, -5 \cdot 80, 5 \cdot 22, -4 \cdot 70, 4 \cdot 23, -3 \cdot 81, ...\}$ ●

7.2.2   Convolution

Having discussed the unit-pulse response of a process, we now start to make use of it. Consider the problem portrayed in figure 7.4. We have a process which is described by its unit-pulse response $g(n)$, and we wish to calculate the output signal sequence $y(n)$ when the input signal is $x(n)$. The input signal can be of any form, an information-bearing signal.

Now we require to calculate the output signal sample at a time corresponding to the index n. If this signal has resulted from an analog-to-digital conversion process, then the nth sample would occur at a time nT_S sec.

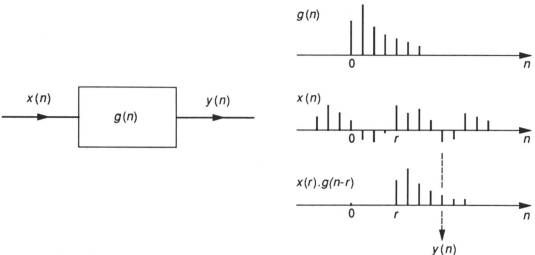

Figure 7.4 Calculation of output signal

Consider the effect of one sample from the input signal $x(r)$, which occurs before or at the time when we calculate the output signal sample. Thus $r \leqslant n$. At an arbitrary time n, this input sample $x(r)$ generates a scaled unit-pulse response $x(r) \cdot g(n-r)$.

Thus at $n = r$ the output sample value is $x(r) \cdot g(0)$

        at $n = r + 1$ the output sample value is $x(r) \cdot g(1)$

        at $n = r + 2$ the output sample value is $x(r) \cdot g(2)$

        and so on.

The actual value of the output sample $y(n)$ will be the sum of the effects of all such input signal samples, summed in the general case from $r = -\infty$ to $r = n$. Hence

$$y(n) = \sum_{r=-\infty}^{n} x(r) \cdot g(n-r) \tag{7.11}$$

This expression is known as the *Convolution* operation, and is often written in shorthand form as

$$y(n) = x(n) * g(n) \tag{7.12}$$

The convolution operation is crucial since it allows us to calculate the output signal knowing the input signal and the unit-pulse response of the process or system, and we have noted above that the unit-pulse response is simply measured. In many practical cases however, it is not that easy to measure, since it can be of extremely small amplitude, but the principle is there. As a theoretical tool it is invaluable.

We shall shortly evaluate an example, and show how simple it is to carry out this operation with discrete-time signals, but first note some extremely important principles:

125

a) The convolution operation is a mathematical structure, and hence the physical significance of the two functions $x(n)$ and $g(n-r)$ is not preserved. It may be written equally well in the following form:

$$y(n) = \sum_{r=0}^{\infty} g(r) \cdot x(n-r) \tag{7.13}$$

assuming that $g(r)$ is non-zero only for $0 \leqslant r < \infty$.

A given output signal may also be generated by exchanging the characteristics of the input signal and the process. Signals and processes are often interchangeable, giving great flexibility in digital signal processing and in signal design.

b) In equation 7.11 we have chosen an upper limit for the index r of n, assuming that the unit-pulse response is causal, only commencing when the input pulse is applied. However, to be more general, we could re-write the equation 7.11 as

$$y(n) = \sum_{r=-\infty}^{\infty} x(r) \cdot g(n-r) \tag{7.14}$$

$$y(n) = \sum_{r=-\infty}^{\infty} g(r) \cdot x(n-r) \tag{7.14a}$$

The expression is the general one for convolution. The number of terms in the summation is limited in practice by zero values in the unit-pulse sequence $g(n)$, and/or in the input signal sequence $x(n)$.

● **Example 7.3** A certain process is described by its unit-pulse response $g(n)$, and has an input signal $x(n)$ suddenly applied. Calculate the output signal $y(n)$.

$$\{g(n)\} = \{4, 3, 2, 1, 0, 0, 0 \ldots\} \qquad n \geqslant 0$$
$$\{x(n)\} = \{1, 2, 0, -1, -2, -1, 0 \ldots\} \qquad n \geqslant 0$$

In order to calculate the convolution, we set out the table below which shows the successive sequences $x(r) \cdot g(n-r)$, and then sums the entries for each value of n.

n	0	1	2	3	4	5	6	7	8	9	10	11
$x(n)$	1	2	0	-1	-2	-1	0	0	0	0	0	0
$x(r) \cdot g(n-r)$	4	3	2	1								
		8	6	4	2							
			0	0	0	0						
				-4	-3	-2	-1					
					-8	-6	-4	-2				
						-4	-3	-2	-1			
$y(n) = \Sigma x(r) \cdot g(n-r)$	4	11	8	1	-9	-12	-8	-4	-1	0	0	0

●

● **Example 7.4** In order to show the symmetry in the convolution expression, reverse the places of the signal $x(n)$ and the unit-pulse response $g(n)$.

$$\{g(n)\} = \{1, 2, 0, -1, -2, -1, 0, 0 \ldots\} \qquad n \geqslant 0$$
$$\{x(n)\} = \{4, 3, 2, 1, 0, 0, 0 \ldots\} \qquad n \geqslant 0$$

We repeat the calculation above:

n	0	1	2	3	4	5	6	7	8	9	10	11
$x(n)$	4	3	2	1	0	0	0	0	0	0	0	0
$x(r) \cdot g(n-r)$	4	8	0	-4	-8	-4						
		3	6	0	-3	-6	-3					
			2	4	0	-2	-4	-2				
				1	2	0	-1	-2	-1			
$y(n) = \Sigma x(r) \cdot g(n-r)$	4	11	8	1	-9	-12	-8	-4	-1	0	0	0

This result is the same as in Example 7.3, demonstrating that convolution is indeed symmetrical, and that signals and processes may be interchanged at will.   ●

Convolution then, for discrete-time systems, is a simple arithmetic process and may be programmed easily for many different types of digital processor. Once the desired unit-pulse response is known, the convolution sum can be calculated by the processor to generate the output signal samples.

● **Example 7.5** By way of a simple example, consider the design of a differentiator. Its response $g(n)$, to a unit-pulse input $p(n)$, may be approximated to

$$\{g(n)\} = \{1, -1, 0, 0, 0, 0 \ldots\}$$

Thus, the unit-pulse response defines a process which reflects changes in the input between one sample and its neighbour.

Working from the convolution equation 7.14, the output signal $y(n)$ of this differentiator will be

$$y(n) = x(n) - x(n-1)$$

This difference equation can then be programmed into a digital signal processor.

In operation, this simple differentiator takes the difference between successive input signal samples, and so approximates the rate-of-change of any input signal.

Let the input signal be

$$\{x(n)\} = \{0, 0, 1, 1, 1, 0, 0 \ldots\}$$

Then the output signal is

$$\{y(n)\} = \{0, 0, 1, 0, 0, -1, 0, 0 \dots\}$$

Alternatively, let the input signal be a linear ramp:

$$\{x(n)\} = \{0, 2, 4, 6, 8, 10, 12 \dots\}$$

Then $\{y(n)\} = \{0, 2, 2, 2, 2, 2, 2 \dots\}$ ●

● **Example 7.6** Programming the convolution equation into a digital processor is fairly simple. We illustrate the principle below with a BASIC program stub.

The unit-pulse response is stored in array $g(n)$, and has M values.

The present input value, and the last $(M-1)$ input values, are stored in an array $x(n)$.

```
10   DIM g(M), x(M)
20   REPEAT
30     INPUT newx
40     GOSUB 100 : REM shuffle x( ) array, and insert new value
50     GOSUB 200 : REM execute convolution
60     PRINT y    : REM output new value
70   FOREVER

100  REM shuffle x( ) array, and insert new value
110    FOR i = (M − 1) TO 1 STEP −1
120    x(i) = x(i − 1)
130    NEXT
140  x(0) = newx
150  RETURN

200  REM execute convolution
210  y = 0
220    FOR r = 0 TO (M − 1)
230    y = y + g(r) * x(r)
240    NEXT
250  RETURN
```

7.2.3 Frequency response and the DFT

The unit-pulse response $g(n)$ has been used to calculate the output signal of a process in response to a general input signal, but we know also that when the input is a phasor, the output signal can be calculated using the frequency response $\mathbf{H}(\omega)$. The unit-pulse response may be used for any kind of input signal via the convolution operation, but the frequency response is used only for the restricted case of phasor inputs. We now discover how to link these two approaches.

For a sampling interval of T_S, and a single-frequency phasor input, the input and output signals are of the form:

$$x(n) = \exp j(\omega T_S n) \qquad (7.15)$$

$$y(n) = \mathbf{H}(\omega) \cdot \exp j(\omega T_S n) \qquad (7.16)$$

Working from the convolution equation 7.14a, we get

$$y(n) = \sum_{r=-\infty}^{\infty} g(r) \cdot x(n-r) \qquad (7.17)$$

Therefore

$$\mathbf{H}(\omega) \cdot \exp j(\omega T_S n) = \sum_{r=-\infty}^{\infty} g(r) \cdot \exp j[\omega T_S(n-r)] \qquad (7.18)$$

or

$$\mathbf{H}(\omega) = \sum_{r=-\infty}^{\infty} g(r) \cdot \exp j(-\omega T_S r) \qquad (7.19)$$

This is a most important result, relating the phasor or frequency response to the unit-pulse response. It is sometimes known as the *Discrete-Fourier-Transform (DFT)* relationship, and is the discrete-time signal form of the normal *Fourier Transform*. (Chapter 13 investigates this relationship in more detail, while appendix A3 outlines the Fourier transform and appendix A5 discusses the DFT.) As such, it provides a relationship between a function of time (the unit-pulse response) and a function of frequency (the phasor response) for a linear system or process.

Conversions between these two domains are frequently made in signal processing design, and in the next section we shall explore the relationship between them.

If the unit-pulse response of a process is given, then equation 7.19 can be used to find the corresponding frequency response, which we shall illustrate below with two examples.

Alternatively, if the desired frequency response of the filter is known, then equation 7.19 can be worked backwards in order to find the corresponding unit-pulse response, and thence via equation 7.8 to select the coefficients of a simple digital filter to carry out this task. This is the filter design problem, which we shall not pursue at the moment. However, the procedure is very similar to that used to find the coefficients in the Fourier series, as shown in appendix A2.

● **Example 7.7** A certain discrete-time process has a unit-pulse response $g(n)$ and we are to find the corresponding frequency response.

$$\{g(n)\} = \{1, 1, 0, 0 \ldots\} \qquad n \geqslant 0$$

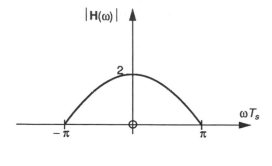 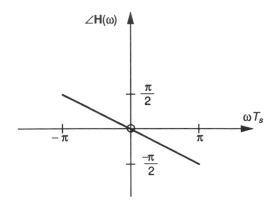

Figure 7.5 Frequency response

Applying equation 7.19 directly,

$$\mathbf{H}(\omega) = 1 + \exp \, j(-\omega T_S)$$
$$= [\exp \, j(\omega T_S/2) + \exp \, j(-\omega T_S/2)] \, . \exp \, j(-\omega T_S/2)$$
$$= [2 \, . \cos(\omega T_S/2)] \, . \exp \, j(-\omega T_S/2)$$

Figure 7.5 shows the magnitude and the phase of this response, which has been expressed in the form *(magnitude) . exp j(phase)*. (Note the standard trick of extracting exp j$(-\omega T_S/2)$ from both terms, in order to form a cosine (or sine) function.)　●

● **Example 7.8**　Repeat the calculation with the following unit-pulse response:

$$\{g(n)\} = \{2, 3, 0, 3, -2\} \qquad n \geqslant 0$$

Applying equation 7.19 directly, we discover that

$$\mathbf{H}(\omega) = 2 + 3 \, . \exp \, j(-\omega T_S) + 3 \, . \exp \, j(-3\omega T_S) - 2 \, . \exp \, j(-4\omega T_S)$$

This equation is in a form which is quite common in discrete-time systems, and can be simplified by taking out a common factor of exp j$(-2\omega T_S)$. The equation then becomes

$$\mathbf{H}(\omega) = [(2 \, . \exp \, j(2\omega T_S) + 3 \, . \exp \, j(\omega T_S) + 3 \, . \exp \, j(-\omega T_S)$$
$$- 2 \, . \exp \, j(-2\omega T_S)] \, . \exp \, j(-2\omega T_S)$$

This simplifies to

$$\mathbf{H}(\omega) = [6 \, . \cos(\omega T_S) + j4 \, . \sin(2\omega T_S)] \, . \exp \, j(-2\omega T_S)$$

So in general, the frequency response for a discrete-time system boils down to a collection of sin/cos terms, together with an exp j$(-\omega\beta)$ term which gives a phase-shift proportional to frequency. Figure 7.6 illustrates the modulus of this particular response function, which demonstrates a simple 'notch' or bandstop filter response. Observe that this conclusion would not easily be seen from the unit-pulse response $g(n)$.　●

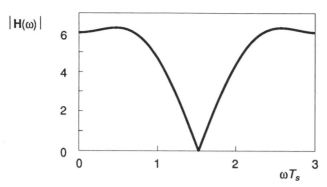

Figure 7.6 Frequency response

7.2.4 Properties of these responses

We now have two general equations. The first is the convolution expression which relates the output signal to the input signal and the unit-pulse response:

$$y(n) = \sum_{r=-\infty}^{\infty} x(r) . g(n-r) \qquad (7.20)$$

The second is the Discrete-Fourier-Transform (DFT) which relates the phasor response to the unit-pulse response:

$$\mathbf{H}(\omega) = \sum_{m=-\infty}^{\infty} g(m) . \exp \mathrm{j}(-\omega T_s m) \qquad (7.21)$$

Let us now explore some general properties of these two responses, which in fact are intimately related.

7.2.4.1 Delay

First, think of a very simple unit-pulse response. Let

$$g(n) = p(n) \qquad (7.22)$$

The unit-pulse response of this simple process is actually a unit-pulse. Each input signal sample is reflected exactly at the output without modification. Thus

$$y(n) = x(n)$$

The corresponding frequency response is then

$$\mathbf{H}(\omega) = \sum_{m=-\infty}^{\infty} p(m) . \exp \mathrm{j}(-\omega T_s m) \qquad (7.23)$$

Since $p(m) = 0$ except at $m = 0$, by definition (equation 7.6), then as may be expected:

$$\mathbf{H}(\omega) = 1 \qquad (7.24)$$

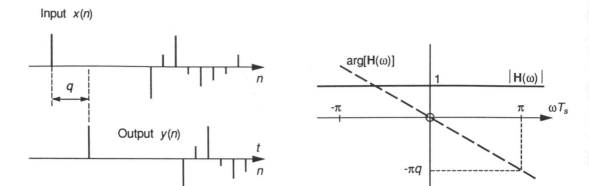

Figure 7.7 Discrete-time system with delay

Now let the unit-pulse response be the delayed unit-pulse:

$$g(n) = p(n - q) \tag{7.25}$$

Here, the input samples are just delayed by q sample times, so that

$$y(n) = x(n - q) \tag{7.26}$$

Applying the same argument as before, we see that

$$\mathbf{H}(\omega) = \exp \mathrm{j}(-\omega T_S . q) \tag{7.27}$$

A signal delay of qT_S sec therefore corresponds to a frequency response whose magnitude is constant, but whose phase changes by q radians per unit ωT_S. When ωT_S changes through an angle π, then the phase has increased by πq radians. This corresponds to a phase gradient of qT_S sec, against a frequency scale of rad/sec (*note that the units are actually rad/(rad/sec)*). An alternative description is a slope of $360q/f_S$ °/Hz, when considered against a frequency scale in Hz.

Figure 7.7 illustrates these properties with a simple diagram. Examples 7.7 and 7.8 illustrate this property too, since the unit-pulse responses of these filters have their centres of symmetry at delays of $T_S/2$ and $2T_S$ respectively.

● **Example 7.9**   A certain discrete-time process has a frequency response whose phase changes linearly by $14 \cdot 4°$ in 100 Hz, and the sampling frequency is 10 kHz. Determine the signal delay defined by this characteristic.

The phase gradient is therefore $144°/kHz$, and since the sampling frequency f_S is 10 kHz, the parameter q is $144 \times 10/360 = 4$.

The delay is therefore qT_S which is 400 μs.

Alternatively, a change of $14 \cdot 4°$ corresponds to $0 \cdot 25$ rad, and this change takes place within a frequency range of $0 \cdot 2\pi$ krad/sec.

The phase gradient is therefore $0 \cdot 25/0 \cdot 2\pi$ or $0 \cdot 4$ msec.

Consequently, $q = 0 \cdot 4/0 \cdot 1 = 4$.   ●

● **Example 7.10** Any digital system which interfaces to the analog world by A-D and D-A converters gives an overall delay of at least T_S sec. Both converters are clocked simultaneously, so even if there is no internal processing, the D-A output value must refer to the *previous* A-D input value.

Thus, without any additional processing, and ignoring the S-H effect (section 6.3.2), the overall frequency response is

$$\mathbf{H}(\omega) = \exp j(-\omega T_S)$$

The phase gradient is therefore T_S sec or $360 T_S^\circ/$Hz. For a sampling frequency of 10 kHz for example, this amounts to $0\cdot1$ msec or $0\cdot036^\circ/$Hz.   ●

Measurement of frequency-response phase is often difficult, particularly if the phase is changing rapidly with frequency. The overall delay is more often obtained by measuring the rate-of-change of the phase characteristic $\varphi(\omega)$, which is known as the *Group Delay*, τ_D:

$$\tau_D(\omega) = -d\lfloor\varphi(\omega)\rfloor/d\omega \tag{7.28}$$

Notice that for a simple process like that defined by equation 7.27, the group delay is simply qT_S sec. However in most practical signal processes and filters, the phase gradient is non-linear, and signals are not only delayed but are distorted too; see example 7.8. Any departures from linearity in the phase gradient are highlighted by variations in the group delay, which are then referred to as *group-delay distortion*.

/ ? 4.2 Frequency response shape

Certain basic properties of the frequency response for a discrete-time process can be seen from equation 7.21. Thus

a) If the unit-pulse response of the process is real (rather than complex), then it follows that the response has these symmetries:

> real$[\mathbf{H}(\omega)]$ has even symmetry about $\omega = 0$
> imag$[\mathbf{H}(\omega)]$ has odd symmetry about $\omega = 0$.

b) Since the coefficients of the unit-pulse response are not functions of ω, it follows that as ω increases the response $\mathbf{H}(\omega)$ will be periodic, since the expression for $\mathbf{H}(\omega)$ is in the form of a Fourier series (see appendix A2). Repetition is reached when $\omega T_S = 2\pi \cdot n$ (where n is any integer), so that the period is $2\pi/T_S$ rad/sec, or the sampling frequency. (See section 6.3.1 which arrives at a similar conclusion via the sampling theorem.)

As an example of this property, in figure 7.8 we plot the frequency response calculated in Example 7.8 but over a wider frequency range.

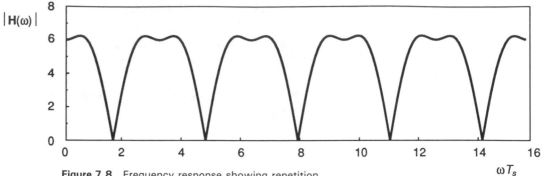

Figure 7.8 Frequency response showing repetition

7.2.4.3 Reciprocal spreading

We make one further deduction regarding these responses, which has extremely wide application and represents a fundamental limitation to all physical processes and signals, of whatever kind. It is known as the principle of *Reciprocal Spreading*.

This principle can best be illustrated for discrete-time signals by taking an ideal lowpass filter characteristic $\mathbf{H}(\omega)$, defined by a bandwidth $\pm\Omega$ rad/sec, and by calculating the corresponding unit-pulse response $g(n)$ (see figure 7.9).

In order to calculate $g(n)$, we must use the inverse form of equation 7.21, enabling $g(n)$ to be calculated if $\mathbf{H}(\omega)$ is known. The classical derivation is to multiply each side of the equation by $\exp j(\omega T_S n)$, and then take the average of each side of the equation with respect to ωT_S. We shall not be pursuing this relationship in detail here, so leave the derivation for appendix A2. The result is

$$g(n) = \frac{1}{2\pi} \int_{-\pi}^{\pi} \mathbf{H}(\omega) \cdot \exp j(\omega T_S n) \cdot d(\omega T_S) \qquad (7.29)$$

Using this expression, we calculate the unit-pulse response for this case, which is

$$g(n) = \frac{\sin(\Omega T_S n)}{\pi n} \qquad (7.30)$$

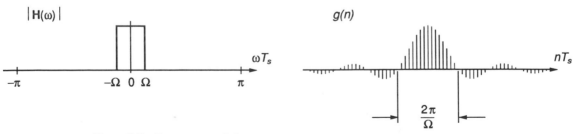

Figure 7.9 Frequency and time responses

134

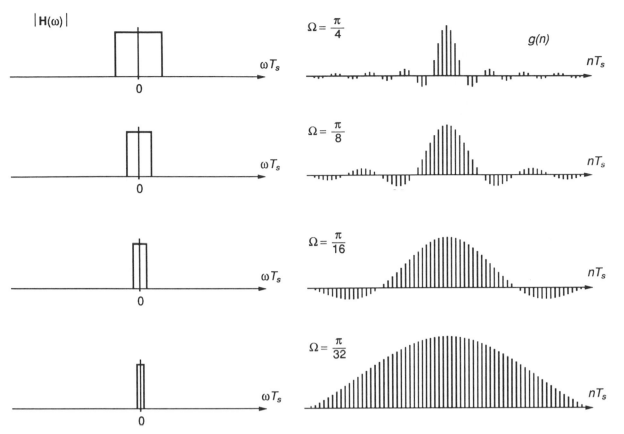

Figure 7.10 Illustration of reciprocal spreading

Figure 7.10 shows the form of this function for various values of Ω, together with the corresponding frequency responses.

Now the *width* of the frequency response is clearly defined as 2Ω rad/sec, and we could take a measure of the *width* of the unit-pulse response as the distance between the first two zeros of the function, which is $2\pi/\Omega$ sec. The product of these two widths is a constant 4π. The value is not important, but the fact that it is a constant is significant.

Clearly then, the width of the frequency response varies reciprocally with the width of the unit-pulse response as the bandwidth Ω is varied. As the frequency response bandwidth is increased, so the unit-pulse response becomes more narrow, and vice versa. A wideband frequency response implies a sharp unit-pulse response, while a narrowband frequency response implies a widespread unit-pulse response.

It is most important to keep this general principle in mind. It can be verified for frequency responses of any kind of course, but does not lend itself to a simple numerical measure, except in very simple cases such as we have just considered. In our later discussion of filter designs, we shall endeavour to keep track of both the frequency and the time performance of each filter.

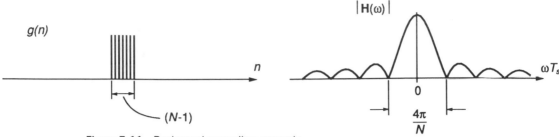

Figure 7.11 Reciprocal-spreading example

● **Example 7.11** A certain discrete-time process has a unit-pulse response $g(n)$:

$$g(n) = A \qquad 0 \leqslant n \leqslant N - 1$$
$$= 0 \qquad \text{elsewhere}$$

Using equation 7.21, we find the corresponding frequency response:

$$|\mathbf{H}(\omega)| = \frac{A \,.\, |\sin(\omega T_S N/2)|}{\sin(\omega T_S)}$$

This is sketched in figure 7.11. It is a similar function to that derived before in equation 7.30, and may be considered to have a *width* of $2\omega_S/N$ rad/sec, or $4\pi/N$ when referred to the ωT_S scale. Thus, the reciprocal spreading principle is again verified:

(Frequency response width) × (Unit-pulse response width) = constant ●

7.3 Continuous-time systems

The argument and relationships for continuous-time systems are very similar to those derived for discrete-time systems, with the following exceptions:

a) Differential equations are used in place of difference equations.
b) Integrals are used in place of summations.

It is not our intention to go through a rigorous derivation of these relationships, merely to point out their significance and applications; so we exploit their similarity with the discrete-time systems discussion in section 7.2. However, the similarity can be pressed only so far, since discrete-time and continuous-time systems are of *different kinds*, and bear only a superficial similarity to one another.

7.3.1 Impulse response

The continuous-time analogy to the *unit-pulse* is the *impulse*. A discrete-time signal of limited length can be built up from a finite number of unit-pulse functions. A continuous-time signal of limited length is built up from an infinite number of impulse functions. The impulse function $\delta(t)$ is therefore more difficult to visualise. It is defined indirectly as follows:

$$\delta(t) = 0 \qquad t \neq 0$$

and

$$\int_{-\infty}^{\infty} \delta(t)\, dt = 1 \qquad\qquad (7.31)$$

It follows from these two conditions that the impulse function has a magnitude of infinity at $t = 0$. It therefore is a strictly non-physical function, a mathematical abstraction!

However, approximations to this function can be made. Consider a short pulse of magnitude A and width $1/A$. As $A \to \infty$, so this approaches the characteristics of an impulse function. In practice a short pulse is used as an approximation to an impulse.

When an impulse is input to a continuous-time signal process, then the corresponding output is known as the *impulse response h(t)*. The impulse response $h(t)$ completely defines the process, in a similar way to that in which the unit-pulse response $g(n)$ defines a discrete-time system.

In mechanical systems, the impulse is well-known as a short sharp shock, like a hammer blow. The system then responds in a time governed by the inertia and dynamics of the system. The electrical impulse behaves in a similar manner.

Derivation of the impulse response from a system differential equation, such as equation 7.5, is clearly a more cumbersome task than finding the unit-pulse response in the discrete-time case (section 7.2.1). In practice this is normally calculated using the *Laplace Transform*, which we are not going to study here since our main target is discrete-time systems. However, the properties of the Laplace Transform are outlined in appendix A4 for reference.

● **Example 7.12** Consider the simple first-order lowpass circuit which we first met in section 3.3.1, when we calculated its frequency response. The circuit is shown again in figure 7.12, and its differential equation, in terms of the signals used here, is

$$y(t) + \tau \frac{dy(t)}{dt} = x(t)$$

To find the impulse response we make $x(t) = \delta(t)$, and then $y(t) = h(t)$. Now after $t = 0$, the input $x(t)$ is zero, so

$$h(t) = -\tau \frac{dh(t)}{dt}$$

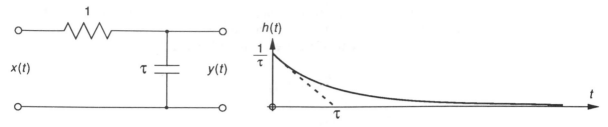

Figure 7.12 Impulse response for first-order lowpass circuit

and $h(t) = A \cdot \exp\left(-\dfrac{t}{\tau}\right)$

where A is an arbitrary constant.

In order to evaluate the arbitrary constant, we note that for an impulsive input, unit charge is placed on the capacitor in zero time.

So $A = 1/\tau$

Hence $h(t) = \dfrac{1}{\tau} \cdot \exp\left(-\dfrac{t}{\tau}\right)$

This also is illustrated in figure 7.12. ●

7.3.2 Convolution

Following the general argument presented in section 7.2.2, except that the signals are now continuous rather than discrete, we arrive at the *convolution integral*:

$$y(t) = \int_{-\infty}^{\infty} x(u) \cdot h(t - u) \cdot \mathrm{d}u \tag{7.32}$$

While the evaluation of this integral may often be unwieldy, its general principle is vital for an understanding of time-frequency relationships in continuous-time systems. We often write the operation in shorthand form:

$$y(t) = x(t) * h(t) \tag{7.33}$$

Notice that in order to make equation 7.32 consistent, the impulse response $h(t)$ must have dimensions of $(\text{sec})^{-1}$. This may seem odd, since the impulse response is apparently a function of time and can be observed in the laboratory. However, remember that the closest we can get to actually generating the impulse response is to apply a short pulse to the input $x(t)$, and then observe the voltage output $y(t)$ which is calculated via the convolution integral of equation 7.32. At best then, we can observe only a *scaled* form of the impulse response.

● **Example 7.13** Following on from example 7.12, we can calculate the

138

output of this circuit for any input, using the convolution integral:

$$y(t) = \frac{1}{\tau} \int_{-\infty}^{t} x(u) . \exp\left[\frac{-(t-u)}{\tau}\right] du$$

Notice that the upper limit of the integral is t rather than ∞, because the impulse response $h(u)$ is zero for $u < 0$.

This expression shows that in order to calculate the output signal at a time t, the impulse response is reversed in time, positioned at $u = t$, and then multiplied by the input signal $x(u)$. The value of the output at this time is then the area under this product. A further illustration occurs in section 10.3.2.1.   ●

7.3.3 Frequency response

Applying a phasor input to the continuous-time process:

$$x(t) = \exp j(\omega t) \tag{7.34}$$

and

$$y(t) = \mathbf{H}(\omega) . \exp j(\omega t) \tag{7.35}$$

Combining this information with the convolution equation, we deduce that

$$\mathbf{H}(\omega) = \int_{-\infty}^{\infty} h(u) . \exp j(-\omega u) . du \tag{7.36}$$

This is known as the *Fourier Integral*, and defines uniquely the frequency response of a continuous-time system, given the impulse response $h(t)$. Appendix A3 gives further details of this calculation.

Reciprocal spreading can be seen more clearly here, by applying the *scaling* principle. If $\mathbf{H}(\omega)$ is the frequency response corresponding to an impulse response $h(t)$, then, from equation 7.36, $\mathbf{H}(\omega/a)/a$ corresponds to $h(at)$.

● **Example 7.14** A sample-and-hold process (section 6.3.2) is essentially a continuous-time process, and has an impulse response which is defined by

$$\begin{aligned} h(t) &= 1/T_S & 0 \leqslant t \leqslant T_S \\ &= 0 & \text{elsewhere} \end{aligned}$$

Any new input value is always held for T_S sec.

Using equation 7.36,

$$\mathbf{H}(\omega) . T_S = \int_{0}^{T_S} \exp j(-\omega u) . du$$

Hence $$\mathbf{H}(\omega) = \frac{\sin(\omega T_S/2)}{\omega T_S/2} . \exp j(-\omega T_S/2)$$

139

Compare this result with equation 6.17 and figure 6.14. Note that working from the impulse response is about the only way to calculate the frequency response of a hybrid process such as the sample-and-hold. (Some people would say that even to speak of the impulse response of such a system is stretching credibility!)   ●

STUDY QUESTIONS

1   Examine these signal processes, and discover whether or not they are linear.

   a) $y = a.x_1 - b.x_2$
   b) $y = a.x + b$
   c) $y = a.\sqrt{x}$
   d) $y = a.(dx/dt)$

2   Draw the signal flow graph for the following difference equation, and calculate its unit-pulse response, $g(n)$.

$$y(n) = \tfrac{1}{2}.x(n) + \tfrac{1}{2}.x(n-1) + y(n-1)$$

What mathematical operation does it perform approximately?

3   Write down the Convolution equation for discrete-time signals. Calculate the output sequence from the process defined in question 2, for an input sequence $x(n)$:

$$\{x(n) = \{1, 2, 4, 0, -2, -2, 0, 0, 0, 0 \dots\}$$

4   Apply the simple differentiator $\{g(n)\} = \{1, -1\}$ to the signal of question 3.

5   Derive the frequency response $\mathbf{H}(\omega)$ from the unit-pulse response $g(n)$, for a general discrete-time system. Then calculate the frequency response for

   a) $\{g(n)\} = \{-1, 2, 0, -2, 1\}$
   b) The process defined in question 2

   Note: $\displaystyle\sum_{r=0}^{\infty} a^r = \frac{1}{(1-a)}$

6   a) What inherent delay is shown by the process of question 5a?
      b) A certain discrete-time signal is delayed by $8T_S$ sec, when the sampling frequency is 12 kHz. What additional phase is added to the frequency response of this signal?

7   State and illustrate the principle of Reciprocal Spreading.

8 Linear Filtering: the Z-transform

OBJECTIVES

To introduce the benefits of using transforms in signal processing system analysis. In particular to

a) Show why transforms are necessary.
b) Introduce the Z-transform for discrete-time systems.
c) Show how the Z-transform links up with Convolution and Frequency Response.

COVERAGE

We commence from the need to calculate the output signal of a system, given its input signal and some measure of its characteristics. The concepts of difference equation, frequency response and convolution are reviewed, and their limitations spelled out. Signal transforms are designed to overcome these limitations, and to provide links between convolution and frequency response.

The Z-transform is then introduced in a logical but non-rigorous fashion, and its major properties are deduced. When applied to discrete-time linear filtering, it expresses the difference equation in convenient form, enables output sequences to be calculated, and characterises the filtering process by a generalised 'frequency response', as well as enabling the true frequency response to be calculated.

8.1 Review and introduction

Our whole objective in studying signal processing has been to develop a set of mathematical and practical tools which will allow us to make full use of the opportunities offered by the burgeoning art of the digital system; of which the microprocessor is only one form. Signals are readily represented

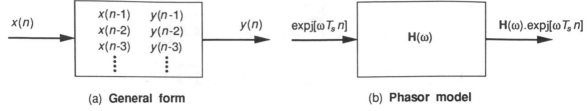

(a) **General form** (b) **Phasor model**

Figure 8.1   Signal-process models

by sequences of numerical sample values (see chapter 6 for instance), and can easily be processed by digital arithmetic.

The paramount task when designing and implementing a signal processing system is to calculate the output signal from a process, given

  a) The input signal sequence.
  b) Adequate information about the process, like the frequency response or the unit-pulse response.

In a discrete-time system, signals exist as sequences of numerical sample values, and the process is executed by means of a *difference equation*. At any given point in time, the output signal sample value is calculated as a linear sum of the input signal sample value and

  a) the past $(N-1)$ input sample values
and/or   *b*) the past M output sample values.

Thus the difference equation in general form is

$$y(n) = \sum_{r=0}^{N-1} a_r . x(n-r) + \sum_{q=1}^{M} b_q . y(n-q) \tag{8.1}$$

where a_r, b_q are the coefficients which define the process.

Figure 8.1a shows our overall view of the signal process model, and figure 8.1b shows how we view it for phasor inputs.

The difference equation describes the *microscopic* behaviour of the signal process on a sample-by-sample basis, but it says nothing about the long-term behaviour of the process, for example its response to a step input or to a sinusoidal input at a certain frequency.

● **Example 8.1**   The following difference equation describes a simple 'running average' filter:

$$y(n) = \tfrac{1}{2}[x(n) + x(n-1)]$$

This filter then retains a memory of the last input sample as well as the present one, and the output sequence may easily be calculated:

n	0	1	2	3	4	5	6	7
$x(n)$	50	66	22	-16	-34	-8	0	0
$y(n)$	25	58	44	3	-25	-21	-4	0

Thus we are able to calculate the output sequence to an arbitrary input sequence, but we have little idea as to what it means, or what is the general purpose of the filter.   ●

We need also a *macroscopic* view of the process, in order to be able to design it to perform some given task. The *frequency response* **H**(ω) gives us such a view, but only for phasor input signals, or for signals which can be modelled by a collection of phasors (see figure 8.1b and chapters 3, 4, 5). We really require a generalised 'frequency response' which will cope with other forms of input signals, such as transient signals, step-like signals, and for realistic random signals like speech for example. Some of these signals are shown in figure 8.2.

Only rarely are signals in sinusoidal form, since information is carried by changes in signal parameters, like amplitude and phase. How then are we to calculate the response of a signal process to some arbitrary input signal, other than by a term-by-term calculation?

The convolution equation does enable us to calculate the output sequence for an arbitrary input sequence, but again it is a term-by-term calculation, equation 7.14 yielding only one sequence value per summation. It is

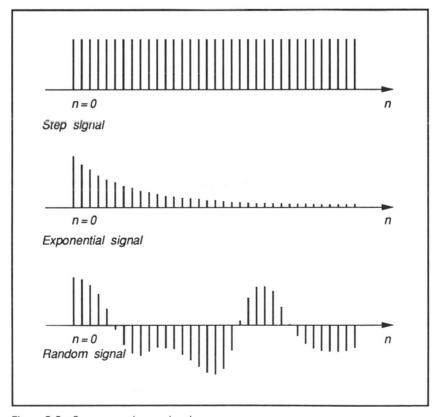

Figure 8.2 Some non-phasor signals

143

generally messy to carry out, and its overall effect is difficult to understand. A set of functional relationships is required, which can then be manipulated algebraically. It is for this reason that transforms have been developed.

8.2 The Z-transform introduced

8.2.1 Transforms in perspective

A signal transform is a mathematical operation which *maps* a signal into a different *domain* or *regime*. The mapping or conversion is chosen so that certain operations can be carried out more simply than was originally the case.

Such a vague description is not much help, so consider a simple example. We wish to calculate numerically the non-linear process:

$$y = x^b \tag{8.2}$$

Now this is a *difficult* operation if the exponent b is not an integer, and would normally be done by using logarithms, as indicated in the diagram of figure 8.3, where nodes represent signals and arrows represent operations.

The calculation of x^b from x is either impossible or at best very difficult, so we take an alternative route by *transforming* the variable x into the *domain* or *regime* of logarithms. In this transformed domain, our difficult operation is replaced by simple multiplication, and when operated on by the inverse transform, we finish up with the desired output variable.

We have already used a form of signal transform in chapter 4, in order to calculate the output of a system when the input signal is modelled by a set of phasors. The alternative, difficult route is of course the convolution operation which we met in chapter 7. Referring back to section 4.1.3 we note that periodic signals can be modelled by a sum of phasor components whose frequencies are multiples of a *fundamental* frequency ω_0, the Fourier series. Thus from equation 4.10, such a signal may be expressed as

$$x(t) = \sum_{k=-N}^{N} \mathbf{C}_k \cdot \exp j(k\omega_0 t) \tag{8.3}$$

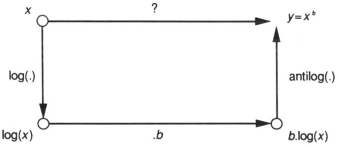

Figure 8.3 A simple transformation

144

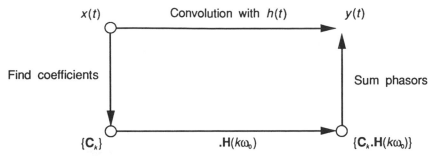

Figure 8.4 Fourier series as a transformation

Thus, given the fundamental frequency ω_0, the original signal $x(t)$ is described completely by the Fourier series set of complex coefficients $\{C_k\}$. This series representation is a transformation into a domain where the system output signal is determined by multiplying each coefficient by the appropriate value of the frequency response $\mathbf{H}(\omega)$:

$$y(t) = \sum_{k=-N}^{N} \mathbf{H}(k\omega_0) . \mathbf{C}_k . \exp \mathrm{j}(k\omega_0 t) \tag{8.4}$$

The diagrammatic form is given in figure 8.4.

So in order to avoid convolution, we have to carry out a seemingly complicated sequence of operations. However, this transformation route is easier than convolution in practice, and when expressed in discrete-time form, the transformation and inverse transformation are efficiently carried out by the *Fast Fourier Transform (FFT)* (chapter 13). It is therefore a practical alternative to convolution.

The Z-transform is just one step forward of the Fourier series example that we have just quoted. We now deal with discrete-time signals, and are not limited to periodic signals of the type which can be modelled by phasors. Our objective again is to calculate the overall characteristics of the output signal from a discrete-time system. The difficult route is the discrete convolution operation, which effectively evaluates the difference equation. The easier route is accomplished by transforming the input signal into the Z domain, $x(n) \Rightarrow X(z)$, where convolution is replaced by a multiplication of the generalised response $H'(z)$. Figure 8.5 shows this operation in diagrammatic form.

Having established the overall view of the Z-transform as far as signal processing is concerned, we can now start to examine the meaning of the transformed variable $X(z)$ and the nature of the generalised response $H'(z)$.

In passing, we note that in the continuous-time regime, the *Laplace transform* corresponds to the Z-transform in the discrete-time world. Both transforms offer a greater insight into the operation of signal processing as well as shortcutting the convolution operation, as we shall see in chapter 9 through using the Z-transform.

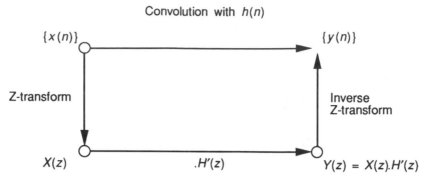

Figure 8.5 The Z-transform in operation

8.2.2 The Z-transform of a signal

We define the Z-transform and then follow with a discussion of what it means, together with an explanation of some of its properties. At this level we are regarding the Z-transform merely as a useful tool which is very important in this business of digital signal processing. In order to prove and verify formally the various properties of the Z-transform, we would need to draw upon some rigorous mathematics, but that is not our intention here.

We begin by regarding the variable z as a *delay operator*, so that z^{-1} represents unit delay.

A discrete-time signal $x(n)$ consists of the set of numerical sample values $\{x(n)\}$, commencing with $n = 0$. We can associate the nth member of the set with a delay of n intervals relative to the start of the sequence, so that we could write it as $x(n) \cdot z^{-n}$.

● **Example 8.2** Example 8.1 gives two sequences $x(n)$ and $y(n)$. When expressed in terms of the z operator they become

$$X(z) = 50 + 66z^{-1} + 22z^{-2} - 16z^{-3} - 34z^{-4} - 8z^{-5} \ldots$$
$$Y(z) = 25 + 58z^{-1} + 44z^{-2} + 3z^{-3} - 25z^{-4} - 21z^{-5} \ldots \qquad ●$$

This then is the simplest expression of the Z-transform, which represents a sequence of numerical values by a polynomial in the variable z. In general form, the Z-transform may be written as

$$X(z) = \sum_{n=0}^{\infty} x(n) \cdot z^{-n} \qquad (8.5)$$

Note that this is a *single-sided* transform; it is defined here only for positive values of the time index n.

The sequence of input values in examples 8.1 and 8.2 is arbitrary, and the Z-transforms of the sequences do not yield much additional understanding of the nature of the signals they represent, although we shall see in section 8.3.1 that this transformed signal can be used for numerical calculations.

When the sequence of values is described by an analytical function, then the Z-transform can produce a compact representation of the signal.

For instance, consider an exponential series of values:

$$x(n) = c^n \tag{8.6}$$

where $c = \exp(-\alpha)$.

Then the Z-transform of this signal is

$$X(z) = \sum_{n=0}^{\infty} c^n \cdot z^{-n} \tag{8.7}$$

Now this is a geometric series, of the form: $\Sigma\, a^n: \{0 \leqslant n \leqslant \infty\}$, for which the sum to infinity is $1/(1 - a)$. Thus

$$X(z) = \frac{1}{1 - cz^{-1}}$$

or

$$X(z) = \frac{z}{z - c} \tag{8.8}$$

We are now regarding the variable z as if it were a *complex variable*, which it is. So in using this summation formula, we are restricted by the condition that the series converges only for $|z| > c$.

Thus the Z-transform has converted a signal which is a sequence of numerical values into a compact algebraic function of z, which can now be used in system calculations. Later, in section 8.3 we shall see the benefit of this move.

● **Example 8.3** Consider the case of the signal $x(n) = c^n$ where $c = 0 \cdot 8$, then

$$\{x(n)\} = \{1 \cdot 0, 0 \cdot 8, 0 \cdot 64, 0 \cdot 51, 0 \cdot 41, 0 \cdot 33 \dots\}$$

and $X(z) = \dfrac{z}{z - 0 \cdot 8}$ ●

8.2.3 Poles and zeros

Having transformed the signal sequence into a rational function of z, we may now make some generalised deductions using features which are called *poles* and *zeros*.

The zero of a function needs little explanation; it is defined by the value of z which makes the function zero. In the function of equation 8.8, the *zero* occurs at $z = 0$. If the numerator contained a factor $(z - b)$ say, then a zero would occur when $z = b$.

A *pole* of the function occurs when the denominator is zero, giving an

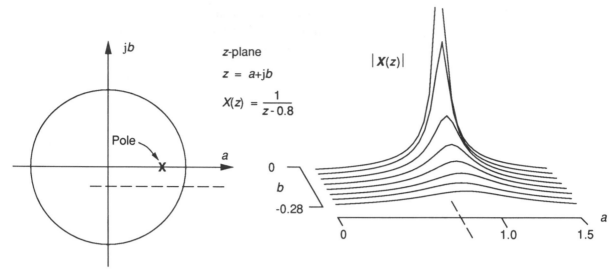

Figure 8.6 Illustrating the influence of a pole

infinite value for the function. Thus, in equation 8.8, a pole occurs when $z = c$ and the value of $X(z) \to \infty$.

The locations of these features give valuable clues regarding the behaviour of the function. In practice, such a function as $X(z)$ is completely defined by the positions of its poles and zeros, and by a multiplicative constant. Notice that z is generally complex and allowed to range in value all over a plane, the z-plane, rather than being constrained to a single line.

Figure 8.6 illustrates the idea of a pole, by working out the value of function $X(z)$ from equation 8.8 along several locii parallel to the horizontal axis. Notice that values of z near the position of the pole give rise to sharply increasing values of $|X(z)|$, while $|X(z)|$ changes much more slowly away from the pole.

We now take some simple examples of discrete-time signals, showing the signal sequence and the corresponding positions of poles and zeros in the z-plane. This is just to develop a feel for the link between these features in the z-plane and the corresponding signal sequence; detailed analysis can wait for later. Figure 8.7 shows some examples for different values of c.

On the basis of this pictorial evidence, we can say that

- *a*) A pole outside the unit circle corresponds to a growing-exponential signal sequence.
- *b*) A pole inside the unit circle denotes a decaying-exponential signal sequence.

This conclusion can be verified rigorously by studying the convergence of the Z-transform series in equation 8.7, and is always true.

We need not be so inhibited as to choose always real values for the pole position c. Indeed, if $c = \exp j(\theta)$ then the signal sequence $x(n)$ represents

148

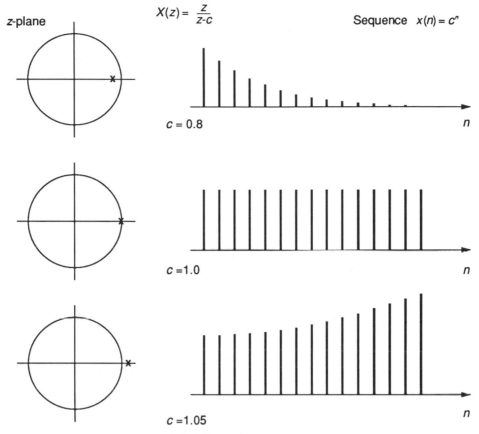

Figure 8.7 Some sequences and their Z-transforms

a discrete phasor, $x(n) = \exp j(\theta n)$, so this is an important choice to consider. *Note, however, that there is a subtle difference between our previous use of the phasor, and this present one. In the present context the signal commences at $n = 0$, whereas in chapter 3, etc. the phasor has always existed and $-\infty < n < \infty$.*

So, under this condition, the pole occurs at $z = \exp j(\theta)$. Figure 8.8 illustrates the effect of this pole for various angles θ. The sequence values are now complex, and are represented in two separate streams, real and imaginary:

$$x(n) = x_R(n) + j x_I(n) \tag{8.9}$$

Examining these diagrams, we can deduce that

a) A signal of constant amplitude corresponds to a pole on the unit circle.

b) The higher the frequency of the signal, the larger the angle of the corresponding pole. In fact, a pole angle of θ rad corresponds to a sinusoid having $2\pi/\theta$ samples per period, or in other words to a frequency of $\theta f_S/2\pi$ Hz.

149

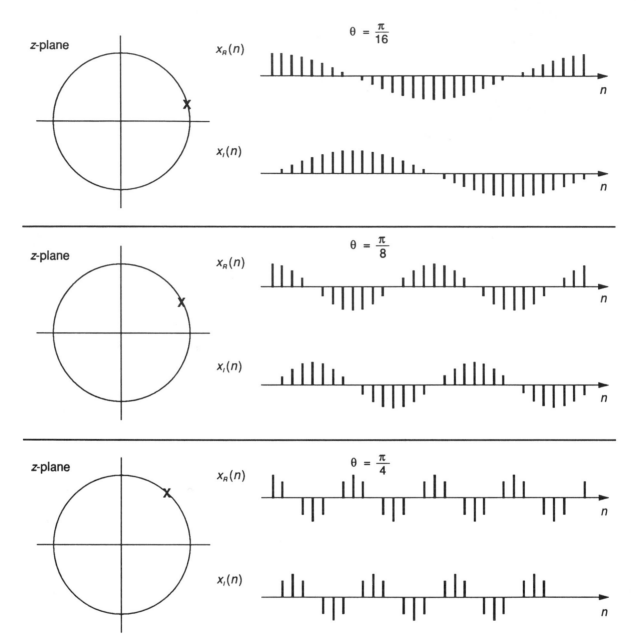

Figure 8.8 Effect of a complex pole

c) Since $\theta = \omega T_S$ for a phasor signal, we see again that for a discrete-time system, the range of valid signal frequencies is limited to $\pm \pi / T_S$ rad/sec or $\pm f_S/2$ Hz.

We can combine conclusions from figures 8.7 and 8.8 to produce a more general example like that in figure 8.9. Here $c = r \cdot \exp j(\theta)$, where r is the radius of the pole and θ is its angle, and the corresponding signal sequence has properties drawn from both figures 8.7 and 8.8.

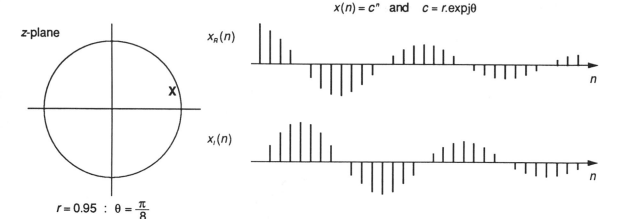

$$x(n) = c^n \quad \text{and} \quad c = r.\exp j\theta$$

$r = 0.95 \;:\; \theta = \dfrac{\pi}{8}$

Figure 8.9 A complex pole, off the unit circle

● **Example 8.4** Find the Z-transform of the sinusoidal signal:
$x(n) = \cos(\theta n)$.

Express $x(n)$ as two exponentials, then

$$x(n) - \tfrac{1}{2}[\exp\; j(\theta n) + \exp\; j(-\theta n)]$$

Transforming each term:

$$2X(z) = \frac{z}{z - \exp\; j(\theta)} + \frac{z}{z - \exp\; j(-\theta)}$$

$$X(z) = \frac{z(z - \cos(\theta))}{z^2 - 2\cos(\theta).z + 1} \qquad ●$$

Z-transform expressions can be produced in similar fashion for many signals, and table 8.1 shows a few of the simpler ones.

Table 8.1 *Some useful Z-transforms*

$x(n)$	$X(z)$
c^n	$\dfrac{z}{z - c}$
n	$\dfrac{z}{(z-1)^2}$
$a^n . \cos(\theta n)$	$\dfrac{z[z - a.\cos(\theta)]}{z^2 - 2a.\cos(\theta).z + a^2}$
$a^n . \sin(\theta n)$	$\dfrac{a.\sin(\theta).z}{z^2 - 2a.\cos(\theta).z + a^2}$

151

8.3 The Z-transform applied to linear filtering

After the discussion in section 8.2, the relationship between the original signal sequence and its Z-transform should now be a little clearer. The Z-transform converts a sequence of numerical values into a rational function of z, and the positions of poles and zeros in the z-plane provide a visual aid to understanding the general form of the overall function $X(z)$. This is parallel to the manner in which the phasor acts as a model of a periodic signal, or a signal-flow graph acts as a pictorial view of a linear equation.

Most benefit can be obtained from the Z-transform through its application to discrete-time linear filters, which gives us a compact way of describing their performance in both time and frequency.

8.3.1 Time response of a filter

We shall now apply the Z-transform to the difference equation of a filter, but as preparation, consider first the Z-transform of a signal delay process:

$$y(n) = x(n-1) \tag{8.10}$$

Hence

$$Y(z) = \sum_{n=0}^{\infty} x(n-1) . z^{-n}$$

$$= z^{-1} \sum_{n=0}^{\infty} x(n-1) . z^{-(n-1)}$$

$$= z^{-1} . X(z) \tag{8.11}$$

since $x(-1) = 0$ by definition.

Evidently then, the effect of a delay of r intervals is to multiply the Z-transform of the signal by z^{-r}, or

$$Z[x(n-r)] = z^{-r} . X(z) \tag{8.12}$$

where $Z[.]$ denotes the Z-transform.

Applying this to a discrete-time filter, we take first a Finite-Impulse-Response (FIR) filter, where the output is calculated as a linear sum of present and previous inputs. Figure 8.10 shows the signal-flow graph for such a difference equation, indicating unit delay by z^{-1}. Then

$$y(n) = \sum_{r=0}^{N-1} a_r . x(n-r) \tag{8.13}$$

and

$$Y(z) = \sum_{r=0}^{N-1} a_r . z^{-r} . X(z) \tag{8.14}$$

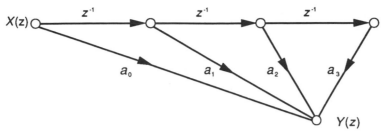

Figure 8.10 Signal-flow graph for FIR filter

Equation 8.14 gives a relationship between the Z-transform of the input signal $X(z)$, and the Z-transform of the output signal $Y(z)$. In broad terms, the ratio of system output to input is known as the *transfer function* of the process, and is a kind of generalised frequency response. In this case, we write

$$Y(z) = H'(z) \, . \, X(z) \tag{8.15}$$

and

$$H'(z) = \sum_{r=0}^{N-1} a_r \, . \, z^{-r} \tag{8.16}$$

Compare equation 8.15 with the convolution equation 7.12, where

$$y(n) = g(n) * x(n) \tag{8.17}$$

The Z-transform has enabled us to replace the convolution summation with a multiplication of Z-transforms; and also defines the transfer function $H'(z)$, which does for general signals what the frequency response $\mathbf{H}(\omega)$ does for phasor signals.

We can relate $H'(z)$ directly to the unit-pulse response $g(n)$. Consider an input signal $x(n)$ which is the unit-pulse sequence $p(n)$ (see section 7.2.1). Then

$$x(n) = p(n) \quad \text{and} \quad X(z) = 1 \tag{8.18}$$

since $p(n) = 0$ except at $n = 0$.

Consequently, when we apply the unit-pulse sequence to the input of a linear filter process, the output sequence $y(n)$ is the unit-pulse response $g(n)$, and the Z-transform of the output $Y(z)$, is the Z-transform of the unit-pulse response:

$$Z[g(n)] = H'(z) \tag{8.19}$$

● **Example 8.5** Consider the case of the simple 'running average' filter used in example 8.1. Then

$$y(n) = \tfrac{1}{2}[x(n) + x(n-1)]$$

153

The unit-pulse response is $\{g(n)\} = \{\tfrac{1}{2}, \tfrac{1}{2}\}$
The transfer function is $H'(z) = Z[g(n)]$
$$= \tfrac{1}{2}(1 + z^{-1})$$
Now we note that the output signal is given by

$$Y(z) = H'(z) \cdot X(z)$$

From example 8.2,

$$X(z) = 50 + 66z^{-1} + 22z^{-2} - 16z^{-3} - 34z^{-4} - 8z^{-5} \ldots$$

so

$$Y(z) = \tfrac{1}{2}(1 + z^{-1}) \cdot (50 + 66z^{-1} + 22z^{-2} - 16z^{-3} - 34z^{-4} - 8z^{-5} \ldots)$$
$$= 25 + 58z^{-1} + 44z^{-2} + 3z^{-3} - 25z^{-4} - 21z^{-5} \ldots \quad \bullet$$

We see from this example that the Z-transform allows us to calculate in algebraic form:

a) The generalised filter response $H'(z)$
b) The output of the filter process.

This sort of calculation is also applicable to recursive structures, giving Infinite-Impulse-Response (IIR), as the next example shows.

● **Example 8.6**   A certain smoothing filter has a recursive difference equation:

$$y(n) = x(n) + 0 \cdot 5 y(n - 1)$$

Take the Z-transform of this equation:

$$Y(z) = X(z) + 0 \cdot 5 z^{-1} \cdot Y(z)$$

and $\quad H'(z) = \dfrac{Y(z)}{X(z)} = \dfrac{1}{1 - 0 \cdot 5 z^{-1}}$

The Z-transform of the output sequence $Y(z)$ can therefore be calculated by $H'(z) \cdot X(z)$ as before, except that this now requires an algebraic long-division sum. While being cumbersome with pen and paper, polynomial multiplication and division can be done quite easily in digital form. For a simple filter like this, the arithmetic is probably easiest done using a tabular calculation of the difference equation, thus:

n	0	1	2	3	4	5	6	7
$x(n)$	50	66	22	-16	-34	-8	0	0
$y(n)$	50	91	$67 \cdot 5$	$17 \cdot 8$	$-25 \cdot 1$	$-20 \cdot 6$	$-10 \cdot 3$	$-5 \cdot 1$
$= x(n) + 0 \cdot 5 y(n-1)$								

●

8.3.2 Frequency response

The frequency response $\mathbf{H}(\omega)$ of a linear filtering process may be obtained quite easily from the Z-transform transfer function $H'(z)$.

Consider first the relationship worked through in section 7.2.3, enabling the frequency response $\mathbf{H}(\omega)$ to be determined from the unit-pulse response $g(n)$:

$$\mathbf{H}(\omega) = \sum_{n=-\infty}^{\infty} g(n) . \exp \text{j}(-\omega T_S n) \tag{8.20}$$

Now we have shown in equation 8.19 above that the Z-transform of the unit-pulse response $g(n)$ is the generalised response $H'(z)$, so

$$H'(z) = \sum_{n=0}^{\infty} g(n) . z^{-n} \tag{8.21}$$

Comparing these two equations, we note that they are very similar, and since $g(n) = 0$ for $n < 0$, the summation limits are effectively the same. Consequently we can say that the frequency response $\mathbf{H}(\omega)$ is given by the generalised response $H'(z)$ when $z = \exp \text{j}(\omega T_S)$. In formal terms:

$$\mathbf{H}(\omega) = H'(z) \big|_{z=\exp \text{j}(\omega T_S)} \tag{8.22}$$

Observe that setting $z = \exp \text{j}(\omega T_S)$ constrains the variable z to run around a circle of unity radius, which we call the *unit circle*. Figure 8.11 shows this condition, and also points out the significance of the cardinal points around the circle.

Some significant values of $\exp \text{j}(\omega T_S)$ are given in table 8.2.

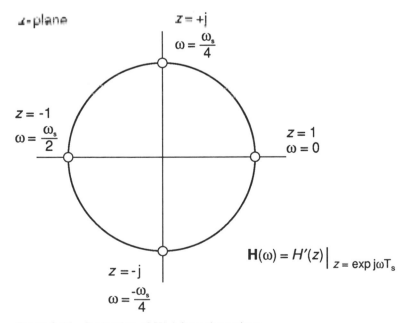

Figure 8.11 Evaluation of $\mathbf{H}(\omega)$ from the z-plane

155

Table 8.2 *Significant values of exp j(ωT_S)*

ω	0	$\omega_S/8$	$\omega_S/4$	$3\omega_S/8$	$\omega_S/2$
ωT_S	0	$\pi/4$	$\pi/2$	$3\pi/4$	π
exp j(ωT_S)	1	$(1+j)/\sqrt{2}$	j	$(-1+j)/\sqrt{2}$	-1

● **Example 8.7** Consider the simple 'running average' filter:

$$y(n) = \tfrac{1}{2}[x(n) + x(n-1)]$$

Then $H'(z) = \tfrac{1}{2}(1 + z^{-1})$

and $\mathbf{H}(\omega) = \tfrac{1}{2}[1 + \exp j(-\omega T_S)]$
$$= \cos(\omega T_S/2) \cdot \exp j(-\omega T_S/2)$$

Now suppose we wish to calculate quickly various point values on the frequency response. Then

ω	0	$\omega_S/4$	$\omega_S/2$
z	1	j	-1
$\|\mathbf{H}(\omega)\|$	1·0	$1/\sqrt{2}$	0

Similar calculations may also be done at frequencies of $\omega_S/8$, etc. but these usually require a calculator.   ●

● **Example 8.8** Repeat the above exercise for the recursive discrete-time system used in example 8.6:

$$y(n) = x(n) + 0\cdot5\,y(n-1)$$

Then $H'(z) = \dfrac{1}{1 - 0\cdot5z^{-1}}$

and $\mathbf{H}(\omega) = \dfrac{1}{1 - 0\cdot5\,\exp j(-\omega T_S)}$

ω	0	$\omega_S/4$	$\omega_S/2$
z	1	j	-1
$\|\mathbf{H}(\omega)\|$	2·0	0·89	0·67

●

We see that having obtained the Z-transform response of a discrete-time system, which we can do quite easily from the difference equation or from the unit-pulse response, we can then simply substitute exp j(ωT_S) for z to obtain the frequency response $\mathbf{H}(\omega)$. Further, certain frequencies like 0 and $\omega_S/2$ yield exceptionally easy calculations for frequency response. It is these properties that are our main motivation for considering the Z-transform at all, and these are the ones that are most commonly used in simple signal processing.

8.3.3 Combinations of filters

The Z-transform response properties of discrete-time signal processes carry through into systems which contain several components. Figure 8.12 gives the signal-flow graphs for two simple structures.

Complicated filtering operations are frequently built up from several processes in cascade, as in figure 8.12a. The overall response for these filters is

$$H'(z) = H'_1(z) \cdot H'_2(z) \cdot H'_3(z) \dots \tag{8.23}$$

We shall see in chapter 9, which considers the design of such filters for specific uses, that the poles and zeros of one section contribute to the poles and zeros of the whole structure. For continuous-time signals, the Laplace transform performs exactly the same function, defining poles and zeros which influence the frequency and time response of the system output signal (see appendix A4).

Figure 8.12b shows a typical feedback connection, and the resulting Z-transform response can be seen to be

$$H'(z) = \frac{H'_1(z)}{1 + H'_2(z)} \tag{8.24}$$

Thus, if both responses are characterised by zeros only, the response $H'_1(z)$ contributes to the zeros of the final response, while $H'_2(z)$ contributes to the poles of the final response. Most practical filters of the IIR variety are in fact made in this way.

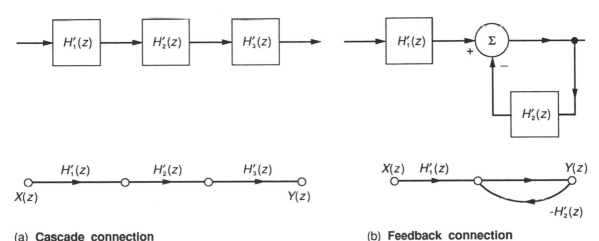

(a) **Cascade connection**

(b) **Feedback connection**

Figure 8.12 Combinations of processes

● **Example 8.9** Find the total response if two simple 'running average' filters are placed in cascade.

$$H'_1(z) = H'_2(z) = \tfrac{1}{2}(1 + z^{-1})$$

The overall response is therefore:

$$H'(z) = (1/4)(1 + z^{-1})(1 + z^{-1})$$
$$= (1/4)(1 + 2z^{-1} + z^{-2})$$

ω	0	$\omega_S/4$	$\omega_S/2$
z	1	j	-1
$\lvert\mathbf{H}(\omega)\rvert$	$1\cdot0$	$0\cdot5$	0

●

STUDY QUESTIONS

1   A simple differentiator has the following difference equation:

$$y(n) = x(n) - x(n - 1)$$

Use this equation to calculate the output sequence when the input sequence is

$$\{x(n)\} = \{0, 1, 2, 3, 4, 4, 2, 0, -2, -1, 0, 0\ldots\}$$

2   Define the Z-transform of a sequence $x(n)$. Write down the Z-transforms of

   a) The input and output signals of question 1
   b) $x(n) = 0\cdot9^n$
   c) $x(n) = 1\cdot1^n$
   d) $x(n) = 0\cdot8^n . \exp j(\pi n/8)$
   e) $\{x(n)\} = \{1, 0, 0, 0, \ldots\}$

3   Derive the Z-transform for the sequence $x(n) = \sin(\theta n)$.

4   A simple predictor has the difference equation:

$$y(n) = 2 . x(n) - x(n - 1)$$

Write down the unit-pulse response and generalised transfer function. Use the Z-transform technique to calculate the output sequence when the input sequence is as in question 1.

5   Determine the frequency response for each of the processes in questions 1 and 4; P1 and P4 respectively. Sketch the modulus, using the cardinal values of z (see section 8.3.2).

6 Determine the overall response, and sketch the frequency response of

 a) P1 and P4 in cascade.
 b) Connected as in figure 8.12b, with P1 in feedback loop and P4 in series.
 c) Connected as in figure 8.12b, with P4 in feedback loop and P1 in series.

9 Linear Filtering: Poles and Zeros

OBJECTIVES

To apply the principles discovered in chapters 7 and 8 to the design of simple filters. In particular to

- *a*) Investigate the properties of response zeros
- *b*) Investigate the properties of response poles
- *c*) Apply these ideas to the design of a pair of bandpass filters.

COVERAGE

After reviewing the frequency response function for discrete-time systems, the nature of simple and complex response zeros is discussed and illustrated. By combining the effects of several zeros, a lowpass filter can be produced.

The properties of poles are then studied in similar fashion, and the resonance-like characteristics are emphasised. Relationships are given which enable a digital resonator to be designed for a particular frequency and Q-factor.

The telecontrol signalling example, first introduced in the Introduction, is then used to illustrate how these design principles may be used. A complementary pair of bandpass filters is designed, first using zeros, then poles and finally both combined.

9.1 Review

In chapters 7 and 8 we have brought together a number of functions to describe linear filtering operations. We now use some of these results in the design of simple filters, and investigate a particular design example.

Staying for the present with discrete-time systems, the *unit-pulse response g(n)* can be deduced easily from the *difference equation* which describes the process. The output sequence $y(n)$ of a linear filter can then be calculated from the input sequence $x(n)$ by the *convolution* equation:

$$y(n) = \sum_{r=-\infty}^{\infty} g(r) \cdot x(n-r) \tag{9.1}$$

The frequency response of a system $\mathbf{H}(\omega)$ describes how the system behaves when the input is a phasor, $x(n) = \exp \mathrm{j}(\omega T_S n)$, and is related to the unit-pulse response by this equation:

$$\mathbf{H}(\omega) = \sum_{n=-\infty}^{\infty} g(n) \cdot \exp \mathrm{j}(-\omega T_S n) \tag{9.2}$$

Frequency response is the way we most frequently define a linear filtering process, and process design is a matter of finding a difference equation which produces the desired frequency response. Valuable insight can be gained by using the Z-transform which expresses the frequency response in a general form. Thus we note that the Z-transform of the unit-pulse sequence $H'(z)$ is related to the phasor frequency response $\mathbf{H}(\omega)$ by

$$\mathbf{H}(\omega) = H'(z) \big|_{z = \exp \mathrm{j}(\omega T_S)} \tag{9.3}$$

Notice that for a *causal* system, that is one where the output occurs *after* the input and not before it, the Z-transform of the unit-pulse sequence possesses only negative powers of z, and is

$$H'(z) = \sum_{n=0}^{\infty} g(n) \cdot z^{-n} \tag{9.4}$$

Since $g(n)$, $\mathbf{H}(\omega)$ and $H'(z)$ are all intimately related, and each defines the filter from its own point of view, we can use for design or analysis, whichever form is most convenient at the time. The Z-transform $H'(z)$ enables the response to be described in terms of *poles* and *zeros*, features which give valuable insight into the design procedure; and this is where we concentrate for the time being.

Our target, then, is to produce a filter with a specific $\mathbf{H}(\omega)$, and the poles and zeros of $H'(z)$ enable this to be done. Notice that for discrete-time systems poles and zeros are defined in the complex z-plane, whereas, for continuous-time systems, poles and zeros are defined in the complex s-plane. The frequency response $\mathbf{H}(\omega)$ is influenced equally by poles and zeros in both domains, but placing of them is much easier for discrete-time systems than for continuous-time systems. Hence the following discussion will be in terms of discrete-time systems and digital processing, rather than continuous-time systems and analog circuit components. Appendix A4 introduces the corresponding ideas for continuous-time signals, incorporating the Laplace transform instead of the Z-transform.

After introducing the importance of poles and zeros, we show that in simple cases they can be used to tailor the frequency response in order to suit a particular application. Zeros are first considered, and then attention is turned to poles, drawing out their distinctive features. Then a design example, taken from the telecontrol system of the Introduction, is tackled with only zeros, only poles and then with poles and zeros combined.

Practical design techniques are more sophisticated than those introduced here, which are just for illustration of what is possible, and to establish certain fundamental principles. There are many textbooks which detail these design techniques, and many computer programs are available for automatic design of filters for all kinds of duty.

9.2 Zeros

9.2.1 Simple zero

A zero is that value of z which makes $H'(z) = 0$. So if this value is r say, then

$$H'(z) = z^{-1} \cdot (z - r) \tag{9.5}$$

Observe that a z^{-1} factor has been introduced so that this Z-transform has all-negative powers of z, corresponding to a delayed output and not an advanced one.

If the zero occurs on the unit-circle, then $\mathbf{H}(\omega)$ is also zero at this point (see equation 9.3). The only possible position for a simple zero is therefore at $r = \pm 1$, corresponding to $\omega = 0$ or $\omega_S/2$.

● **Example 9.1** Consider the design of a discrete-time system which approximates to differentiation. In the frequency domain, the first requirement is that the process should have zero response to zero-frequency signals.

That is $\mathbf{H}(\omega) = 0$ at $\omega = 0$

and $H'(z) = 1 - z^{-1}$

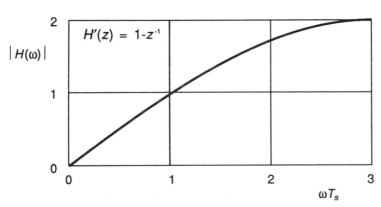
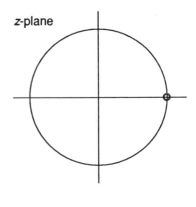

Figure 9.1 Frequency response of simple differentiator

while $y(n) = x(n) - x(n-1)$

Consequently $|\mathbf{H}(\omega)| = 2 . \sin(\omega T_S/2)$ from equation 9.3.

Figure 9.1 plots this frequency response, showing the zero quite clearly. This is not a particularly good approximation to a differentiator, but is one that is often used. More of that in chapter 11. ●

9.2.2 Pair of complex zeros

A simple zero can eliminate a signal with a frequency of 0 or $\omega_S/2$ rad/sec, but more frequently we wish to remove components at other frequencies. For example, consider the system diagram of figure 9.2, which shows the procedure for measuring the mean-power value of a sinusoidal signal, using discrete-time processing.

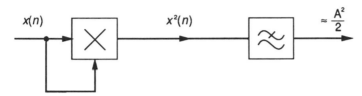

Figure 9.2 System diagram for power measurement

Let the input signal be $x(n) = A . \cos(\omega_1 T_S n)$. Then after squaring we get

$$x^2(n) = (A^2/2) . [1 + \cos(2\omega_1 T_S n)] \tag{9.6}$$

$x^2(n)$ is the instantaneous power, but the mean-power is given by the average value (see chapter 10), which is $A^2/2$. In order to observe this as a measured value, the $2\omega_1$ frequency component must be eliminated, which can be done by using a lowpass filter which has zeros at $\pm 2\omega_1$ rad/sec.

Now let $\theta = 2\omega_1 T_S = 4\pi f_1/f_S$. Then we may form the response function to have two zeros at angles of $\pm\theta$ in the z-plane, as follows:

$$H'(z) = z^{-2} . [z - \exp \mathrm{j}(\theta)] . [z - \exp \mathrm{j}(-\theta)]$$
$$= z^{-2} . [z^2 - 2 . \cos(\theta) . z + 1] \tag{9.7}$$

So this expression gives the general form for forcing a pair of conjugate zeros at angles $\pm\theta$ in the z-plane, corresponding to frequencies $\pm\theta f_S/2\pi$ Hz.

● **Example 9.2** It is required to eliminate a frequency of $1 \cdot 2$ kHz in a discrete-time processing system with a sampling frequency of 8 kHz.

First discover the corresponding z-plane angle for the zeros.

$$\theta = 2\pi f/f_S = 0 \cdot 94 \text{ rad}$$

The coefficient of z in equation 9.7 then becomes

$$2 . \cos(0 \cdot 94) = 1 \cdot 18$$

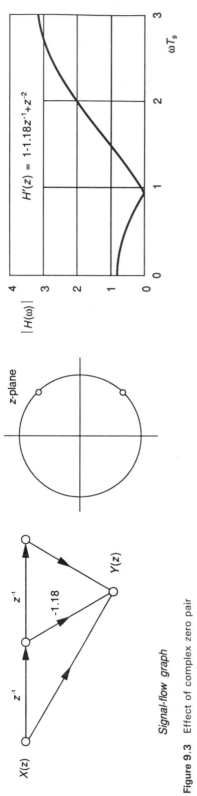

Signal-flow graph

Figure 9.3 Effect of complex zero pair

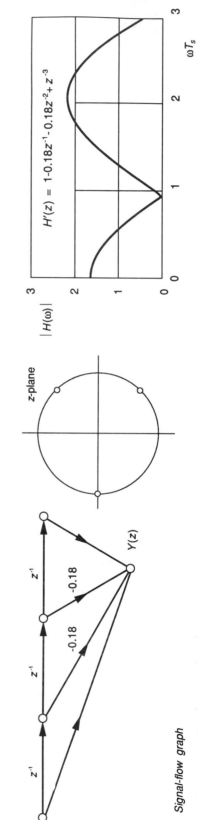

Signal-flow graph

Figure 9.4 Frequency response of revised filter

Hence $H'(z) = z^{-2} \cdot (z^2 - 1 \cdot 18z + 1)$

$$= 1 - 1 \cdot 18z^{-1} + z^{-2}$$

and $y(n) = x(n) - 1 \cdot 18x(n - 1) + x(n - 2)$

The frequency response is

$$H(\omega) = 1 - 1 \cdot 18 \cdot \exp j(-\omega T_S) + \exp j(-2\omega T_S)$$

The magnitude of this response is shown in figure 9.3, together with the zero positions, and the signal-flow graph of the process. ●

Systems with several zeros can be constructed by simply multiplying together the factors which define the individual zeros.

● **Example 9.3** Suppose that the system in example 9.2 is also to have zero frequency response at 4 kHz.

4 kHz is the Nyquist frequency, so this requirement means a simple zero at $z = -1$.

Then $H'(z) = z^{-3} \cdot (z^2 - 1 \cdot 18z + 1) \cdot (z + 1)$

$$= z^{-3} \cdot (z^3 - 0 \cdot 18z^2 - 0 \cdot 18z + 1)$$

$$= 1 - 0 \cdot 18z^{-1} - 0 \cdot 18z^{-2} + z^{-3}$$

and $y(n) = x(n) - 0 \cdot 18x(n - 1) - 0 \cdot 18x(n - 2) + x(n - 3)$

$$H(\omega) = 1 - 0 \cdot 18 \cdot \exp j(-\omega T_S) - 0 \cdot 18 \cdot \exp j(-2\omega T_S) + \exp j(-3\omega T_S)$$

This response is plotted in figure 9.4. ●

9.2.3 Zeros in general

We have established the effect of simple zeros, and of a complex pair of zeros. Since these have been on the unit-circle, with a radius of $1 \cdot 0$, they have influenced the frequency response directly, forcing zeros of response.

This idea may be developed to represent the response of a certain class of linear filters in general terms:

$$H'(z) = A \cdot z^{-N} \prod_{i=1}^{N} (z - z_i) \tag{9.8}$$

This equation tells us that the filtering process has N zeros, placed at positions $\{z_i\}$. When all the factors are multiplied out, they form an Nth degree polynomial in z, which therefore works back to define a difference equation having N delay elements in it. In many practical filters, N may well be in the order of several hundred.

The multiplicative constant A just sets the overall gain of the filter, for instance setting $H(0) = 1$.

In equation 9.8, the zeros themselves are shown singly, but complex zeros

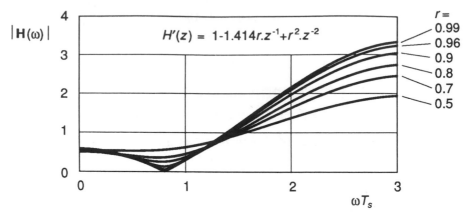

Figure 9.5 Effect of altering the radius of a zero-pair

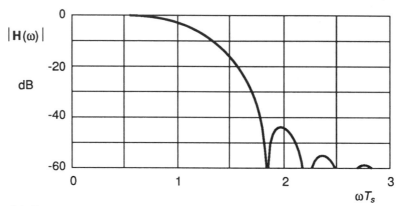

(a) **Frequency response**

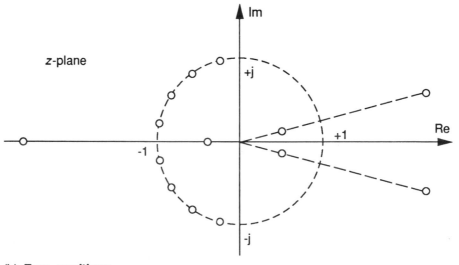

(b) **Zero positions**

Figure 9.6 An all-zero lowpass filter

always occur in conjugate pairs for real filters, so that terms like that in equation 9.7 will also be used in representing the response. Further, zeros are not necessarily constrained to lie on the unit-circle, their radii may be greater or less than unity. Figure 9.5 shows the effect of altering the radius of a conjugate pair of zeros which have an angle of $\pi/4$ rads.

These concepts are brought together and illustrated in figure 9.6, which shows a lowpass filter containing only zeros; those in the stopband being on the unit-circle, while those in the passband are off the unit-circle. Zeros in the stopband force the response to be very low in this region, so that the attenuation is high.

Unfortunately this simple type of filter design shows a very gradual change between the high gain in the passband and the low gain in the stopband, and sharper transitions are generally required in practice.

The techniques of design whereby these zero positions are found are not our concern here, but may be followed up in one of the many textbooks on the subject. Design procedures are not often derived by considering the positions of response zeros directly, and these usually emerge as a result of having carried out some algebraic manipulation of the overall response function.

9.3 Poles

A zero causes the system frequency response to go to zero, or to dip in its vicinity. A pole is the opposite; in the neighbourhood of a pole, the response function peaks, and at the pole position itself the function goes to infinity. Hence, a zero allows a signal at a certain frequency to be removed or rejected, while a pole allows a signal at that frequency to be enhanced or magnified.

9.3.1 Simple pole

Let us take the case of a pole at a certain location r, analogous to the simple zero in section 9.2.1. The Z-transform response is therefore

$$H'(z) = \frac{z}{z - r} \tag{9.9}$$

Translating this into the difference equation for input signal x and output signal y, we remind ourselves first that

$$H'(z) = \frac{Y(z)}{X(z)} \tag{9.10}$$

Hence $Y(z) = X(z) + r \cdot z^{-1} Y(z)$

and $y(n) = x(n) + r \cdot y(n-1)$ \hfill (9.11)

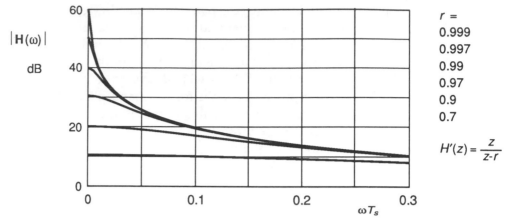

Figure 9.7 Frequency response for a simple pole

This expression therefore defines the recursive difference equation process that we met before in sections 3.3.2, 7.1.2 and 7.2.1.

Examining the frequency response with the tools that we now have available, we note that the frequency response gain at $\omega = 0$ (ie. $z = 1$) is $1/(1 - r)$, while the frequency response gain at $\omega = \omega_s/2$ (ie. $z = -1$) is $1/(1 + r)$ (see section 8.3.2). The overall expression for frequency response is

$$\mathbf{H}(\omega) = \frac{1}{1 - r \cdot \exp\, \mathrm{j}(-\omega T_S)} \tag{9.12}$$

The magnitude of this response is plotted in figure 9.7 for various values of the coefficient r. Notice how

a) The response peaks at $\omega = 0$ ($z = 1$), which is the point on the unit-circle nearest to the position of the pole.
b) The magnitude of the peak response increases as the value of r approaches unity. In fact $|\mathbf{H}(0)| = 1/(1 - r)$.
c) The response becomes very sharp as the pole radius r approaches unity. Extremely selective filters can therefore be made in digital form.

If the pole was actually on the unit-circle ($r = 1$), then clearly the response would be infinite, which is not to be allowed! There is a definite restriction therefore that

$$r < 1 \tag{9.13}$$

● **Example 9.4** Consider the example 9.2, taken above in section 9.2.2 to show power measurement with a sinusoidal signal. The complex zero was selected so as to eliminate completely the twice-frequency component at $1 \cdot 2$ kHz with a sampling frequency of 8 kHz. The simple-pole recursive filter that we have here acts as a lowpass filter but amplifies the zero-frequency component more than the sinusoidal component.

The zero-frequency component is the 'wanted' signal component, while the $1 \cdot 2$ kHz signal component is 'unwanted'.

Before filtering $\quad \dfrac{\text{Wanted signal}}{\text{Peak unwanted signal}} = 1$

After filtering, this ratio becomes $\quad |H(0)| / |H(\omega)|$

Now $\quad |H(0)| = 1/(1-r)$

and $\quad |H(\omega)|^2 = 1/[1 + r^2 - 2r \cdot \cos(2\pi \cdot 1 \cdot 2/8)]$
$$= 1/[1 + r^2 - 1 \cdot 18r]$$

So, after filtering,

$$\dfrac{\text{Wanted signal}}{\text{Peak unwanted signal}} = \dfrac{\sqrt{(1 + r^2 - 1 \cdot 18r)}}{(1-r)}$$

$$= 90 \cdot 1 \text{ or } 39 \text{ dB} \quad \text{if } r = 0 \cdot 99$$

This value for r therefore improves the discrimination of the system between signals at zero-frequency and $1 \cdot 2$ kHz, by emphasising the zero-frequency component. As $r \to 1 \cdot 0$, so the selection improves. ●

● **Example 9.5** The expressions derived in example 9.4 may also be used to discover the BANDWIDTH of this filtering process. Taking the modulus of equation 9.12, we are required to find the value of the angle $\theta = \omega T_S$, so that the gain at this frequency is 3 dB less than the zero-frequency gain.

Thus $\quad |\mathbf{H}(\theta)|^2 = |\mathbf{H}(0)|^2/2$

Re-arranging the expressions above, we discover that

$$\cos(\theta) = \dfrac{4r - 1 - r^2}{2r} \tag{9.14}$$

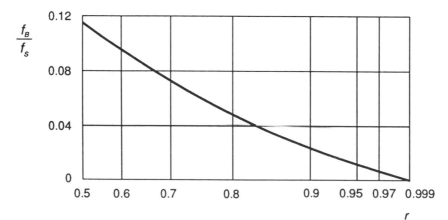

Figure 9.8   3 dB bandwidth of simple pole, versus pole radius

169

This relationship is fairly complicated, and not to be memorised, but can be illustrated by the graph in figure 9.8, which shows how the 3 dB bandwidth varies with the pole radius r. Notice the highly non-linear horizontal axis which distorts the curve!   ●

Before we leave the simple pole, note that the unit-pulse response of this process is (section 8.2.2):

$$g(n) = r^n \tag{9.15}$$

As $r \to 1 \cdot 0$, so the unit-pulse response sequence dies away more slowly, but the bandwidth of the filter response gets narrower. This is a further illustration of the principle of *Reciprocal Spreading*, which was introduced in section 7.2.4.

9.3.2 Pair of complex poles

A simple pole will emphasise the frequency response in the neighbourhood of $\omega = 0$ or $\omega = \omega_S/2$, depending upon whether the pole is along the positive or negative real axis. To emphasise a frequency response at some intermediate frequency, we need to establish a conjugate pair of poles at angles $\pm \theta$.

Then we may form a response function which has a pair of poles at angles of $\pm \theta$ in the z-plane, and at radius r as follows:

$$H'(z) = \frac{z^2}{[z - r \cdot \exp \text{j}(\theta)] \cdot [z - r \cdot \exp \text{j}(-\theta)]}$$

$$= \frac{z^2}{z^2 - 2r \cdot \cos(\theta) \cdot z + r^2} \tag{9.16}$$

This is the general form for forcing a pair of conjugate poles at angles $\pm \theta$ in the z-plane, corresponding to frequencies $\pm \theta f_S/2\pi$ Hz.

● **Example 9.6** Using similar figures to those of example 9.2, we require to emphasise a frequency of $1 \cdot 2$ kHz in a discrete-time processing system with a sampling frequency of 8 kHz.

First discover the corresponding z-plane angle for the poles:

$$\theta = 2\pi f/f_S = 0 \cdot 94 \text{ rad}$$

The coefficient of z in equation 9.16 then becomes

$$- 2r \cdot \cos(0 \cdot 94) = - 1 \cdot 18r$$

Choose, somewhat arbitrarily, $r = 0 \cdot 9$; then

$$H'(z) = z^2/(z^2 - 1 \cdot 06z + 0 \cdot 81)$$
$$= 1/(1 - 1 \cdot 06z^{-1} + 0 \cdot 81z^{-2})$$

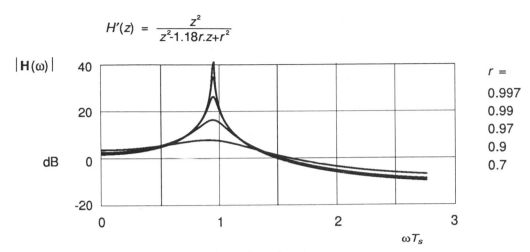

$$H'(z) = \frac{z^2}{z^2 - 1.18r.z + r^2}$$

$r =$
0.997
0.99
0.97
0.9
0.7

Figure 9.9 Frequency response of complex pole-pair

and $y(n) = x(n) + 1 \cdot 06\,y(n-1) - 0 \cdot 81\,y(n-2)$

The frequency response is

$$\mathbf{H}(\omega) = 1/[1 - 1 \cdot 06 \,.\, \exp\, j(-\omega T_S) + 0 \cdot 81 \,.\, \exp\, j(-2\omega T_S)] \qquad \bullet$$

Selection of the pole radius r in the above example was arbitrary, but in order to design useful filters we need to show how the response is influenced by this radius. Figure 9.9 shows how the frequency response varies when the pole radius r is changed, for the pole angle θ as calculated in example 9.6.

This response is very similar to the resonant circuit in analog systems, and so the structure is often called a *digital resonator*. As the pole radius increases and the pole position nears the unit-circle where frequency response is calculated, so the response gets more peaky, the gain at resonance increases and the bandwidth decreases. The bandwidth is often expressed in the context of (resonant frequency)/(bandwidth) which is known as the *Quality Factor* or *Q-factor* of the resonator.

The calculation of these parameters is tedious, but working from equation 9.16 we can deduce that for $r \rightarrow 1 \cdot 0$,

$$\text{Gain at resonance} \simeq \frac{1}{2(1-r)\,.\, \sin(\theta)} \tag{9.17}$$

$$Q \simeq \frac{\theta}{2(1-r)} \tag{9.18}$$

where $r > 0 \cdot 8$.

These properties are illustrated in figure 9.10, which shows the set of resonances at different pole angles θ, for a constant pole radius $r = 0 \cdot 95$. Notice how the gain at resonance varies according to $1/\sin(\theta)$, and that the resonator bandwidth stays approximately constant as the angle is increased.

171

$$H'(z) = \frac{z^2}{z^2 - 1.94 \cos\theta z + 0.941}$$

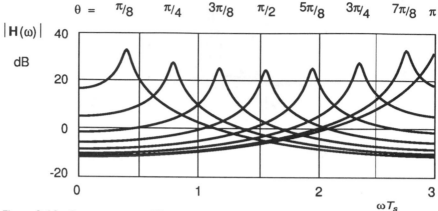

Figure 9.10   Resonances at different frequencies

Now regarding the corresponding unit-pulse response, section 8.2.3 has discussed the relationship between pole position and signal sequence. From table 8.1 we can see that a pair of conjugate poles such as we are studying here, gives rise to a unit-pulse response of the form:

$$g(n) = r^n . \cos(\theta n) \tag{9.19}$$

Thus, as the pole radius is increased towards $1\cdot0$,

   a) The bandwidth decreases—equation 9.18
   b) The unit-pulse response dies away more slowly—equation 9.19.

The *Reciprocal Spreading* principle is vindicated yet again!

● **Example 9.7**   Let us explore the example 9.6 in more detail. We require to emphasise a frequency component of $1\cdot2$ kHz when the sampling frequency is 8 kHz. However, in order to give a more rational reason for choosing the pole radius *r*, suppose that we require the bandwidth of this resonator to be 100 Hz.
   From example 9.6 we already have that $\theta = 0\cdot94$ rad.
   The Q-factor is to be $1200/100$ or 12.
   Working from equation 9.18 $(1 - r) = \theta/2Q = 0\cdot039$
   Hence $r \simeq 0\cdot96$

The maximum gain of the resonator is approx.

$$1/[2 . 0\cdot039 . \sin(0\cdot94)] = 15\cdot9 \quad \text{or} \quad 24 \text{ dB}$$

Then $H'(z) = z^2/(z^2 - 1\cdot13z + 0\cdot92)$
$$= 1/(1 - 1\cdot13z^{-1} + 0\cdot92z^{-2})$$

and $\quad y(n) = x(n) + 1 \cdot 13\, y(n-1) - 0 \cdot 92\, y(n-2)$

The frequency response is

$$\mathbf{H}(\omega) = 1 / [1 - 1 \cdot 13 \exp \mathrm{j}(-\omega T_S) + 0 \cdot 92 \exp \mathrm{j}(-2\omega T_S)] \qquad \bullet$$

9.4 A design example

9.4.1   A telecontrol system

The telecontrol requirements were first outlined in section I.4.1 of the Introduction. This example system controls a switch at a remote point, for example an electrical supply circuit-breaker or a railway point machine. In order to carry out this control function, we need three items of information to be communicated over a cable system:

a) ON/OFF control of the switch
b) ON/mid-way/OFF indication of the state of the switch
c) Telemetered quantity such as current.

In section I.4.1 we evaluated the information content of these three items, and in section 5.1.4 we allocated some channel frequencies which would enable these three signals to share the same cable.

Having spent some time developing tools to analyse linear filtering, we may now use this skill to design a simple receiver to suit this application. Our design will not be realistic since there are several interacting requirements to be met in the practical case, but we choose a simplified approach in order to illustrate how the design process might proceed and what trade-offs there are in the design of linear filters

The sub-set problem that we are to tackle is the $1 \cdot 5$ kHz demodulator in figure 5.4. This is placed at the Outstation, and must receive and act unequivocally on the control information to open or close the switch. We shall use two frequencies:

1 kHz tone—denotes that the switch is to open
2 kHz tone—denotes that the switch is to close

The transmission channel carries a signal which is $1 \cdot 5 \pm 0 \cdot 5$ kHz. By using two distinct signals, and requiring that only one of them should be acted upon, we reduce the possibility of false operation due to interference on the cable circuit.

So, figure 9.11a shows the schematic of the receiver structure, which consists of two filters, followed by two detectors and a decision process. Figure 9.11b shows broadly how this receiver would be realised in hardware terms. We now require to discover the filter coefficients which may be programmed into the digital signal processor (DSP) device.

The detectors assess the signal level in each receiver channel, and do so

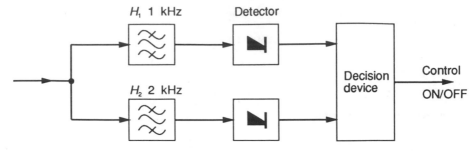

(a) **Schematic of receiver structure**

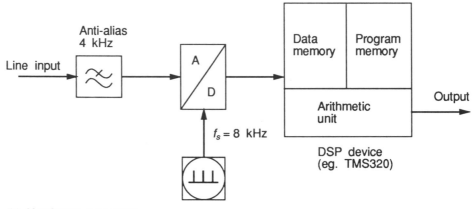

(b) **Hardware structure**

Figure 9.11   Telecontrol design problem

by rectifying the signal and then averaging the result so as to obtain a reliable estimate. This process is discussed in detail in chapter 10. Having obtained estimates of signal power for each channel, these are then compared to see which channel contains the greater power and hence the most likely signal. Without any noise or interference signals, the decision process is extremely simple, but in practice there must be some uncertainty about the result because of disturbances during transmission of the signal.

The two filters are to be *complementary* or *orthogonal*, since the filter which receives the 1 kHz signal must reject the 2 kHz signal, and vice-versa.

In this case, as well as the two frequencies that constitute the signalling options, there is a third frequency of 3 kHz which arises from the indication channel (see table 5.1 and figure 5.4). This appears as unwanted interference and must be removed. Summarising these requirements, we can specify the two filters as follows:

Filter 1   To pass the frequency of 1 kHz, and to minimise the response at frequencies of 2 and 3 kHz.

Filter 2   To pass the frequency of 2 kHz and to minimise the response at frequencies of 1 and 3 kHz.

Since these are to be bandpass filters, it is desirable to make each response symmetrical about its centre frequency. For simplicity in calculation, we choose a sampling frequency of 8 kHz.

9.4.2 Solution using zeros

For all-zero filters, we concentrate on removing unwanted signal components by forcing zeros to occur at these frequencies.

Consider filter 1, which is to eliminate frequencies of 2 and 3 kHz. A zero will also be forced at 0 kHz, so as to make the response symmetrical about the desired frequency of 1 kHz and hence form a peak in the response at this frequency. Table 9.1 lists the attributes of each zero, and builds up the factors which will form the generalised response $H'(z)$ using the general terms of equations 9.5 and 9.7.

The response for filter 1 is therefore

$$H_1'(z) = z^{-5}(z-1)(z^2+1)(z^2+1\cdot41z+1)$$
$$= 1 + 0\cdot41z^{-1} + 0\cdot59z^{-2} - 0\cdot59z^{-3} - 0\cdot41z^{-4} - z^{-5}$$

Hence

$$y(n) = x(n) + 0\cdot41x(n-1) + 0\cdot59x(n-2)$$
$$- 0\cdot59x(n-3) - 0\cdot41x(n-4) - x(n-5) \quad (9.20)$$

Figure 9.12a shows the signal-flow graph for implementing this filter, and figure 9.13 shows the frequency response.

Filter 2 can be designed in similar fashion (see Table 9.2).
The response is therefore

$$H_2'(z) = z^{-6}(z-1)(z^2 - 1\cdot41z + 1)(z^2 + 1\cdot41z + 1)(z+1)$$
$$= 1 - z^{-2} + z^{-4} - z^{-6}$$

Hence

$$y(n) = x(n) - x(n-2) + x(n-4) - x(n-6) \quad (9.21)$$

Figure 9.12b shows the corresponding signal-flow graph, and figure 9.13 the frequency response.

Notice that the filters do accomplish the design spec. which was laid out.

Table 9.1 *Zeros for filter 1*

Frequency	Angle	Type	Factor
0 kHz	0 rad	simple	$z - 1$
2	$\pi/2$	complex	$z^2 + 1$
3	$3\pi/4$	complex	$z^2 + 1\cdot41z + 1$

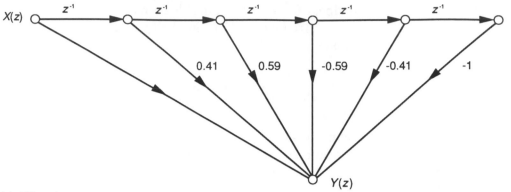

(a) **Filter 1**

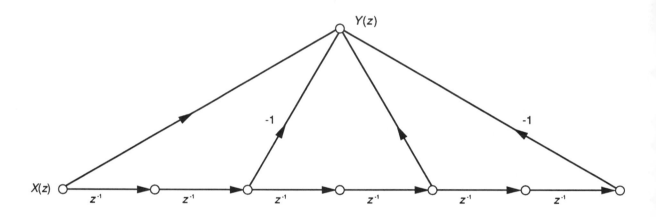

(b) **Filter 2**

Figure 9.12 Signal-flow graphs of all-zero filters

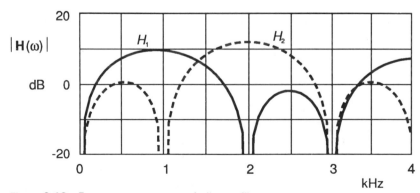

Figure 9.13 Frequency responses of all-zero filters

176

Table 9.2 *Zeros for filter 2*

Frequency	Angle	Type	Factor
0 kHz	0	simple	$z - 1$
1	$\pi/4$	complex	$z^2 - 1\cdot41z + 1$
3	$3\pi/4$	complex	$z^2 + 1\cdot41z + 1$
4	π	simple	$z + 1$

However, notice too:

a) Filter 1 has a large response in the region of 4 kHz. This might be an embarrassment, depending upon what other interference signals are expected. An additional simple zero at 4 kHz would control this.

b) The filters have different gains at their centre frequencies, which would be taken into account when comparing signal levels in the two receiver channels.

c) Filter 2 has a particularly simple difference equation, requiring only addition and subtraction and no multiplications. In general this is desirable, since arithmetic multiplication is often a time-consuming and intricate operation, although modern DSP devices are designed for fast multiplication. There are not many filters which can be made without any multiplications, but multiplication by a power of 2 is easily accomplished by shifting the digital word to the right or left.

9.4.3 Solution using poles

Zeros were used to eliminate unwanted frequency components. Poles can be used to enhance the frequencies that we want, and in general can do the job more economically.

For filter 1, we require a pole in the region of 1 kHz, that is a pole angle of $\pi/4$ rad. In order to get good discrimination against unwanted frequency components, let us choose a Q-factor of say 20, which gives a bandwidth of 50 Hz. Then, from equation 9.18 we determine that $r = 0\cdot98$.

Hence using the standard form for a complex pole pair (equation 9.16), we see that

$$H_1'(z) = \frac{z^2}{z^2 - 1\cdot39z + 0\cdot96}$$

and

$$y(n) = x(n) + 1\cdot39y(n - 1) - 0\cdot96y(n - 2) \tag{9.22}$$

Figure 9.14a shows the signal-flow graph for this filter, and its frequency response is shown in figure 9.15.

Filter 2 requires a pole in the region of 2 kHz, giving a pole angle of $\pi/2$ rad, and hence $\cos(\theta) = 0$. Choose a similar bandwidth to that for filter

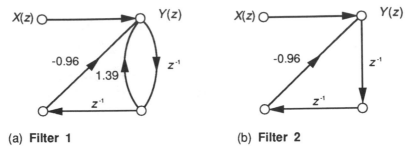

(a) **Filter 1**　　　　　　　　　　　　　　(b) **Filter 2**

Figure 9.14　Signal-flow graphs of all-pole filters

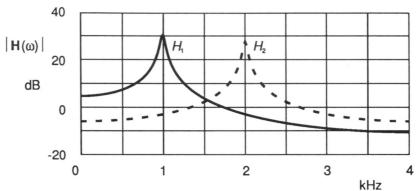

Figure 9.15　Frequency responses of all-pole filters

1, that is 50 Hz. The Q-factor in this case is therefore 40, and the pole radius is still $r = 0 \cdot 98$. Hence

$$H_2^\prime(z) = \frac{z^2}{z^2 + 0 \cdot 96}$$

and

$$y(n) = x(n) - 0 \cdot 96 y(n-2) \tag{9.23}$$

Figure 9.14b shows the signal-flow graph and figure 9.15 shows the frequency response.

Notice:

a) The peaky response around the selected frequency.
b) That the difference equation for filter 2 is again simpler than that for filter 1, due to its response being symmetrical about $z = \pm j$.
c) The maximum gains of each filter are not the same; this must be taken into account when comparing the signals in the two receiver channels.
d) The unwanted frequencies are not now eliminated entirely, but are just attenuated. It remains to be seen whether the attenuation will be adequate. Table 9.3 list the figures for attenuation relative to the maximum gain, taken from the graph.

Table 9.3 *Filter attenuation of unwanted signals*

	Frequency	Attenuation
Filter 1	0 kHz	26 dB
	2 kHz	34 dB
	3 kHz	40 dB
Filter 2	0 kHz	34 dB
	1 kHz	31 dB
	3 kHz	31 dB

9.4.4 Comparison of time responses

Both our examples, using zeros and using poles, fulfil the simple design requirements placed on these filters. However, figures 9.13 and 9.15 do show considerable differences in the bandwidths of the two sets of filters.

The all-pole filters of figure 9.15 show very narrow bandwidths, which might be a good thing if there is much noise and interference about on the cable circuit. On the other hand, the principle of reciprocal spreading tells us that a narrow bandwidth implies a slow time response, so that the wider bandwidth might be preferred if the signal is to change quickly with time.

Without making heavy weather of the calculation, we now calculate the corresponding time responses, and then show in figure 9.16 the step responses that are obtained.

Since these filters are bandpass, we are concerned to know how they behave in response to a step input at frequency ω_0, the centre frequency of the filter. A discrete-time signal of frequency ω_0 is represented by its Z-transform (table 8.1):

$$x(n) - \cos(\theta n) \qquad 0 \leqslant n$$

$$X(z) = \frac{z[z - \cos(\theta)]}{z^2 - 2\cos(\theta).z + 1} \tag{9.24}$$

Now if we take filter 2, $\theta = \pi/2$ and $\cos(\theta) = 0$. This will then give simplified results, and the step-sinusoidal input signal is given by

$$X(z) = \frac{z^2}{z^2 + 1} \tag{9.25}$$

Taking the all-zero filter, the output sequence in response to a step input at 2 kHz is

$$Y(z) = X(z).H_2'(z)$$

$$Y(z) = \frac{z^2(1 - z^{-2} + z^{-4} - z^{-6})}{z^2 + 1} \tag{9.26}$$

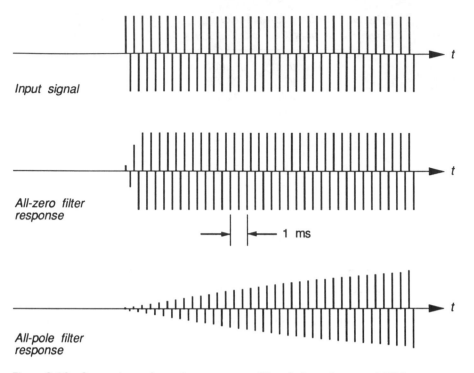

Input signal

All-zero filter
response

1 ms

All-pole filter
response

Figure 9.16 Comparison of transient responses (filter 2, input burst at 2 kHz)

Hence

$$Y(z) = 1 - 2z^{-2} + 3z^{-4} - 4z^{-6} + 4z^{-8} - 4z^{-10} \ldots$$

and

$$\{y(n)\} = \{1, 0, -2, 0, 3, 0, -4, 0, 4, 0, -4, 0, 4, 0, -4, 0, 4, \ldots\}$$

So the initial step transient is over within 6 intervals, and the steady-state response is maintained from then on.

Taking the all-pole filter, the output sequence will be

$$Y(z) = X(z) . H_2^1(z)$$

$$Y(z) = \frac{z^4}{(z^2 + 1)(z^2 + 0 \cdot 96)} \tag{9.27}$$

Hence

$$Y(z) = 1 - 1 \cdot 96z^{-2} + 4 \cdot 80z^{-4} - 7 \cdot 53z^{-6} + 10 \cdot 15z^{-8} - 12 \cdot 66z^{-10} \ldots$$

and

$$\{y(n)\} = \{1, 0, -1 \cdot 96, 0, 4 \cdot 80, 0, -7 \cdot 53, 0, 10 \cdot 15, 0, -12 \cdot 66, 0 \ldots\}$$

This transient shows no sight of settling out, and in fact runs for a long time, as figure 9.16 shows. The two output transients are compared, for the all-zero filter and for the all-pole filter. The lengths of the two transients reflect the inverses of the bandwidths, as predicted by the reciprocal spreading principle.

In this particular application there is no requirement for the signal to change rapidly with time, so either filter set might do; although the all-pole filter may well be preferred, since it will prevent short spikes of interference from operating the detector at the wrong time.

9.4.5 Using both poles and zeros

In practical filter design, poles and zeros are combined to provide the best compromise for filter characteristics. Figure 9.17 shows the frequency responses of two filters, which combine both the zeros and the poles of the filters designed previously. We appear to get the best of both worlds, since the passband is narrow and peaky, leading to good selectivity, but the attenuation of filter 1 at 2 kHz is infinite and so on.

These filters are an illustration of the cascade principle which was introduced in section 8.3.3. The responses of an all-zero filter and an all-pole filter are multiplied together, so the individual zeros and poles appear directly in the combined response. The overall signal-flow graph is drawn simply by cascading the individual signal-flow graphs, and the transient responses are dominated by those of the all-pole filters.

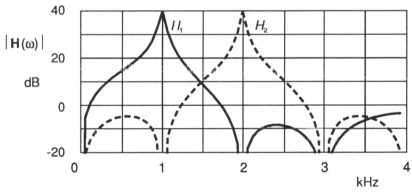

Figure 9.17 Frequency responses of combined filters

STUDY QUESTIONS

1 Design a discrete-time filter which has simple zeros at $\omega = 0$ and at $\omega = \omega_S/2$. Give the difference equation, the signal-flow graph and the frequency response.

2 A discrete-time filter is to have a response zero at $4 \cdot 4$ kHz, and the sampling frequency is 12 kHz. Find the difference equation and sketch the frequency response.

3 Design an all-zero filter which is to have response zeros at 2·5, 3·5 and 5·0 kHz. The sampling frequency is 10 kHz.

4 Design a simple recursive filter which has a pole at the Nyquist frequency, of radius 0·9. Sketch the response.

5 A certain filter like the one in question 4 operates with a sampling frequency of 10 kHz, and has an input signal $x(n)$. The input signal has a wanted component of magnitude 2·0 at 5 kHz, and an unwanted component of magnitude 0·5 at 3 kHz. Calculate the wanted-to-unwanted signal ratio, before and after the filter.

6 Design a digital resonator for a frequency of 100 Hz with a Q-factor of 20. The sampling frequency is 1 kHz. Sketch the frequency response.

7 Add zeros at 0 and 500 Hz to the resonator in question 6, and sketch the result.

10 Signal Processes: Averaging

OBJECTIVES

To introduce averaging as a practical signal process. In particular to

- a) Illustrate applications where averaging is applied
- b) Classify the types of practical averagers
- c) Analyse the effect of non-ideal averaging
- d) Illustrate analog and digital averaging systems.

COVERAGE

Situations are described where an average is needed. These include true-rms measurement of waveforms, spectral analysis, correlation and radio agc.

A signal model is introduced for the analysis of averaging, consisting of a mean value and a fluctuating component. This is then used to investigate short-term and running averages, and the idea of signal-to-noise ratio. Sinusoidal fluctuations are described in some detail, and random fluctuations are dealt with briefly.

Practical systems which carry out these processes are then described: the analog integrator and the discrete-time integrator.

Averaging is one of the most common signal processes in use and yet often is misunderstood. Classical averages of waveforms are fairly easy to grasp and have been used in these notes from chapter 2 onwards. The ideal average of a signal waveform is obtained by dividing the area under the waveform by the averaging time T, where T is the signal period for a periodic signal, or is infinite time for a non-periodic power signal (section 2.3.1).

When taking the average of a practical signal, neither of these two conditions can be met exactly. If the signal is periodic, the period may not be known; and clearly integrating for infinite time is not a practical option!

A further practical restriction arises since most signals have an average value which is changing with time, and hence there is only a limited time in which to make an estimate of the average value. Questions immediately spring to mind, such as how accurate can the estimate be, and how long should we average to obtain a given precision? We give some guarded answers in this chapter, with a fuller treatment of averaging random disturbances in chapter 15.

10.1 When averages are needed

First, we survey some application areas where averages must be taken, in order to show that this is indeed an important signal process. Understanding the limitations of this process will be of great help in understanding some of the more exotic signal processing applications to be covered in future studies. Note that we are merely using these examples to establish that signal averaging is a widely used process; details of the examples are not our immediate concern, they just generate signals which need to be averaged.

A simple example will illustrate the fundamental dilemma posed by the deceptively simple question, *"What is the average value of this waveform?"*

Consider the record of temperature in a building. A central-heating controller will be required to measure the average temperature over intervals of the order of a minute, in order to exercise close control over the temperature fluctuations. However, for the comfort of personnel working in the environment, an average temperature taken over a working day might be more significant. An average evaluated over a week or a month might be required in order to assess the thermal efficiency of the building, while a yearly average is necessary perhaps in order to evaluate the capacity required of a solar heating system. An 'average' therefore is defined with respect to its application, there is no such thing as a simple, single 'average' for a practical signal.

10.1.1 True-rms measurement

The mean-square value, or mean power, of a signal $x(t)$ is given by

$$P = \mathrm{avg}[x^2(t)] \tag{10.1}$$

The rms value is simply $\sqrt{P}$.

For a sinusoidal signal,

$$x(t) = A \cdot \cos(\omega_0 t) \tag{10.2}$$

and

$$x^2(t) = (A^2/2)[1 + \cos(2\omega_0 t)] \tag{10.3}$$

$$P = A^2/2 \tag{10.4}$$

RMS value is frequently measured because it is directly related to power, and gives a more fundamental measure than rectified-mean which is a simpler alternative. A true-rms measuring instrument therefore consists of a squaring device followed by an averager (equation 10.1).

The squaring device might be a magnetic attraction system as in the *moving-iron* meter, or it may be a thermal device measuring the temperature of a resistive element, or it may be an electronic multiplier.

In a conventional moving-pointer meter, the average value of the squared signal is indicated by the pointer movement, because the inertia of the moving parts will allow a response only to fluctuations with a period of the order of 1 second or greater.

In an electronic instrument such as a *digital-voltmeter (DVM)*, the intrinsic process can respond very rapidly, so the averaging must be done electronically, and must be designed carefully.

10.1.2 Spectral analysis

There is a wide range of applications where it is necessary to measure the signal power resident in a defined frequency band. A band-separating filter is used to isolate a narrow range of frequencies, and then the mean-power is estimated, either by the true-rms method as in section 10.1.1, or by a measurement of the rectified-signal mean which is a simpler alternative. Either way, it is necessary to estimate the mean value of a signal, be it squared or normal.

Figure 10.1 shows an example of a speech waveform, and illustrates nicely the time-varying property of the average. The figure shows several syllables of an utterance, ignoring the fine structure which has frequency components of the order of a kHz or so, and emphasises the variation in mean power between adjacent syllables.

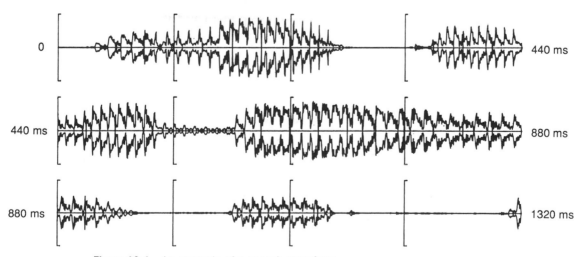

Figure 10.1   An example of a speech waveform

185

In order to model this waveform, and hence compare it with other similar waveforms, we require to estimate the frequencies and magnitudes of the components which construct it. This we can do by taking a series of bandpass filters, each passing a narrow range of frequencies around a different centre frequency, and then assessing the signal power in each one of these bands. Figure 4.7 shows the kind of structure that we have in mind, although that example is for a coarse subdivision of the overall frequency band, suitable for a small set of flashing disco lights.

An example of quite a different kind, but which nevertheless uses the same principle, is that of the telecontrol system introduced in section I.4.1 and discussed further in section 5.1.4 and section 9.4. Consider the channel which transmits control information for the switch, where any one out of two possible signals is to be sent. A simple but effective signalling scheme is to transmit one of two possible frequencies for each of the two control states. Since the whole signal is to be in a band centred on 1·5 kHz, let the two frequencies be 1·0 and 2·0 kHz respectively.

A suitable detector would consist of two bandpass filters, one centred on 1·0 kHz and the other on 2·0 kHz (see figure 9.11). The average signal power is then measured for each of these two outputs, and a *switch ON* command is accepted if signal power is present in the 1·0 kHz filter but not in the 2·0 kHz filter. This is also quite a safe signalling system, as befits a secure application like control of a sub-station circuit-breaker.

A further adaptation of the same principle is used in radio receivers. The radio-frequency signal power picked up by an antenna varies enormously as the radio is moved about, particularly if the radio set is in a moving vehicle like a car. It is necessary to control the amplification or *gain* of the receiver

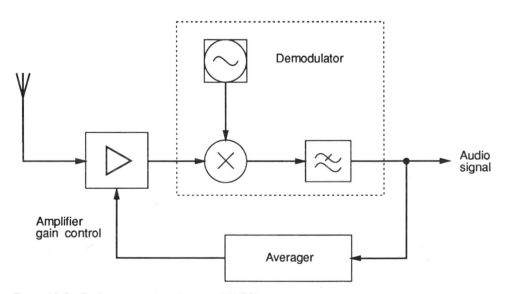

Figure 10.2 Radio automatic-gain-control (AGC)

in order to compensate for this effect, and figure 10.2 shows an outline system diagram of the *Automatic-Gain-Control (AGC)* which is used.

Figure 10.2 is of course extremely simplified. We use the very simple idea of demodulation of an amplitude-modulated (AM) signal introduced in section 5.1.3 to illustrate the receiver structure. The part we have added here starts with taking the average value of the demodulated AM signal, which is then an estimate of the radio-frequency signal strength. This is used to control the gain of the radio-frequency amplifier so as to compensate partially for the change in signal strength at the antenna.

The average process must ignore the modulation, the audio signal being carried by the radio system, while following accurately any fluctuations in radio-frequency signal strength as the antenna is moved. So the design of this averager is not straightforward, although the technique is used very successfully for broadcast receivers, and without this facility a receiver would be extremely sensitive to its physical position.

10.1.3 Correlation

A distinctive class of operations where averaging is used is typified by the correlation process, as introduced in section 1.3.2. A common application is where an unknown input signal is to be identified. It may be a speech signal, an utterance which is to be recognised, or it may be a received data signal. In either case we restrict the number of possible signals which can be identified, to a sct $W = \{w(m, t)\}$, where m denotes the mth member of that set. The input signal is $x(t)$, and is to be compared with a replica of each of the members of the set W, to see which one it is most like. We define the measure of similarity, or *correlation*, by the following relationship:

$$r(m) = \text{avg}[x(t) . w(m, t)] \tag{10.5}$$

Recognition or identification of the input signal in terms of one member of the set of recognisable signals is given by that element $r(m)$ which has the maximum value.

Clearly this operation is an order of magnitude more complicated than those we have considered so far, but it is a basic process and is widely used. We shall pursue some applications at a later date, but for now we note that successful *correlation* depends upon a correct measure of an average value.

10.2 Principles

We now start to investigate the principles which lie behind practical or non-ideal averaging. These principles are very simple, but of extreme importance.

10.2.1 A signal model

Consider a simple model of the kind of signal with which we are concerned:

$$x(t) = x_m + g(t) \tag{10.6}$$

This signal has a mean value of x_m, which is masked by a fluctuation about the mean, which we shall call $g(t)$, the *garbage* component. Figure 10.3 illustrates these two signal components. $g(t)$ may consist of sinusoidal component(s), or may be random.

In order to measure the mean value of this signal, we pass it to a signal process which evaluates an *estimate of the mean*, $\hat{x}_m$. This estimate may be expressed as

$$\hat{x}_m = x_m + \varepsilon \tag{10.7}$$

where ε is an error.

The fractional error is then ε / x_m, but this is not always a useful measure of how well the average is being determined; for instance the case where $x_m = 0$ yields infinity for this ratio.

We can express the effectiveness of this operation more usefully, by using the improvement in *signal-to-noise power ratio*. In this context the 'signal' is the mean value x_m, while the 'noise' is the fluctuation $g(t)$. Thus, before the averaging process commences, the *signal-to-noise-power ratio* SNR_i is given by

$$\mathrm{SNR}_i = \frac{x_m^2}{\mathrm{avg}\,[g^2(t)]} \tag{10.8}$$

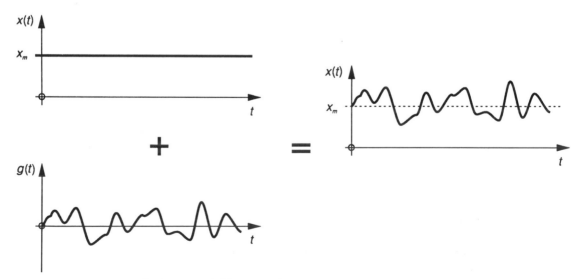

Figure 10.3 Signal model for averaging

After the averaging process the signal-to-noise power ratio is given by

$$\text{SNR}_o = \frac{x_m^2}{\text{avg}\,[\varepsilon^2]} \tag{10.9}$$

The ratio of these two ratios will give the measure of improvement imparted by the averaging process. Initially the mean value is heavily masked by the fluctuation $g(t)$, but after the averaging process we hope that the residual error ε is small. Thus

$$\frac{\text{SNR}_o}{\text{SNR}_i} = \frac{\text{avg}\,[g^2(t)]}{\text{avg}\,[\varepsilon^2]} \tag{10.10}$$

Notice that avg[ ] denotes an ideal or true or theoretical average, and that the two averages mentioned in equation 10.10 are subtly different from one another. However, this need not concern us yet, since we are looking at broad principles.

● **Example 10.1** A certain signal is described by

$$x(t) = 5 + 3\,.\,\cos(1000t)$$

Thus, $x_m = 5$ and $g(t) = 3\,.\,\cos(1000t)$.

Before averaging, $\text{SNR}_i = 25/(9/2) = 5\cdot56$ or $7\cdot45$ dB.

After averaging, suppose that the 1000 rad/sec component is reduced in magnitude by 100 times. Then the output signal of the averaging process is

$$y(t) = 5 + 0\cdot03\,\cos(1000t)$$

Thus, the output signal-to-noise power ratio becomes

$$\text{SNR}_o = 5\cdot56 \times 10^4 \text{ or } 47\cdot45 \text{ dB}$$

This particular averager has improved the SNR by 10 000 times, or 40 dB.   ●

10.2.2 Types of average

It may be thought that an average is a straightforward operation, and that there cannot be more than one type of average. While this is true for the ideal long-term average, when considering an average taken over a finite time, several options open up.

A *short-term* estimate of the average may be defined thus:

$$\hat{x}_m(T) = \frac{1}{T} \int_0^T x(t)\,.\,\mathrm{d}t \tag{10.11}$$

This short-term average amounts to integrating the waveform function $x(t)$ over a limited time interval T sec, and hence will be a function of T, since it only forms the true average when $T \rightarrow \infty$. Consequently we can write

$$\text{avg}[x(t)] = \lim_{T \to \infty} [x_m(T)] \qquad (10.11a)$$

This short-term average can easily be implemented as we see later in section 10.3, but it has the disadvantage that a new estimate of the mean value is available only at intervals of T sec. The output is discrete-time with an interval of T sec. In certain applications this is completely satisfactory, but in others it is necessary to have a continuous estimate of mean value. For these applications we use the *running average*:

$$\hat{x}_m(t) = \frac{1}{T} \int_{t-T}^{t} x(t) \, . \, dt \qquad (10.12)$$

The running average gives the short-term average as estimated over the previous T sec, at all times. It is therefore a genuine measure of changes in the mean value of the signal, as they occur.

10.2.3 Averaging with sinusoidal fluctuation

Now we take a simple sinusoidal form of fluctuation, representing an unwanted signal component, or 'noise', and discover how it is affected by short-term averaging.

Returning to our signal model, we define

$$x(t) = x_m + b \, . \, \cos(\omega t + \varphi) \qquad (10.13)$$

so

$$g(t) = b \, . \, \cos(\omega t + \varphi) \qquad (10.14)$$

Taking the definition of short-term average from equation 10.11, we see that the residual error (equation 10.7) becomes

$$\varepsilon = \frac{1}{T} \int_{0}^{T} g(t) \, . \, dt \qquad (10.15)$$

Applying our current definition of the fluctuation $g(t)$ from equation 10.14, we see that

$$\varepsilon = (b/\omega T) \, . \, [\sin(\omega T + \varphi) - \sin(\varphi)] \qquad (10.16)$$

As the frequency ω is varied, this error cycles in value, reaching zero when ωT is a multiple of 2π, or when fT is an integer. Local maxima occur between these zero points, and reach a maximum with respect to the arbitrary phase angle φ, when φ is a multiple of $\pi/2$.

So, we may put a maximum bound on the value of error ε:

$$|\varepsilon| \leqslant 2b/\omega T \qquad (10.18)$$

This error which occurs in determination of the mean value is plotted in figure 10.4, assuming that $\varphi = \pi/2$.

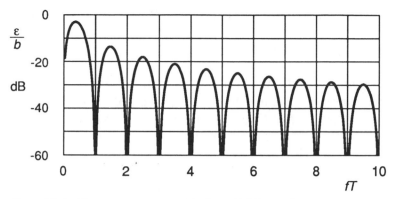

Figure 10.4 Measurement error with sinusoidal fluctuation

So much for the algebra, but what significance does this result have for practical averaging? The first application that we mention is one which makes use of the fact that the error is zero when $\omega T = 2\pi n$, where n is an integer.

Digital voltmeters commonly use the integration principle to convert the input analog quantity into a digital reading. Because of the large amount of mains-supply wiring in any laboratory, workshop or factory, there is a high likelihood that signal wires will contain a 50 Hz interfering signal, which is called 'pick-up' or 'hum'. If the integration time T is selected to be a multiple of 20 ms, the 50 Hz signal period, then errors due to mains pick-up are eliminated.

● **Example 10.2** A certain signal is run over a long cable, and during transmission picks up an interfering signal with a frequency of 100 Hz. The signal mean is to be measured with a DVM. What integrating time would ensure that the effect of the interference is eliminated?

The period of the interference is 10 ms. An integrating time which is $10n$ ms would be suitable, where n is integer.

Suppose that a figure of $n = 4$ is chosen, giving an integrating time of 40 ms. Then any interfering signal having a frequency which is a multiple of 25 Hz will be eliminated from the measurement.   ●

● **Example 10.3** A certain rms-value measurement commences by squaring a sinusoidal signal, giving a signal $x(t)$ which is to be averaged, of the form:

$$x(t) = A\,[1 + \cos(\omega t)]$$

The unwanted fluctuation has a frequency which is in the region of 100 Hz, but which is not known exactly. For how long must integration proceed, if the residual error is to be less than 2% of the mean?

191

The fractional error is ε/x_m which is $(2/\omega T)$ in this case.
Hence $(2/\omega T) \leqslant 0 \cdot 02$ or $fT \geqslant 15 \cdot 92$.
Since $f \simeq 100$ Hz, then $T \geqslant 160$ ms. ●

● **Example 10.4** When assessing a speech signal, it is frequently necessary to measure the power in each of a number of narrow frequency bands. Since the filters have narrow bandwidths, the output of each filter can be considered to be almost sinusoidal. The averaging process described in example 10.3 is therefore applied to the output of each filter.

Suppose that such an analysis is carried out over a frequency range of 300–3400 Hz, and that an estimate of the mean power is required to better than say 5%.

Following a similar argument to example 10.3, we deduce that $fT \geqslant 6 \cdot 37$.

Now the lowest frequency band is at 300 Hz, and when squared yields a frequency component at 600 Hz. Consequently $T \geqslant 10 \cdot 6$ ms.

Notice that this type of calculation defines a fundamental limitation on measurement uncertainty. ●

10.2.4 Averaging with random fluctuation

Although many measurements take place in the presence of sinusoidal fluctuations or interference, other measurements have to contend with random fluctuations. Random interference can arise from a collection of many sinusoidal components, from modulated signals, and from physical mechanisms like electronic noise.

The mathematics of random signals clearly has certain features additional to those that we have considered so far, but we can draw simple conclusions about the effect of random fluctuations, using a naive argument. A more realistic argument will be found in chapters 14 and 15.

It will be best to use a discrete-time signal and the discrete-time version of integration, when considering random fluctuations. Thus

$$x(n) = x_m + g(n) \tag{10.19}$$

The integration process is equivalent to a summation of samples, hence

$$\hat{x}_m = \frac{1}{N} \sum_{n=0}^{N-1} x(n) \tag{10.20}$$

The measurement error is given by

$$\varepsilon = \frac{1}{N} \sum_{n=0}^{N-1} g(n) \tag{10.21}$$

When averaged over a sufficiently large number of samples N, the error tends to zero, since random samples are equally likely to be positive as negative in sign. This is intuitive and is the justification for measuring the average with random interference.

However, when N is finite, then the error is finite and random, and can be measured by its *variance* or its *standard deviation*. Suppose that the original random fluctuation $g(n)$ has a variance (or power) of σ^2. Then, if the individual samples of $g(n)$ are independent, it can be shown that the variance of the sum of $g(n)$ is

$$V[\varepsilon] = \sigma^2 / N \tag{10.22}$$

Referring back to equation 10.10, we see that the improvement in signal-to-noise power ratio is N times.

● **Example 10.5**   A certain discrete-time signal has a mean value of $3 \cdot 0$ V and a random fluctuation about the mean which has a standard deviation of $0 \cdot 5$ V.

Determine the number of samples which must be averaged in order to establish a signal-to-noise power ratio of 1000 (or 30 dB).

The fluctuation power after averaging is to be

$$3 \cdot 0^2 / 1000 = 0 \cdot 009 \text{ V}^2$$

However, the fluctuation power before averaging is $0 \cdot 25$ V^2. The number of samples to be averaged is thus: $N \geqslant 0 \cdot 25 / 0 \cdot 009$ or $27 \cdot 8$.   ●

This calculation has not been presented with any precision, since random signals require further tools, but the result is generally correct and may be used to obtain a first-estimate of the effectiveness of a given discrete-time averaging operation. Chapter 15 gives a more complete discussion.

10.3   Practical processors

Having discussed the need for averaging, and the principles which lie behind the operation, we now present several practical averagers. They are quite simple items of hardware or software, but a proper understanding of how they perform can only be gained by first grasping the principles given above.

We commence with a study of the classical *analog integrator*, which has been a fundamental building block of signal processing for the past 40 to 50 years. The *RC averager* is also used extensively, and some of its properties are an improvement on the integrator, while its performance may be analysed from either the time or the frequency domain points of view. Discrete-time versions exist for both these processes.

10.3.1 Analog integrator

This classic circuit is shown in figure 10.5. The clue to the operation of this circuit lies in the amplifier, which has a very high voltage-gain $-A$, and a very high input resistance R_i. An amplifier with these parameters is called an *operational amplifier*.

The output voltage $v_o(t)$ is given by

$$v_o(t) = - v_i(t) . A \tag{10.23}$$

Since the voltage-gain is very high, usually greater than 10^5, the amplifier input voltage $v_i(t)$ is negligible compared with $v_o(t)$. $x(t)$ and $v_o(t)$ are in the same order, so $v_i(t)$ is also small compared with the source voltage $x(t)$. Consequently, $v_i(t)$ may be set approximately to zero, which simplifies enormously the analysis of the circuit. The input current to the amplifier is insignificant too, so

$$\frac{x(t)}{R} \simeq - C \frac{d[v_o(t)]}{dt} \tag{10.24}$$

or

$$v_o(t) \simeq \frac{-1}{CR} \int x(t) . dt \tag{10.25}$$

The output voltage is therefore proportional to the integral of the input voltage. If an integral over a fixed time is required, then the integrating capacitor starts by being discharged, and after the integration time, holds a voltage proportional to the integral of the input signal.

In practice the approximation involved in making the analysis is extremely small, and the integration is nearly ideal. Operational amplifiers like the one used in this circuit are readily available.

Although this circuit takes the average value of a signal by simple integration as in equation 10.11, it is necessary to spell out the algebra for a practical understanding of its operation. This we do for a sinusoidal

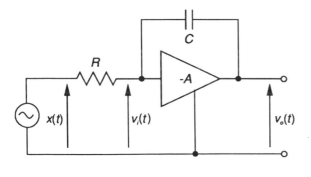

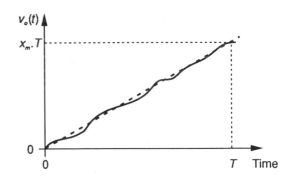

Figure 10.5 Analog integrator

fluctuation, using the signal of equation 10.13:

$$v_o(T) = \frac{x_m T}{CR} + \frac{b}{\omega CR} \cdot [\sin(\omega T + \varphi) - \sin(\varphi)] \qquad (10.26)$$

Thus although the output voltage increases linearly with the integration time T, the fluctuation component does not, and the ratio of mean-value output and maximum-fluctuation output remains the same as declared in section 10.2.3.

10.3.2   RC averaging circuit

A simpler analog circuit which is frequently used for averaging is the first-order lag or lowpass filter circuit shown in figure 10.6.

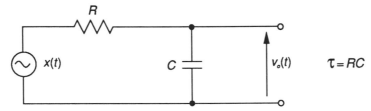

Figure 10.6   *RC averaging circuit*

10.3.2.1   Running average

We may view this circuit first as a running average, albeit a weighted one, by combining the impulse response with the convolution integral (section 7.3).

Thus, for a time constant τ, the impulse response of the circuit in figure 10.6, as discussed in example 7.11, is

$$h(t) = \frac{1}{\tau} \cdot \exp - (t/\tau) \qquad t \geqslant 0 \qquad (10.27)$$

The output of the circuit $y(t)$ is given by the convolution of the input signal $x(t)$ and the impulse response $h(t)$:

$$y(t) = \int_{-\infty}^{\infty} x(u) \cdot h(t - u) \, du$$

$$= \frac{1}{\tau} \int_{-\infty}^{t} x(u) \cdot \exp - \left(\frac{t - u}{\tau} \right) du \qquad (10.28)$$

Equation 10.28 is very similar to the running average expression of equation 10.12, except that the signal $x(t)$ which is being averaged is first multiplied by an exponential weighting function. Signal values in the recent past are weighted more heavily than those in the distant past, as illustrated in figure 10.7.

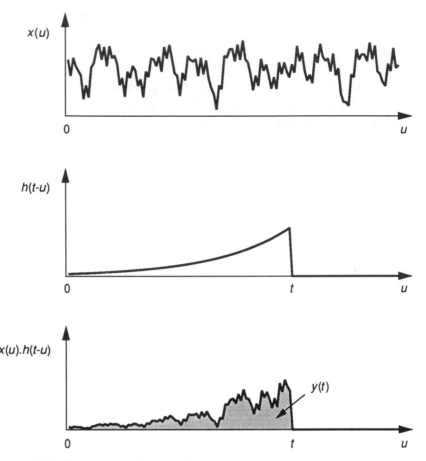

Figure 10.7 Exponential weighting of running average

Although this is a useful way of looking at the operation of the circuit, it is generally a tedious calculation to carry out. Consequently, for noise or interference which can be expressed in terms of sinusoidal components, the frequency domain analysis is normally used.

10.3.2.2 Frequency analysis

We have already analysed the phasor response of this circuit in section 3.3.1, but we re-examine it now in order to show that it functions as an averager by reducing the magnitude of the fluctuation around the average.

Consider a signal where the fluctuation $g(t)$ consists of a number of sinusoidal components:

$$g(t) = \sum_{r=1}^{M} b_r . \cos(\omega_r t) \tag{10.29}$$

There are M separate components making up $g(t)$, each having a different frequency ω_r and a different magnitude coefficient b_r. The power in these fluctuation components is then

$$\text{avg}\,[g^2(t)] = \sum_{r=1}^{M} b_r^2/2 \qquad (10.30)$$

Now the general expression for the phasor or frequency response of this lowpass filter is

$$\mathbf{H}(\omega) = \frac{H_0}{1 + j\omega\tau} \qquad (10.31)$$

where $\tau = RC$.

Thus at the output of the filter, the zero-frequency component (the mean value) will be $(x_m . H_0)$.

Each of the other frequency components will now have a magnitude $b_r.|\mathbf{H}(\omega_r)|$, and the mean-square value of the remaining fluctuation or error component will be

$$\text{avg}\,[\varepsilon^2(t)] = \sum_{r=1}^{M} b_r^2.|\mathbf{H}(\omega_r)|^2/2 \qquad (10.32)$$

Since $|\mathbf{H}(\omega_r)|$ diminishes as ω_r increases (equation 10.31), so the effect of the various fluctuation components is reduced by this filtering process.

Examining the expression for filter frequency response (equation 10.31), we notice that for large frequencies such that $\omega\tau \gg 1$, then

$$|\mathbf{H}(\omega_r)| \simeq H_0/\omega_r\tau \qquad (10.33)$$

Since we are normally trying to achieve significant reduction of any fluctuation signal, this condition applies, and we might express it more conveniently as

$$|\mathbf{H}(\omega_r)| \simeq H_0 . f_0/f_r \qquad (10.34)$$

where $f_0 = 1/2\pi\tau$.

Figure 10.8 shows the form of this relationship, and enables a quick

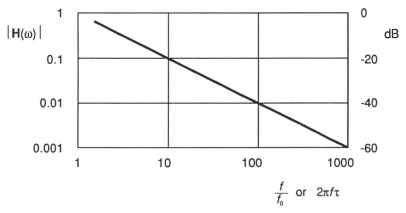

Figure 10.8 Performance of *RC* averager

estimate to be made of the attenuation afforted by an RC averager to inputs of various frequencies. The peak magnitude of the output signal with a frequency ω_r is

$$\max[y(t)] = b_r \cdot H_0 \cdot \frac{f_0}{f_r} \qquad (10.35)$$

For a zero-frequency 'signal' and a single frequency disturbance, the signal-to-noise ratio improvement is

$$\frac{\text{SNR}_o}{\text{SNR}_i} = \frac{f_r^2}{f_0^2} \qquad (10.36)$$

● **Example 10.6** Consider a case where there is only one interfering or fluctuation error component of frequency 100 Hz, and having a magnitude $b = x_m$. This is then the classic case of rms measurement with a 50 Hz sinusoidal input signal.

$$x(t) = x_m[1 + \cos(2\pi \cdot 100t)]$$

Determine the value of time constant τ which will reduce the peak error component to less than 2% of the mean value.

Now initially the fluctuation has a peak value of x_m, and we have to reduce it less than $0 \cdot 02 x_m$. Referring to equation 10.34 this means that

$$f_0 \leqslant 0 \cdot 02 \times 100 \text{ Hz}$$

So $f_0 \leqslant 2$ Hz or $\tau \geqslant 79 \cdot 5$ ms.   ●

● **Example 10.7** Take example 10.6 again, but this time the criterion is that the mean-square value of the error will be less than 2% of the mean-square value of the mean.

Initially, the mean-square-value of the mean is x_m^2, and the mean-square value of the fluctuation is $x_m^2/2$.

The mean-square value of the new fluctuation is $(x_m \cdot f_0/f_r)^2/2$.

Hence for this case:

$$(f_0/f_r)^2/2 \leqslant 0 \cdot 02 \quad \text{and} \quad f_0 \leqslant 20 \text{ Hz} \quad \text{or} \quad \tau \leqslant 7 \cdot 95 \text{ ms.} \qquad ●$$

● **Example 10.8** A certain signal consists of the following components:

Frequency	0	0.8	1.2	2.5	kHz
RMS magnitude	$1 \cdot 0$	$0 \cdot 2$	$0 \cdot 4$	$0 \cdot 3$	V

Determine the improvement in mean-square fluctuation, after having been processed by a first-order lowpass filter with time constant $1 \cdot 59$ ms.

For this filter, $f_0 = 100$ Hz or $0 \cdot 1$ kHz. Hence, after the filter, the signal will consist of the following:

Frequency	0	0·8	1·2	2·5	kHz
RMS magnitude	1000	25·0	33·3	12·0	mV

Before

After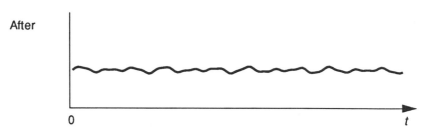

Figure 10.9 Example of signal filtering for determining the mean value

Before filtering, the signal-to-noise power ratio is

$$\mathrm{SNR}_i = 3 \cdot 45 \ (5 \cdot 4 \ \mathrm{dB})$$

After filtering, the signal-to-noise power ratio is

$$\mathrm{SNR}_o = 532 \cdot 5 \ (27 \cdot 3 \ \mathrm{dB})$$

The improvement is therefore $154 \cdot 3$ (or $21 \cdot 9$ dB).

Figure 10.9 illustrates this example, assuming zero phase for all the input frequency components. ●

Notice that several different error criteria have been used without comment, so we need to compare them. The *maximum error* is used when the fluctuation is modelled by a single frequency component, as in the integrator analysis. When the fluctuation is modelled by several frequency components, the *mean-square* or *rms* measure of errors is more meaningful since the peak error depends on the relative phases of the different frequency components. This latter style of assessment has been used in the *RC* averager analysis, but could also be used in the analysis of the integrator.

Analysis of the *RC* averager has concentrated on the phasor response $\mathbf{H}(\omega)$, showing how the various frequency components are reduced in magnitude. However the input signal must be present for a certain time to allow steady-state conditions to be achieved. When the input signal is suddenly applied at $t = 0$, the mean output-signal changes exponentially:

$$\mathrm{avg}\,[v_o(t)] = x_m[1 - \exp(-t/\tau)] \tag{10.37}$$

At a time t after applying a new signal, the mean-output is within a fraction

$\exp(-t/\tau)$ of its final or steady-state value, and this discrepancy must be added to the error ε which has been calculated above. In practice the situation is more complicated, since the mean value of the input signal does not usually change abruptly by such large amounts, and hence the discrepancy is less than given in equation 10.37, being predicted accurately by the convolution integral of equation 10.28. Thus time and frequency descriptions of the operation are intimately related and mutually restricting, as discussed in chapter 7. Calculations must be done for each application, to see the importance of this effect.

10.3.3 Discrete-time integration

In the discrete-time world, the input signal will be in the form:

$$x(n) = x_m + g(n) \tag{10.38}$$

Integration, involving many samples, is easily accomplished by summing the values:

$$y(rN) = \frac{1}{N} \sum_{n=(r-1)N+1}^{rN} x(n) \tag{10.39}$$

This operation is equivalent to the analog integrator in section 10.3.1, and generates a new estimate of the mean value every N input samples.

A running average is also simply achieved, by adding together the most recent N sample values of the input:

$$y(n) = \frac{1}{N} \sum_{r=0}^{N-1} x(n-r) \tag{10.40}$$

For example, for $N = 4$,

$$y(n) = [x(n) + x(n-1) + x(n-2) + x(n-3)]/4 \tag{10.41}$$

This is the difference equation for a 4-stage moving-average digital filter. For large values of N, the process requires many additions in each sample interval, and consumes quite an amount of computing power. This operation is programmed economically by holding the scaled set of the $(N-1)$ previous input values in a storage array $x(\ )$, where $x(r)$ holds the input value from r sample times in the past. The integrated output is y, and the algorithm in summary is then:

Subtract the $(N-1)$th previous input value	$y = y - x(N-1)$
Shuffle down the storage array $x(\ )$	for $i = (N-1)$ to 1 step -1
	$x(i) = x(i-1)$
	next
Store the current input value	$x(0) = \text{input}/N$
Add the new input to the accumulated total	$y = y + x(0)$

(For large values of N, involving many elements in the array shuffle, it will be more economical to maintain a *circular buffer*, leaving the stored values of previous inputs in place, but updating a pointer to the current position in the buffer at each sample time.)

The difficulty in realising the above algorithm is due to the necessity of retaining all recent input samples up to $(N-1)$ intervals ago, before subtracting the most distant value from the current total. A much simpler process is achieved by recognising that we already have an estimate of the mean value $y(n-1)$, which just needs to be updated by the new sample value.

So we have a new input sample $x(n)$ and have to generate a new estimate of the mean value $y(n)$, while making use of the most recent estimate $y(n-1)$. If we were implementing the algorithm of equation 10.40 directly, then we would add a quantity $x(n)/N$ to the running total, and subtract a quantity $x(N-1)/N$. Instead, we can add $x(n)/N$ to a substantial fraction of the most recent estimate: $y(n-1).(N-1)/N$.

So, the new estimate of mean value is updated every sample, but is based on a combination of the new input sample and the previous estimate $y(n-1)$. This is then the discrete-time equivalent of the analog RC averager, the system first introduced in section 3.3.2 and studied further in section 9.3.1. Let $\alpha = (N-1)/N$.

$$y(n) = (1 - \alpha) . x(n) + \alpha . y(n-1) \tag{10.42}$$

where $\alpha \to 1 \cdot 0$.

Knowing the phasor response $\mathbf{H}(\omega)$ of this system from section 3.3.2, we can calculate how this running averager performs:

$$\mathbf{H}(\omega) = \frac{1 - \alpha}{1 - \alpha . \exp \mathrm{j}(-\omega T_S)} \tag{10.43}$$

Unfortunately, there is no simple asymptotic form to this frequency response, as was the case with the analog equivalent. However, the calculations can be done, and we note that

$$|\mathbf{H}(\omega)|^2 = \frac{(1 - \alpha)^2}{(1 + \alpha^2) - 2\alpha . \cos(\omega T_S)} \tag{10.44}$$

The minimum value of the response $|\mathbf{H}(\omega)|$ is $(1 - \alpha)/(1 + \alpha)$, which occurs when $\omega T_S = \pi$, and defines the best possible reduction of a fluctuation component. Figure 10.10 shows a family of response characteristics, for differing values of α.

As in the case of the RC averager, we have used the phasor response $\mathbf{H}(\omega)$ to calculate the amount by which the noise or interference is reduced, but have ignored the transient when a signal is suddenly applied to the discrete-time averager. This transient is exponential in nature (section 9.3.1), and is defined by a time-constant of $-T_S/\ln(\alpha)$.

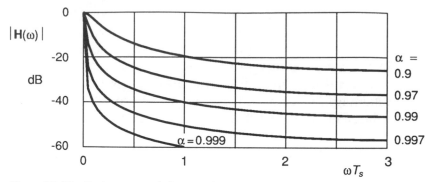

Figure 10.10 Performance of discrete-time averager

● **Example 10.9**   Refer back to example 10.6, and re-work it assuming that we are now in the discrete-time domain. Determine the required sampling frequency, if the value of coefficient α is $0\cdot99$.

$$x(n) = x_m[1 + \cos(\omega T_s n)]$$

where $\omega = 2\pi100$ rad/sec.

We note that $|\mathbf{H}(\omega)|$ is to be $0\cdot02$, and that $\alpha = 0\cdot99$.

Substituting these values in equation $10\cdot44$, we deduce that $\cos(\omega T_s) \leqslant 0\cdot87$ and that $T_s \geqslant 0\cdot82$ ms while $f_s \leqslant 1\cdot22$ kHz.   ●

● **Example 10.10**   Referring back to example 10.7, we note that the criterion for solving this problem is that

$$|\mathbf{H}(\omega)|^2/2 \leqslant 0\cdot02$$

Applying this to equation 10.44, we see that

$$\cos(\omega T_s) \leqslant 0\cdot99375 \quad \text{and} \quad T_s \geqslant 0\cdot17 \text{ ms} \quad \text{while } f_s \leqslant 5\cdot62 \text{ kHz} \qquad ●$$

● **Example 10.11**   We re-work example 10.8, using a discrete-time averager with $\alpha = 0\cdot97$ and a sampling frequency of 10 kHz. The input signal is described by

Frequency	0	0·8	1·2	2·5	kHz
RMS magnitude	1·0	0·2	0·4	0·3	V

First we calculate the equivalent values of ωT_s, and then the relative response of the averager.

Frequency	0	0·8	1·2	2·5	kHz
ωT_s	0	0·50	0·75	1·57	rad
Relative response $|\mathbf{H}(\omega)|$	100	6·14	4·15	2·15	$\times10^{-2}$
Output rms magnitude	1000	12·3	16·6	6·5	mV

Before filtering, the signal-to-noise power ratio is

$$\text{SNR}_i = 3\cdot45 \quad \text{(or } 5\cdot4 \text{ dB)}$$

After filtering, the signal-to-noise power ratio is

$$\text{SNR}_o = 2131 \quad \text{(or } 33 \cdot 3 \text{ dB)}$$

The improvement is therefore 618   (or $27 \cdot 9$ dB).   ●

STUDY QUESTIONS

1   Review the situations where an average is required, and add one or two applications of your own.

2   A certain signal is given by $x(t) = 3 + 2 \cdot \cos(500t)$. Determine the signal-to-noise power ratio of mean value to fluctuation. What time interval should be chosen for integration, so as to eliminate the fluctuation component entirely?

3   Distinguish between
      a) Long-term average
      b) Short-term average
      c) Running average

   Why are three different averages needed?

4   The signal of question 2 is to be integrated in order to measure the mean value. For how long should integration proceed, if the residual error is to be less than $0 \cdot 1\%$ of the mean?

5   A certain signal has a mean value of $1 \cdot 5$ V which is disturbed by a random fluctuation having a variance of $0 \cdot 7$ V^2. Determine the number of sample values to be averaged in order to increase the signal-to-noise power ratio to 40 dB.

6   Explain the working of an analog integrator. The output signal of such an integrator is limited to ± 5 V, and the component values are $R = 10$ kΩ and $C = 0 \cdot 2$ μF. What is the maximum input-signal mean-value which can be allowed for an integration time of 10 ms, ignoring the contribution made by the fluctuation?

7   Choose the components of an RC averaging circuit which, with sinusoidal fluctuation, will give a similar maximum-error to the integrator mentioned in question 6.

8   A signal is described by the following frequency components:

Frequency	0	0·5	2·5	14·0	kHz
RMS magnitude	2·5	0·4	0·3	0·5	V

   Determine the improvement in signal-to-noise power ratio when it is processed by an RC averaging circuit having a time constant of $3 \cdot 18$ ms.

9   Re-work question 4 using a discrete-time averager. Given that $\alpha = 0 \cdot 9995$, calculate the required sampling frequency and the time constant of the measurement.

10 Re-work question 8, assuming a discrete-time averager with a sampling frequency of 30 kHz and coefficient $\alpha = 0 \cdot 95$.

11 Signal Processes: Differentiation and Integration

OBJECTIVES

To show that the familiar mathematical operations of differentiation and integration are important signal processes. To introduce the approximations to these operations which are used in practice, for both continuous-time and discrete-time signals

COVERAGE

For each process, we commence with a few applications to show how important is the principle behind it. Then the response of each process to classical signals like impulse, step, phasor is evaluated. Simple practical implementations using continuous signals and *RC* circuits are then introduced, before investigating the discrete-time form.

Signal processes, like most other things, cannot be classified into neat unique categories. The elementary processes of differentiation, prediction and integration can be described as operations in the *time domain*, but inevitably they have descriptions in terms of *frequency response* as well.

Differentiation and integration are familiar enough as mathematical operations, but we need to consider how they might be achieved in practice, as additional tools in the signal processing armoury. Prediction is a new concept but is a simple extension of differentiation. These operations can precisely be defined for continuous-time signals but have no exact equivalent for discrete-time signals, so the transfer of ideas from continuous-time to discrete-time domains must be done carefully.

In this discussion we shall restrict ourselves to *linear filtering* signal processes. Differentiation and integration can often be accomplished more accurately by numerical algorithms, but processes based on linear filtering are those most frequently used.

11.1 Differentiation

One of the classical uses of differentiation in signal processing is to estimate the gradient or rate-of-change of a variable. Thus differentiating a transducer signal which is proportional to distance covered yields velocity, while differentiating a velocity signal yields acceleration. These concepts are used wherever movement takes place; in a vehicle, on a robot arm or any position-control device. Measurement of distance may be accomplished through rotation of a wheel, by a radar system, or by a linear transducer depending upon variable resistance for instance.

One important application is where rate-of-change of phase denotes frequency, which we first met when considering angle modulation in chapter 4, and which we shall discuss further in chapter 12.

Differentiation is used when sudden changes in signal magnitude are to be detected. One particular example of this application is for video signals, where a TV picture frame consists mostly of flat areas of colour. The amount of information which must be transmitted can be reduced by identifying the *changes* in picture level and then using this information to represent the picture. This is known as *differential coding*, and can result in great savings in transmitted bandwidth, allowing tolerable pictures to be transmitted at 2 Mbit/s or less, in place of the 100 Mbit/s or so which is necessary for the uncompressed picture.

11.1.1 Continuous-time signals

Consider a continuous-time signal $x(t)$. The differentiation process is defined by an output $y(t)$, where

$$y(t) = \frac{d\,[x(t)]}{dt} \tag{11.1}$$

In order to discover what effect this operation has, we apply the process to various classical input signals, and define ideal responses to these stimuli. The effectiveness of various practical forms of the differentiation process can then be assessed by comparing the practical responses with the ideal ones.

11.1.1.1 Classical response

Consider the case where the input signal is a classical *step* function. Then

$$\begin{aligned} x(t) &= 1 \qquad t \geqslant 0 \\ &= 0 \qquad t < 0 \end{aligned} \tag{11.2}$$

and $\quad y(t) = \delta(t) \tag{11.3}$

The differentiation process therefore marks the step change in the input signal, and is zero elsewhere, which represents one range of applications.

If the input signal is now a classical *ramp* function, then

$$x(t) = \alpha t \qquad t \geqslant 0$$
$$\quad\;\; = 0 \qquad t < 0 \tag{11.4}$$

and
$$y(t) = \alpha \qquad t \geqslant 0$$
$$\quad\;\; = 0 \qquad t < 0 \tag{11.5}$$

The differentiation process now measures the slope of the ramp, and so is representative of a further class of applications.

These classical time-domain operations are illustrated in figure 11.1, together with an example of an arbitrary waveform and the effect of differentiation upon it.

The most important classical signal that we have used previously is the *phasor*, and we now investigate the effect of differentiation upon this.

$$x(t) = \exp j(\omega t) \tag{11.6}$$

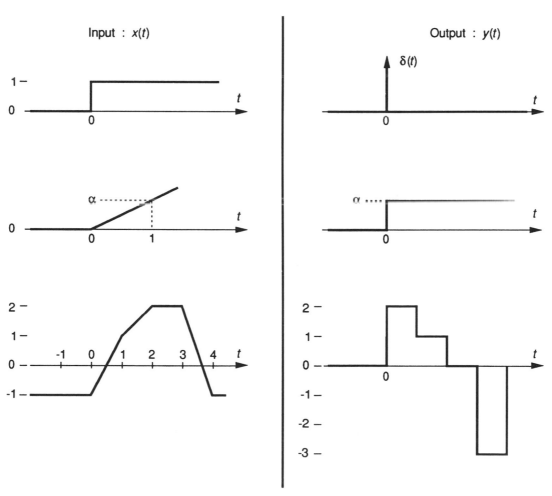

Figure 11.1 Classical differentiation

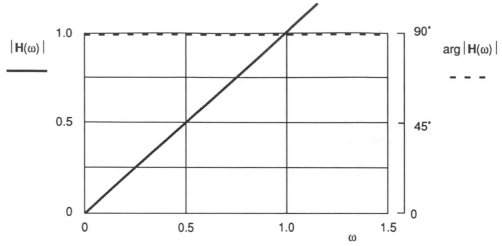

Figure 11.2 Frequency response of ideal differentiator

Then $y(t) = j\omega . \exp j(\omega t)$ (11.7)

Interpreting this operation in terms of *frequency response*, which is the complex number relating the output phasor to the input phasor, we discover that

$$\mathbf{H}(\omega) = j\omega$$ (11.8)

Thus $|\mathbf{H}(\omega)| = \omega$ and $\arg[\mathbf{H}(\omega)] = 90°$. This description of differentiation is the most useful one, since it is in terms of frequency response, which is universally understood in signal processing and can be applied to all kinds of signal. It is illustrated in figure 11.2.

11.1.1.2 Practical process

The simplest practical approximation to differentiation is given by an *RC* circuit as shown in figure 11.3. Simple analysis shows that the frequency response is

$$\mathbf{H}(\omega) = \frac{j\omega\tau}{1 + j\omega\tau}$$ (11.9)

where $\tau = RC$.

Provided that $\omega\tau \ll 1$, then this approximates to differentiation. Since most processed signals are strictly bandlimited, then this condition can usually be met, and the normalised frequency response shown in figure 11.3 enables the degree of approximation to be estimated. For instance, if the phase shift must be at least $80°$, then $\omega\tau < 0 \cdot 18$. Notice that the amplitude response is a better approximation to the ideal than the phase response.

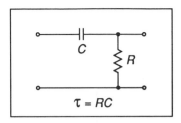

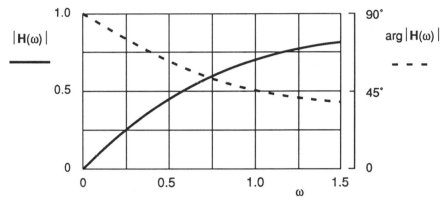

Figure 11.3 *RC* differentiator frequency response

● **Example 11.1** A signal having a bandwidth of 3 kHz is to be differentiated using a simple *RC* circuit. The gain error at 3 kHz is to be less than 5%.

Referring to equation 11.9, the error criterion is satisfied by

$$1 + (2\pi \cdot 3\tau)^2 \leqslant (1 \cdot 05)^2$$
$$\tau \leqslant 0 \cdot 017 \text{ ms or } 17 \ \mu\text{s}$$

Now check on the phase error at this frequency, which from equation 11.9 is

$$\arctan[\omega\tau] = \arctan[2\pi \cdot 3 \cdot 17 \cdot 10^{-3}] = 17 \cdot 8° \qquad ●$$

Time domain operations similarly can be evaluated by simple analysis. Thus the time responses are

Unit step input $\qquad y(t) = \exp(-t/\tau) \qquad\qquad t \geqslant 0 \qquad$ (11.10)

Ramp input, α V/sec $\quad y(t) = \alpha\tau[1 - \exp(-t/\tau)] \qquad t \geqslant 0 \qquad$ (11.11)

These are illustrated in figure 11.4 and show that for a good approximation to the ideal response, τ should be small.

A differentiator can also be constructed using an *operational amplifier* such as employed in the integrator of section 10.3.1. Figure 11.5a shows the

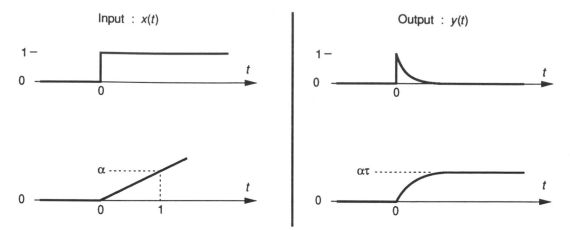

Figure 11.4 *RC* differentiator time responses

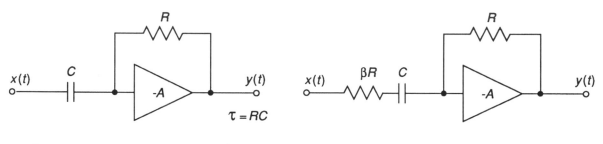

(a) **Basic circuit**   (b) **Gain-limited circuit**

Figure 11.5   An operational differentiator

basic circuit. In terms of the amplifier gain A,

$$\mathbf{H}(\omega) = -\frac{j\omega\tau}{1 + j\omega\tau/A} \tag{11.12}$$

This circuit is therefore a good approximation to an ideal differentiator provided that $\omega\tau \ll A$, a condition which is easily met since the operational-amplifier gain is normally $> 100\,000$. The transient response however, tends to be limited more by the bandwidth of the amplifier, than by this gain condition.

One potential problem with any differentiator is the rising gain at high frequencies, since any noise outside the signal band W will be amplified to become of greater significance than it ought. This effect is simply controlled by adding an extra resistor of value βR as in figure 11.5b. The frequency response now becomes

$$\mathbf{H}(\omega) = -\frac{j\omega\tau}{1 + j\omega\beta\tau} \tag{11.13}$$

The response now levels out to a gain of $1/\beta$ at frequencies above $\omega = 1/\beta\tau$, and this may be controlled by circuit design.

210

● **Example 11.2** The signal of example 11.1, having a bandwidth of 3 kHz, is now to be differentiated with an operational circuit. A gain of 50 is required at 2 kHz, and the gain error at 3 kHz is to be less than 5%. Design the circuit.

Referring to equation 11.13, the gain at a frequency f will be $2\pi f \tau$. Hence, at 2 kHz and a gain of 50,

$$\tau = 50/(2\pi . 2) = 3 \cdot 98 \text{ ms}$$

The gain error will be due to the denominator term, whose modulus must be less than $1 \cdot 05$ in order to order to give an error of less than 5% at 3 kHz. So

$$1 + (2\pi . 3 \beta \tau)^2 \leqslant (1 \cdot 05)^2$$
$$\beta \leqslant 0 \cdot 0043$$

The maximum gain will therefore be $1/\beta$ which is 234, and is reached at frequencies somewhat greater than $9 \cdot 37$ kHz.

We may check out the necessary amplifier gain by using equation 11.12. At the maximum signal frequency, $A \gg \omega\tau$.

Hence $A \gg 2\pi . 3 . 3 \cdot 98$ or $A \gg 75$

Compared with the RC differentiator in example 11.1, the time constant is larger, leading perhaps to more convenient circuit values. For example,

RC differentiator 100 kΩ, 170 pF
Operational diff. 100 kΩ, 40 nF ●

11.1.2 Discrete-time signals

There is a fundamental difficulty when trying to apply the differentiation process to discrete-time signals, since differentiation is defined only with reference to continuous-time signals.

However, we note from the above discussion that differentiation is a linear operation and is accomplished in practice by normal linear-filtering processes. No doubt a similar approach for discrete-time processes will yield good approximations to differentiation. So we explore first the frequency response of a simple moving-average or all-zero digital filter, and see how it may be made to match the frequency response of the ideal differentiator, which is wholly imaginary (equation 11.8).

11.1.2.1 General principle

In order to draw out the basic principles, we consider a *non-causal* filter, which includes future as well as past input values in its calculation. Such a filter can be used to operate upon stored data, but not of course to process real-time signals. A digital filter of this sort has N stages, and we assume first that N is odd, $N = 2M + 1$. With input sequence $x(n)$ and output sequence $y(n)$, the filter is described by the following difference equation:

$$y(n) = \sum_{r=-M}^{M} a_r \cdot x(n - r) \tag{11.14}$$

This filter has the frequency response $\mathbf{H}(\omega)$:

$$\mathbf{H}(\omega) = \sum_{r=-M}^{M} a_r \cdot \exp \mathrm{j}(-\omega T_S r) \tag{11.15}$$

Note that the frequency response is the sum of a number of complex exponential functions, and that a wholly imaginary response can be obtained if they are combined in conjugate pairs to give a set of *sine* functions of $\omega T_S r$. So the set of coefficients $\{a_r\}$ must have odd symmetry, and we ensure that $a_{-r} = -a_r$. Thus

$$\mathbf{H}(\omega) = -\mathrm{j}2 \cdot \sum_{r=1}^{M} a_r \cdot \sin(\omega T_S r) \tag{11.16}$$

We have set $a_0 = 0$ since a differentiator has zero gain at zero frequency. In order to get a $+90°$ phase shift, the function would have to be multiplied by minus one. The difference equation of the filter is now in the form:

$$y(n) = \sum_{r=1}^{M} a_r \cdot [x(n - r) - x(n + r)] \tag{11.17}$$

Two issues now present themselves. First, the true significance of a non-causal filter, which seems an anachronism; and second the method of calculating the set of M coefficients $\{a_r\}$. Light can be shed on both these difficulties by considering the simplest form of digital differentiator, a first-order system.

11.1.2.2 A first-order differentiator

Let $M = 1$ in equation, 11.17, and hence $N = 3$. Then

$$y(n) = a_1 \cdot [x(n - 1) - x(n + 1)] \tag{11.18}$$

Now the sequence $\{x(n)\}$ represents sample values on a continuous-time function $x(t)$, whose slope is given by

$$\frac{\mathrm{d}x(t)}{\mathrm{d}t} = \lim_{\delta \to 0} \frac{x(t + \delta) - x(t - \delta)}{2\delta} \tag{11.19}$$

In the discrete-time case we have sample values only at intervals T_S sec, so the slope at $n = 0$ is approximately $[x(n + 1) - x(n - 1)]/2T_S$. Since the time interval T_S is a function only of the sampling rate, it is usually regarded as a universal scaling constant as far as discrete-signal processing is concerned, and by comparing equations 11.18 and 11.19 we would say that

$$a_1 = -1/2 \tag{11.20}$$

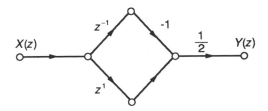

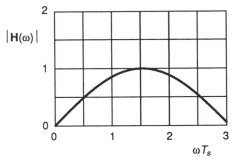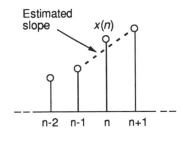

Figure 11.6   A first-order differentiator ($N = 3$)

Consequently the digital filter is described by the following set of equations:

$$y(n) = (-1/2) \cdot x(n-1) + (1/2) \cdot x(n+1) \tag{11.21}$$

$$H'(z) = (1/2) \cdot (z^1 - z^{-1}) \tag{11.22}$$

$$\mathbf{H}(\omega) = j \cdot \sin(\omega T_S) \tag{11.23}$$

Figure 11.6 shows the signal flow-graph, the frequency response and the approximate slope calculation for this simple filter.

Equation 11.22 suggests a technique for turning this non-causal filter into a usable one. Form a new filter with a response $H_1'(z) = z^{-1} \cdot H'(z)$:

$$H_1'(z) = (1/2) \cdot (1 - z^{-2}) \tag{11.24}$$

Then $$\mathbf{H}_1(\omega) = [\exp j(-\omega T_S)] \cdot j \sin(\omega T_S) \tag{11.25}$$

We have therefore converted a *non-causal* filter, requiring future values of the input signal, into a *causal* or usable filter, by inserting a delay of one sampling interval at the input. This additional delay shows up in the

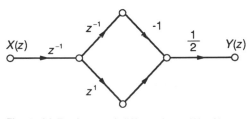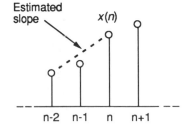

Figure 11.7   A causal differentiator ($N = 3$)

frequency response as a phase term which increases linearly with frequency (see section 7.2.4), and which can therefore be accommodated in any signal processing by keeping an account of accumulated delays. Figure 11.7 shows the revised signal flow-graph, and the new basis for estimating signal slope, which is now centralised at $(n - 1)$.

● **Example 11.3** Following the previous two examples, we consider again the signal with a bandwidth of 3 kHz, with a 5% error criterion at 3 kHz, and apply this simple discrete-time differentiator.

The frequency response of equation 11.25 shows that the only parameter that can be manipulated is the sampling frequency f_S.

Thus, the error criterion is $\omega T_S - \sin(\omega T_S) = 0 \cdot 05 \omega T_S$

Evaluating this numerically, we establish that $\omega T_S \simeq 0 \cdot 55$

Hence $f_S \geqslant 34 \cdot 3$ kHz

Note that the phase response is ideal, except from that due to a T_S delay.   ●

The unit-pulse response of this filter is now

$$\{g(n)\} = \{1/2, 0, -1/2, 0, 0 \ldots\} \qquad k \geqslant 0 \tag{11.26}$$

The step response is

$$\{u(n)\} = \{1/2, 1/2, 0, 0, 0 \ldots\} \qquad k \geqslant 0 \tag{11.27}$$

So, in summary, we have defined a simple first-order differentiator for discrete-time signals, which has the following properties:

a) In its non-causal form the frequency response is totally imaginary.
b) The magnitude of the frequency response is a fair approximation to the ideal, for frequencies less than say $\omega_S/8$.
c) The causal form of the filter is simple, and usable with care.

However, the step response is spread over two sampling times, and we are estimating the slope of the signal over a range of $2T_S$, which is a coarse measurement. Why not estimate the slope using adjacent samples?

11.1.2.3   Half-sample delay

A causal filter which will estimate the slope of the signal over adjacent samples is easily defined, bearing the basic definition of equation 11.19 in mind:

$$y(n) = x(n) - x(n - 1) \tag{11.28}$$

However, this expression lacks an obvious centre point about which a symmetrical structure can be developed, and this will be important later on when we derive some higher-order differentiators. The centre of symmetry of this process is at $(n - \frac{1}{2})$. What are we to make of this?

If z^{-1} represents unit delay, then $z^{-1/2}$ presumably represents a delay of $\frac{1}{2}$. Making use of this revolutionary concept, we may express the Z-transform response of this filter in a form which contains a symmetrical portion and a delay:

$$H'(z) = z^{-1/2} \cdot (z^{1/2} - z^{-1/2}) \tag{11.29}$$

$$\mathbf{H}(\omega) = [\exp \mathrm{j}(-\omega T_S/2)] \cdot \mathrm{j}2 \sin(\omega T_S/2) \tag{11.30}$$

The frequency response is therefore characterised by a phase slope corresponding to a delay of $T_S/2$, which is perfectly consistent with the Z-transform description and with the time-domain operation as illustrated in figure 11.8. (See section 7.2.4 for the relation between delay and phase response.)

The unit-pulse response and the step response are now

$$\{g(n)\} = \{1, -1, 0, 0, 0 \ldots\} \qquad k \geqslant 0 \tag{11.31}$$

$$\{u(n)\} = \{1, 0, 0, 0 \ldots\} \qquad k \geqslant 0 \tag{11.32}$$

The step response is therefore ideal for a discrete-time signal, and the frequency response is considerably better than for the previous case of $N = 3$. For this case, $N = 2$. It is fascinating to speculate on the reason why such a minute alteration to the difference equation of the filter should have a major influence on the final response!

The half-sample delay is apparently a useful tool to have in signal processing, and the general conclusion to draw is that *even-order* differentiators are to be preferred to *odd-order* filters. The only possible difficulty is the effective delay of $T_S/2$, which must always be borne in mind.

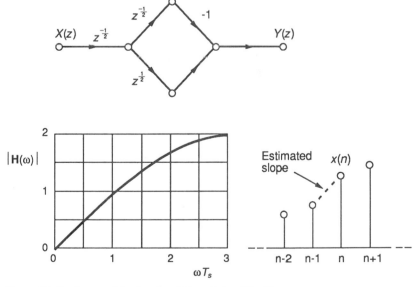

Figure 11.8 A causal first-order differentiator ($N = 2$)

215

● **Example 11.4** Let us apply this new differentiator to the case discussed in previous examples.

Equation 11.30 gives the frequency response, whence the error criterion is that

$$\omega T_S - 2 \cdot \sin(\omega T_S / 2) = 0 \cdot 05 \omega T_S$$

By numerical calculation $\omega T_S \simeq 1 \cdot 1$

Hence the necessary sampling frequency is $f_S \geqslant 17 \cdot 1$ kHz.

Note the improved (i.e. lower) figure over the previous case in example 11.3, and that the phase is again exact apart from that due to a $T_S/2$ delay. ●

11.1.2.4 Higher-order differentiators

In order to improve the operation of a discrete-time differentiator, we must allow more degrees of freedom in the design, and one standard way of doing this is to model the input signal before sampling $x(t)$, by a polynomial in t of degree $(M + 1)$, which of course is differentiable. So in the neighbourhood of the origin $t = 0$, we represent the input signal $x(t)$ by a polynomial:

$$x(t) = \sum_{q=0}^{M+1} \alpha_q \cdot t_q \tag{11.33}$$

The corresponding discrete-time signal will be a sampled form of $x(t)$, and for an even-ordered filter $(N = 2M)$, we consider the samples to be taken at unity spacing, but with a half-sample offset. Thus: $n = \cdots - 3/2, -1/2, 1/2, 3/2, \ldots$.

Forming the sample values in terms of this polynomial expansion for $M = 2$ for instance:

$$x(-3/2) = \alpha_0 - (3/2)\alpha_1 + (9/4)\alpha_2 - (27/8)\alpha_3$$
$$x(-1/2) = \alpha_0 - (1/2)\alpha_1 + (1/4)\alpha_2 - (1/8)\alpha_3$$
$$x(1/2) = \alpha_0 + (1/2)\alpha_1 + (1/4)\alpha_2 + (1/8)\alpha_3$$
$$x(3/2) = \alpha_0 + (3/2)\alpha_1 + (9/4)\alpha_2 + (27/8)\alpha_3 \tag{11.34}$$

Now from equation 11.33, the gradient of $x(t)$ at $t = 0$ is clearly α_1, so the digital process is required to calculate α_1 given a set of $2M$ signal sample values. The solution of this set of simultaneous equations needs only be done once, and Table 11.1 lists some coefficient values $\{a_r\}$ for even-ordered filters, leaving out the intervening algebra which is tedious.

Similar calculations can be done for odd-order filters, using integer values for sampling times, and some of these results are displayed in Table 11.2.

Figure 11.9 shows the frequency magnitude responses of the differentiators in Tables 11.1 and 11.2. The frequency responses show that the behaviour of the odd-order differentiators is generally very poor for

Table 11.1 *Some coefficients for even-order differentiators:* $N = 2M$

M	a_1	a_2	a_3
1	-1		
2	$-9/8$	$1/24$	
3	$-75/64$	$125/1920$	$-3/640$

Table 11.2 *Some coefficients for odd-order differentiators:* $N = 2M + 1$

M	a_1	a_2	a_3	a_4
1	$-1/2$			
2	$-2/3$	$1/12$		
3	$-3/4$	$3/20$	$-1/60$	
4	$-4/5$	$1/5$	$-4/105$	$1/280$

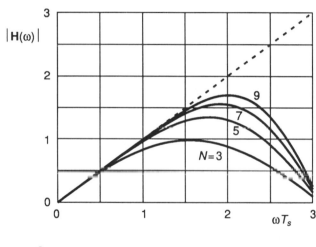

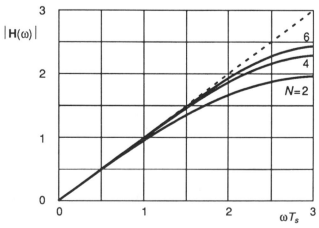

Figure 11.9 Frequency responses for differentiators

frequencies greater than about $\omega_S/4$, while those for even-order differentiators are considerably better.

● **Example 11.5** Let us now compare one of these improved differentiators with those of the last two examples. Take $N = 6$.

Applying the general frequency response equation, 11.16 for instance, after evaluating the coefficients of Table 11.1, we see that

$$|\mathbf{H}(\omega)| = 1\cdot172\,.\,\sin(\omega T_S/2) - 0\cdot065\,.\,\sin(3\omega T_S/2)$$
$$+ 0\cdot0047\,.\,\sin(5\omega T_S/2)$$

Since the error criterion that we have been following is 5% at a specified frequency, this comes down to

$$0\cdot95\omega T_S = |\mathbf{H}(\omega)|$$

Numerical evaluation yields $\omega T_S \simeq 2\cdot25$.

Hence $f_S \geqslant 8\cdot38$ kHz.

So for a given signal, and a given error criterion, this differentiator will operate with a much lower sampling frequency than either of the two simpler ones.

The step response of this differentiator may easily be evaluated by following through the difference equation, and is

$$\{u(n)\} = \{0\cdot0047, -0\cdot060, 1\cdot112, -0\cdot060, 0\cdot0047, 0, 0\ldots\} \qquad ●$$

Note that apart from phase contributions due to the inherent delay in the filter, the phase of each filter response is $90°$. Step responses can easily be calculated from a knowledge of the filter coefficients. The design technique outlined here is not necessarily the best, but it has showed up the essential characteristics of discrete-time differentiators. Other design techniques, based on synthesising the desired frequency response from scratch, can produce very good wideband differentiators, particularly if $N \geqslant 16$ say.

11.2 Prediction

The concept of signal prediction seems more suited to speculation rather than calculation, but it is in fact an important signal process. Clearly, it is not possible to predict, with complete certainty, the future value of a signal which is carrying information, so some statistical uncertainty remains in any predictor. In this section we shall merely introduce the concept using first-order cases for illustration. Extending these ideas to realistic signals is another story.

One obvious application of prediction is for collision-avoidance. Suppose that the positions of two aircraft have been tracked by radar over a period of some minutes; then this data can be used to predict where each aircraft might be at some time in the future, and hence to assess the likelihood of a collision.

There are several applications in signal coding, for voice and for video signals. If a signal waveform is reasonably predictable in the near future, then instead of transmitting every sample value of the signal, a reduced amount of information about the signal is sufficient to enable it to be reconstructed at the receiving end of a transmission link.

An example which is close to home is that of a typical domestic room heating system. Control of the heat input is often through an ON/OFF valve, while room temperature is sensed by a thermostat which commands more heat input when the temperature drops below the desired level. While this might seem to be an adequate control system, the dynamics of the situation often cause the temperature to cycle over a large range. In the first place, the temperature tends to rise exponentially in response to a step heat input. In the second place, there is a transport delay between heat being injected from a radiator and the resulting air temperature being sensed by the thermostat. Consequently, when the thermostat senses that the temperature has reached the set limit and commands the heat input to be turned off, heat continues to flow into the room. The temperature overshoots before being brought under control, and occupants of the room alternately feel cool and hot.

A predictor is required in order to anticipate the rise in temperature and cause the heat input to be switched off before the temperature reaches the set level. We return briefly to this example in section 11.2.3.

11.2.1 A first-order continuous-time predictor

A signal $x(t)$ is known in the past, up to $t - t_0$, and we require to estimate its value for positive values of time. If we model the signal by a polynomial such as that in equation 11.33, then we can evaluate the coefficients by using values taken from the past record of the signal, and then project the polynomial calculation forwards into the region where $t > t_0$. However, the polynomial model might not be the best one to use in practice, and some knowledge of the signal characteristics is necessary in order to choose a suitable model and make future predictions with confidence.

A first-order predictor will illustrate the principle. Let the output of the prediction process be $y(t)$, which is to estimate the future value of the signal at time $(t + T)$. A first-order prediction is based on the signal slope being constant, and

$$y(t) = x(t) + T \cdot \frac{\mathrm{d}x(t)}{\mathrm{d}t} \tag{11.35}$$

The frequency response of this process is given by letting $x(t) = \exp \mathrm{j}(\omega t)$, and

$$\mathbf{H}(\omega) = 1 + \mathrm{j}\omega T \tag{11.36}$$

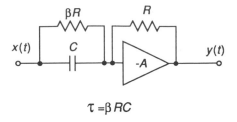

$$\tau = \beta RC$$

(a) **Operational amplifier circuit**

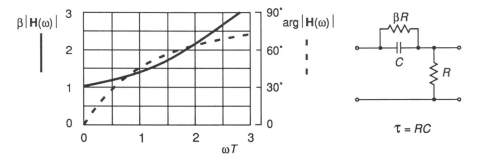

(b) **Frequency response**   (c) **Phase-lead circuit**

Figure 11.10   Predictor circuits

This response may be synthesised by the operational amplifier circuit of figure 11.10a, whose frequency response is shown in figure 11.10b. A passive circuit in figure 11.10c approximates to the required response when $\omega\tau \ll (\beta + 1)/\beta$, and in control circles this is known as a *phase-lead* circuit.

11.2.2   A first-order discrete-time predictor

The first-order discrete-time predictor implements the difference equation corresponding to equation 11.35. Thus, evaluating the current slope of the input signal using an $N = 2$ differentiator (section 11.1.2.3), and attempting to predict the value of the input signal at a time mT_S sec into the future, gives this difference equation:

$$y(n) = (1 + m) \cdot x(n) - m \cdot x(n - 1) \tag{11.37}$$

The corresponding frequency response is given by

$$\mathbf{H}(\omega) = (1 + m) - m \cdot \exp \mathrm{j}(-\omega T_S) \tag{11.38}$$

The magnitude response is plotted in figure 11.11 and shows that the predictor combines a uniform response with a rising response which is characteristic of a differentiator. We do not have the space to discuss this

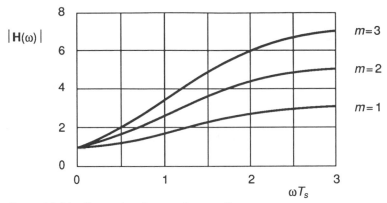

Figure 11.11 First-order discrete-time predictor

operation in any further detail, but before leaving the topic we comment briefly on the heating control system mentioned above.

A simple BASIC program can simulate the behaviour of the system, which incorporates an exponential lag and a delay, and is controlled by a simple predictor. The code given below is not a complete program, but describes each signal process involved in the simulation, and indicates how they are run in sequence.

```
100  REM input signal x, past values stored in array x( )
110  REM feedback signal (heater input) is "fb", scaled by A
120  REM temperature is "y" and predictor output is "z"

140  REM Main program loop
150  FOR i = 0 to N  1
160    GOSUB 200
170    temp(i) = y
180  NEXT
190  STOP

200  REM signal delay by "nd": nd = 3
220    FOR k = nd TO 1 STEP  − 1: x(k) = x(k − 1): NEXT
230    x(0) = fb * A
240  REM exponential system response: alpha = 0·98
244    y = (1 − alpha) * x(nd) + alpha * lasty
250  REM predictor: gamma = nd for optimum condition
254    z = (1 + gamma) * y − gamma * lasty
256    lasty = y
260  REM threshold decision: lu = 50: ll = 47
262    IF fb > 0 AND z > lu THEN fb = 0
264    IF fb = 0 AND z < ll THEN fb = 1
270  RETURN
```

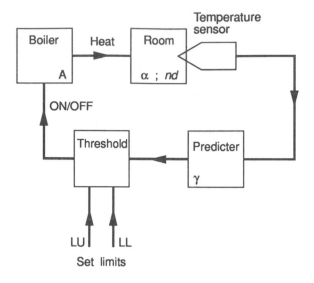

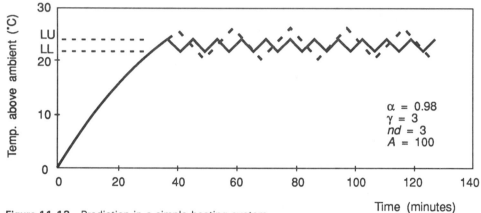

Figure 11.12 Prediction in a simple heating system

Figure 11.12 gives the temperature responses of this idealised system incorporating a small delay. The full curve represents the behaviour when a first-order predictor is used, having a parameter *gamma* equal to the system delay, while the dashed curve is the response when no prediction is included. The overall temperature swing has been reduced by more than half, and the cyclic period has been reduced, leading to a more comfortable environment for the inhabitants.

The predictor does not remove the cyclic variation entirely in this case, since it is not the optimum controller, but it has improved the time response.

11.3 Integration

Integration is the inverse operation of differentiation. In Chapter 10, we have already met the idea of integration as a method of taking a signal average, but in that application it is only necessary to enhance the mean value of the signal at the expense of all other signal frequency components. The integrator behaves like a lowpass filter.

In this discussion we are concerned more with integration as a mathematical operation, and wish it to generate the true integral of the input signal. Many of the applications apply to the equations of motion, where for example if velocity is being measured, the integral of velocity will give distance travelled.

Sometimes a modulated signal is being processed, and the integral of *instantaneous-frequency* will give the absolute phase of the signal.

11.3.1 Continuous-time signals

A continuous-time signal $x(t)$ is applied to an integration process with an output signal $y(t)$. Then

$$y(t) = \int_{-\infty}^{t} x(t) \,.\, \mathrm{d}t \tag{11.39}$$

Such a process is tested by applying the standard classical input signals.

11.3.1.1 Classical response

If the input signal is an *impulse*, then

$$x(t) = \delta(t) \tag{11.40}$$

$$
\begin{aligned}
y(t) &= 1 \quad & t \geqslant 0 \\
&= 0 \quad & t < 0
\end{aligned}
\tag{11.41}
$$

If the input signal is a *step function*, then the output signal is a *ramp function*, ie.

$$
\begin{aligned}
y(t) &= t \quad & t \geqslant 0 \\
&= 0 \quad & t < 0
\end{aligned}
\tag{11.42}
$$

These properties are illustrated in figure 11.13, together with the response to an arbitrary waveform.

The frequency response of this process is easily obtained in a similar manner to that for the differentiation, and

$$\mathbf{H}(\omega) = \frac{1}{\mathrm{j}\omega} \tag{11.43}$$

Notice that it requires a constant phase-shift of $-90°$, but that the magnitude is inversely proportional to frequency and implies infinite gain at zero frequency. This condition will be difficult to meet in practical systems, and may be undesirable. Figure 11.14 illustrates this frequency response.

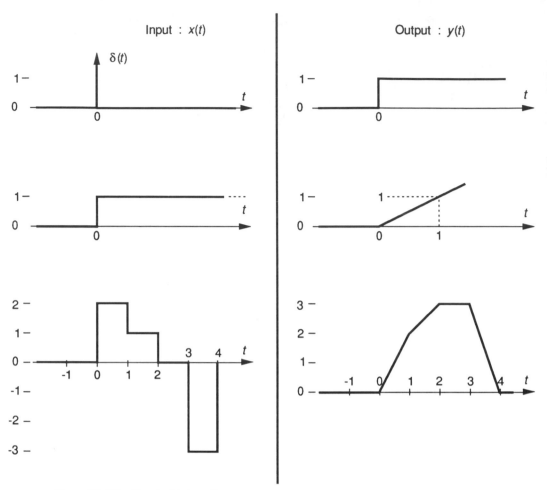

Figure 11.13 Classical integration

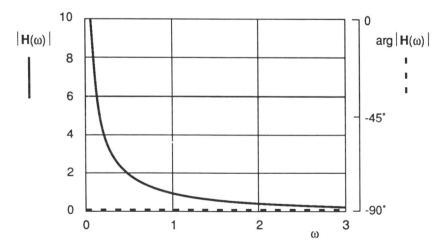

Figure 11.14 Frequency response of ideal integrator

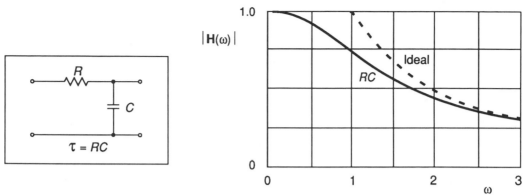

Figure 11.15 *RC* integrator frequency response

11.3.1.2 Practical process

As in the cases of the other processes discussed in this chapter, a simple *RC* circuit can be employed as an approximate integrator, whose frequency response is illustrated in figure 11.15. In this case

$$\mathbf{H}(\omega) = \frac{1}{1 + j\omega\tau} \tag{11.44}$$

Notice that the *RC* circuit gain is finite at zero frequency, whereas the ideal integrator has infinite gain at zero frequency. At higher frequencies, the approximation of the *RC* circuit is adequate provided that $\omega\tau \gg 1$.

● **Example 11.6** Integration is to be carried out by an *RC* circuit, and is to have a frequency response error of less than 5% for frequencies greater than 20 Hz.

We have the error criterion (equation 11.44):

$$1 + (\omega\tau)^2 \leqslant (1 \cdot 05)^2 \cdot (\omega\tau)^2$$
$$\omega\tau \geqslant 3 \cdot 12$$
$$\tau \geqslant 24 \cdot 9 \text{ ms}$$

For example $R = 100 \text{ k}\Omega; C = 6 \cdot 2 \ \mu\text{F}$
Note the large capacitor value! ●

The time responses of this circuit are easily found:

Impulse input $\qquad y(t) = \frac{1}{\tau} \cdot \exp(-t/\tau) \qquad (11.45)$

Step input $\qquad y(t) = 1 - \exp(-t/\tau) \qquad (11.46)$

These approximations are satisfactory if τ is large, but notice that this implies a very low magnitude response in the case of impulse input (equation 11.45).

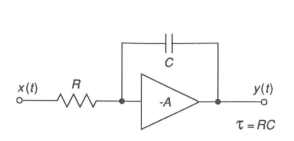

(a) **Basic circuit**

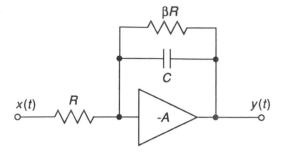

(b) **Gain-limited circuit**

Figure 11.16   An operational integrator

An operational integrator is derived by interchanging the R and C components in the differentiator (figure 11.5). Carrying out a similar analysis, we conclude that the integrator is nearly ideal if $\omega\tau A \gg 1$. The zero-frequency gain is $-A$, which is large for an operational amplifier ($A > 100\,000$). However, this large gain can be an embarrassment, since incidental bias currents, offset voltages and so on are amplified enormously and result in a *drift* of the signal baseline at the output of this type of integrator. In practice the zero-frequency gain is usually controlled by a shunt resistor, as shown in figure 11.16.

For this revised circuit,

$$\mathbf{H}(\omega) = \frac{-\beta}{1 + j\omega\tau\beta} \qquad (11.47)$$

The zero-frequency gain is now β, and a good approximation to the ideal integrator performance is obtained when $\omega\tau\beta \gg 1$, ie. at high frequencies.

● **Example 11.7**   A certain integrator is to have a maximum zero-frequency gain of 100, and is to have a frequency response error of less than 5% for frequencies greater than 20 Hz.

Then $\beta = 100$.

The modulus of the denominator term of equation 11.47 determines the frequency response error. Thus, in this case,

$$1 + (\omega\tau\beta)^2 \leqslant (1 \cdot 05)^2 . (\omega\tau\beta)^2$$
$$\omega\tau \geqslant 0 \cdot 031$$
$$\tau \geqslant 0 \cdot 25 \text{ ms}$$

For example $R = 10 \text{ k}\Omega; \ C = 25 \text{ nF}$
The amplifier gain must satisfy $A \gg \beta$
Notice the benefits in this case, of using the operational amplifier circuit!   ●

11.3.2 Discrete-time signals

Integration turns out to be easier to accomplish with discrete-time signals than differentiation. The frequency response of an ideal integrator requires a $90°$ phase shift (equation 11.43), as did the differentiator (equation 11.8). However, the $1/\omega$ magnitude response does not lend itself to synthesis via the techniques that we used for the differentiator (section 11.1.2), so an approach via the frequency response of the process is not helpful.

To produce a useful integrator using the moving-average type of filter would require a very large number of terms in order to produce a fair approximation to the magnitude frequency response, so we turn to a *recursive* type of linear filter.

Consider a discrete-time integration process, with input sequence $x(n)$ and output sequence $y(n)$. As time progresses, so $y(n-1)$ contains the previous estimate of the signal integral, which must be updated by an increment calculated from the last few values of the input signal. Thus, we have a general difference equation:

$$y(n) = y(n-1) + F[x(n), x(n-1), x(n-2)\,...]\qquad(11.48)$$

where $F[\ ]$ represents an undefined function.

Design of the integrator is carried out therefore in the time domain, and the accuracy of integration depends solely on the estimated increment of area since the last sampling instant. Integration algorithms can be deduced by representing the input signal before sampling, by the polynomial in equation 11.33, and then using terms of increasing order to estimate the incremental area.

The simplest, first-order approximation represents the signal by a polynomial of degree 0. The incremental area is then $T_S \cdot x(n)$, which is known as the *rectangular* approximation, and gives rise to the following difference equation, if we remove the T_S factor by normalisation:

$$y(n) = y(n-1) + x(n)\qquad(11.49)$$

Then

$$H'(z) = \frac{z}{z-1}\qquad(11.50)$$

This filter has a pole at $z=1$, or $\omega = 0$, and so fulfils the criterion that the gain of the integrator should be infinite at zero frequency, and the approximation should be best at low frequencies. However, this feature is difficult to handle in practice, since instability may occur due to vagaries of arithmetic round-off and the like. Consequently, a gain coefficient α is used, which has the effect of reducing the gain at zero frequency to $1/(1-\alpha)$:

$$y(n) = \alpha \cdot y(n-1) + x(n)\qquad(11.51)$$

where $\alpha \rightarrow 1$.

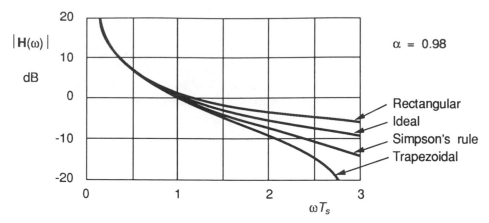

Figure 11.17   Discrete-time integrator frequency responses

$$\mathbf{H}(\omega) = \frac{1}{1 - \alpha \cdot \exp\, j(-\omega T_S)} \tag{11.52}$$

and

$$|\,\mathbf{H}(\omega)\,|^2 = \frac{1}{(1 + \alpha^2) - 2\alpha \cdot \cos(\omega T_S)} \tag{11.53}$$

This is none other than the conventional first-order recursive filter that we met first in section 3.3.2, and then in section 10.3.3 as a discrete-time averager. However, the gain of this version of the filter is different from the ones we met before, since the overall criterion of performance is different.

Figure 11.17 illustrates this magnitude response, together with the ideal response, and one or two other approximations which we shall discuss presently.

The frequency response of this approximate integrator (equation 11.52) is not obviously inversely proportional to frequency as it should ideally be (equation 11.43), but applying small angle approximations to the *cos* and *sin* terms in equation 11.52 produces

$$\mathbf{H}(\omega) \simeq \frac{1}{j\alpha \cdot \omega T_S} \tag{11.54}$$

where $(1 - \alpha) \ll \omega T_S \ll 1$.

So, this circuit is a good approximation to an integrator for low frequencies.

● **Example 11.8**   Consider a discrete-time integrator for the situation in example 11.7. The zero-frequency gain is to be limited to 100, and the frequency response error is to be less than 5% for frequencies less than 20 Hz.

The zero-frequency gain is $1/(1 - \alpha) = 100$, so $\alpha = 0.99$.

Using the frequency response equation 11.53, we set up the error criterion:

$$1 \cdot 05 / \omega T_S = | \mathbf{H}(\omega) |$$

Evaluating equation 11.53 numerically, $\omega T_S \simeq 1 \cdot 03$.

Hence the sampling frequency is $f_S \geqslant 18 \cdot 3$ kHz.

Notice that the phase shift at this frequency, as calculated by equation 11.52, is $-60°$. This is well removed from the ideal of $-90°$.   ●

Reviewing now the time response, we note from the difference equation that the unit-pulse response is

$$\{g(n)\} = \{1, \alpha, \alpha^2, \alpha^3, ..., \alpha^n ...\} \tag{11.55}$$

This is a sampled form of the impulse response of the RC network shown above (equation 11.45), and has a time constant of $\tau = -T_S / \ln(\alpha)$. So this simple discrete-time integrator yields a similar time response to the continuous-time RC integrator, but their frequency responses are similar only at low frequencies.

A better approximation is obtained by using a polynomial of degree 1 to model the signal before sampling. This gives rise to the *trapezoidal* approximation, where the incremental area is estimated as $[x(n) + x(n-1)]/2$ in normalised form, and

$$y(n) = \alpha \cdot y(n-1) + (1/2) \cdot x(n) + (1/2) \cdot x(n-1) \tag{11.56}$$

$$H'(z) - \frac{1}{2} \cdot \frac{z+1}{z-\alpha} \tag{11.57}$$

$$\mathbf{H}(\omega) = \frac{1 + \exp j(-\omega T_S)}{2[1 \quad \alpha \cdot \exp j(-\omega T_S)]} \tag{11.58}$$

The frequency response of this approximation is also shown in figure 11.17, where we note that the approximation gets worse as the Nyquist frequency is approached.

A further possibility for a discrete-time integrator is to employ a quadratic polynomial for estimating the area increment, which is commonly known as *Simpson's rule*. The usual form of this rule calculates the area over *two* discrete intervals, but in line with our definition of this process in equation 11.48, we have used the quadratic estimator to yield the area of only *one* time interval, giving a difference equation:

$$y(n) = \alpha \cdot y(n-1) + (5/12) \cdot x(n) + (2/3) \cdot x(n-1) - (1/12) \cdot x(n-2) \tag{11.59}$$

The frequency response of this approximation is also included in figure 11.17, but its calculation is left to the reader. Simpson's rule is a well known technique for numerical integration, but the frequency response clearly shows an increasing error as the Nyquist frequency is approached.

Comparing the frequency responses in figure 11.17, we note considerable departures from the ideal, so an integrator of this sort cannot easily be designed from the frequency point of view. However the time responses are quite good, and usually these are what matter. The *rectangular* approximation is the simplest, and apparently the most accurate over a range of frequencies up to the Nyquist frequency.

Notice that we have simplified this discussion by ignoring the phase component of frequency response. Whereas the 90° phase shift was guaranteed for the discrete-time differentiators, the phase of the discrete-time integrators varies with frequency.

More elaborate, adaptive algorithms are used for precise numerical integration, but many of these are not linear filters and so cannot be expressed in terms of a frequency response.

STUDY QUESTIONS

1   An *RC* differentiator has a resistance of 10 kΩ and is to give a phase error of no more than 10° at 5 kHz. Calculate the corresponding gain error at this frequency and the capacitor value.

2   The differentiator of question 1 is now to be realised by an operational circuit whose gain is to be limited to 100. Determine the new time constant for a similar performance, and calculate the circuit gain at 5 kHz.

3   A discrete-time first-order differentiator is to be employed on a signal whose bandwidth is 5 kHz. Determine the frequency-response magnitude-error at 5 kHz, for $N = 3$ and $N = 2$ versions of the differentiator. The sampling frequency is 20 kHz.

4   An *RC* operational integrator is to have a zero-frequency gain of 250, and a frequency-response magnitude-error of less than 2% for frequencies above 10 Hz. Design such a circuit, and calculate the phase error at 100 Hz.

5   A discrete-time first-order integrator operates at a sampling frequency of 12 kHz, and with a coefficient α of 0·98. Calculate the zero-frequency gain, and the frequency-response error in magnitude and phase at a frequency of 2 kHz.

12 Signal Processes: Complex Signals

OBJECTIVES

To introduce the idea of complex-signal processing, which is a direct implementation of certain mathematical concepts. In particular to

a) Extend the use of the complex-phasor signal model
b) Show how complex processes can be carried out
c) Demonstrate the importance of this concept by practical example.

COVERAGE

Commencing with the phasor model discussion of chapter 3, we emphasise again the importance of this concept, and then show how complex signals and processes may easily be represented in the real world. Complex signals are represented by I and Q components. The Analytic signal is a generalised form for the complex phasor, and possesses a one-sided frequency spectrum. Analytic signals are formed by using the Hilbert transform, and one implementation of discrete-time Hilbert transform is introduced.

These principles are applied to Doppler frequency measurements, and to single-sideband (SSB) signals.

Complex functions should now be familiar objects to the reader. The frequency response of a system is a complex function of *frequency*. It is defined as the response to a phasor, which is a complex function of *time*.

When we wish to create a mathematical model of a real signal, we use pairs of conjugate phasors, as described in chapters 3, 4 and 5. So the idea of a complex signal has been shown to be useful when modelling real situations, but do complex signals have a part to play in actual signal processing?

We shall show that the concept of a complex signal (ie. a complex function of time) is extremely valuable in practice, and allows some elegant solutions to certain problems. It is a key feature of much discrete-time signal processing.

Since the complex signal is such a central concept, we are able only to introduce and discuss it broadly in this chapter, while giving a few simple applications. Readers who go on to practice signal processing will come across the idea in many different contexts.

12.1 Principles

12.1.1 Phasors re-visited

In chapter 3 we introduced the phasor, a complex function of time $\mathbf{x}(t)$, which has a magnitude A and an initial angle φ:

$$\mathbf{x}(t) = A \cdot \exp \mathrm{j}(\omega t + \varphi) \tag{12.1}$$

Recall that this expression may also be written in the form:

$$\mathbf{x}(t) = \mathbf{A} \cdot \exp \mathrm{j}(\omega t) \tag{12.2}$$

where $\mathbf{A} = A \cdot \exp \mathrm{j}(\varphi)$.

Here, the multiplying constant $\mathbf{A}$ is a *complex magnitude*, and $\exp \mathrm{j}(\omega t)$ is a *unit phasor*. The unit phasor determines the base frequency of the phasor signal, while the complex magnitude $\mathbf{A}$ represents the information carried by it.

The complex phasor signal $\mathbf{x}(t)$ has been expressed above in polar coordinates, but it could equally well be stated in Cartesian coordinates. Thus

$$\mathbf{x}(t) = A \cdot \cos(\omega t + \varphi) + \mathrm{j}A \cdot \sin(\omega t + \varphi) \tag{12.3}$$

These two coordinates represent the *real* and *imaginary* parts of the phasor. Each is a real quantity, but together with the j operator, they describe a complex function of time.

12.1.2 Complex signals—a compact notation

Continuous-time processes use signals which are real functions of time (voltage for instance), and discrete-time processes use real numbers to represent signal samples; whereas the instantaneous value of a complex signal must be described by two parameters, magnitude/angle or real/imaginary.

A complex signal can be written in the general form:

$$\mathbf{x}(t) = x_I(t) + \mathrm{j}x_Q(t) \tag{12.4}$$

The two individual components correspond to the real and imaginary parts of the complex quantity, but in signal processing parlance they are referred to as

$x_I(t)$ In-phase component (corresponds to real part)
$x_Q(t)$ Quadrature component (corresponds to imaginary part).

So we may generate a complex phasor in the real world simply by using two signal components based on the real and imaginary parts of a phasor, as suggested by equation 12.3, and we note that the Q-component is the I-component phase-shifted by $-90°$. This unique relationship maintains the phasor rotation, and clearly holds for a phasor of any frequency. Thus for this case,

$$x_I(t) = A \cdot \cos(\omega t + \varphi)$$
$$x_Q(t) = A \cdot \sin(\omega t + \varphi)$$
$$(12.5)$$

A complex phasor signal can therefore be generated from a real signal by resolving it into in-phase and quadrature components as above.

Suppose that we have a real signal:

$$x(t) = a \cdot \cos(\omega t + \varphi)$$
$$(12.6)$$

The complex phasor corresponding to this signal, is obtained by observing that the I-component will be a scaled version of the signal itself ($A = a/2$), while the Q-component will be the I-component shifted through $-90°$.

Figure 12.1 shows, via a signal flow-graph, how such a complex phasor can be generated from a real input signal.

Notice that the combined operation generates the positive-frequency phasor for this signal, but that if a $+90°$ phase-shift is used to generate the Q-component, the negative-frequency phasor is generated. This complex signal format is therefore capable of representing positive and negative frequencies.

● **Example 12.1** Generate the I and Q components for the complex phasor equivalent of the following signal:

$$x(t) = 10 \cdot \sin(200t + 30°)$$

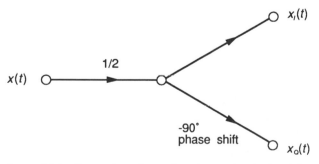

Figure 12.1 Generation of complex phasor signal: signal flow-graph

The I-component is merely the signal, scaled by $\frac{1}{2}$:

$$x_I(t) = 5 . \sin(200t + 30°)$$

The Q-component is shifted by $-90°$:

$$x_Q(t) = 5 . \sin(200t + 30° - 90°)$$
$$= 5 . \sin(200t - 60°) \qquad \bullet$$

Although this discussion has been on the basis of a signal at a single frequency, clearly it also holds for a range of frequencies provided that each one is shifted by $-90°$, and in particular any signal representable by a sum of sinusoidal components like those discussed in chapter 4. Appendix A3 shows that an aperiodic signal possesses a valid frequency spectrum, and chapter 16 shows that random signals do also. Therefore the conversion process that we are considering here applies to any kind of signal, provided that a $-90°$ phase-shift can be achieved at all frequencies within the signal bandwidth.

From here on we shall frequently refer to complex signals in terms of their I and Q components as well as their magnitude and angle.

● **Example 12.2**   Generate the I and Q components for the phasor equivalent of the following signal:

$$x(t) = 4 . \cos(500t + 20°) - 6 . \cos(700t - 45°)$$

Then

$$x_I(t) = 2 . \cos(500t + 20°) - 3 . \cos(700t - 45°)$$
$$x_Q(t) = 2 . \cos(500t - 70°) - 3 . \cos(700t - 135°) \qquad \bullet$$

The operation of phase-shifting by $-90°$ is often known as the *Hilbert Transform*, and the complex phasor representation is called the *Analytic Signal*, whose properties we shall investigate further in section 12.1.3.

The Hilbert transform of a signal $x(t)$ is often written as $\hat{x}(t)$, and consequently the (complex) analytic signal is

$$\mathbf{x}(t) = \tfrac{1}{2} . [x(t) + j\hat{x}(t)] \tag{12.7}$$

Note that achieving a phase shift of $-90°$ over a range of frequencies is not easy in practice, and that its effect on the signal waveform is dramatic and often unexpected. Figure 12.2 shows the example of a bandlimited square wave and its Hilbert transform.

The reason for this waveform distortion of course is that a component at frequency f Hz is delayed effectively by $1/(4f)$ sec when phase-shifted by $-90°$. Each frequency component is therefore delayed by a different amount, and the phase-shifted signal has an entirely different waveform from the original.

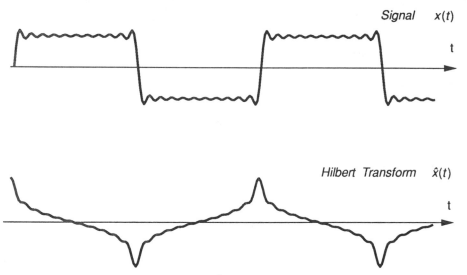

Figure 12.2 Effect of phase-shift by $-90°$ (Hilbert transform)

12.1.3 One-sidedness, the analytic signal

A real signal at a single frequency is represented by two conjugate phasors, leading to a description in terms of both positive and negative frequency. Any real signal therefore has a two-sided frequency spectrum or description, no matter whether it is periodic, aperiodic or random. (See also appendices A2, A3.)

Consider a real signal $x(t)$, which has a frequency description or spectrum $\mathbf{X}(f)$. For this discussion we have no need to specify the nature of $\mathbf{X}(f)$, it is just a complex function of frequency which represents the real signal $x(t)$. Since $x(t)$ is real, it follows that $\mathbf{X}(f)$ has conjugate symmetry, as illustrated in figure 12.3, and

$$\mathbf{X}(f) = \mathbf{X}^*(-f) \tag{12.8}$$

The corresponding analytic signal, however, resolves the real signal into only one phasor-type component, and hence leads to a one-sided frequency description.

We shall often denote the analytic signal corresponding to a real signal $x(t)$ as $\mathbf{x}(t)$. However in this case, in order to distinguish clearly between the original signal and the corresponding analytic signal, we suppose that an analytic signal $\mathbf{y}(t)$ is formed from the real signal $x(t)$, and has a frequency spectrum or description $\mathbf{Y}(f)$. Then

$$\begin{aligned} \mathbf{Y}(f) &= 0 & f < 0 \\ &= \mathbf{X}(f) & f \geqslant 0 \end{aligned} \tag{12.9}$$

and $\quad \mathbf{y}(t) = \tfrac{1}{2} . [x(t) + j\hat{x}(t)]$ (12.10)

This too is illustrated in figure 12.3.

235

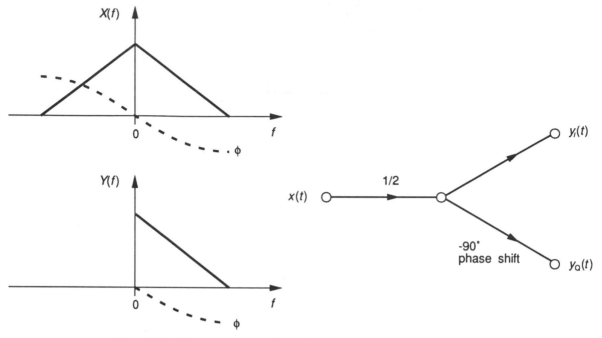

Figure 12.3 Frequency spectra for real and analytic signals

Since the analytic signal is complex, it possesses the usual attributes of magnitude and angle. Thus for a general analytic signal $\mathbf{y}(t)$:

$$\text{mag}[\mathbf{y}(t)] = \sqrt{[y_I(t)^2 + y_Q(t)^2]} \qquad (12.11)$$

$$\text{angle}[\mathbf{y}(t)] = \arctan[y_Q(t)/y_I(t)] \qquad (12.12)$$

The *instantaneous frequency* of this signal is defined as the rate-of-change of angle with time.

In the case where $\mathbf{y}(t)$ is the analytic signal representing a real signal $x(t)$, $y_I(t)$ is $x(t)/2$ and $y_Q(t)$ is $\hat{x}(t)/2$. Notice that these properties of magnitude and angle cannot be deduced from the original signal, but they are often extremely valuable, particularly for modulated signals. Complex signals are not necessarily analytic, although most of those that we shall discuss here will be.

● **Example 12.3** Calculate the magnitude and angle for the analytic signal derived from the real signal:

$$x(t) = 2A \cdot \cos(\omega t + \varphi)$$

Then using equation 12.5:

$$y_I(t) = A \cdot \cos(\omega t + \varphi)$$
$$y_Q(t) = A \cdot \sin(\omega t + \varphi)$$

236

And from equations 12.11 and 12.12:

$$\text{mag}\,[\mathbf{y}(t)] = A$$
$$\text{angle}\,[\mathbf{y}(t)] = \omega t + \varphi$$

So, the analytic signal has separated out the magnitude and angle of the original real signal, which is not surprising since the whole idea is based on the complex phasor. The analytic signal is a generalised form of phasor, extending this simple model to more realistic situations. In this case

$$\mathbf{y}(t) = A \,.\, \exp\, j(\omega t + \varphi) \qquad \bullet$$

We shall be able to outline one or two applications shortly, but a full understanding of the value of this concept comes only with experience.

● **Example 12.4**   Consider the case where the signal of the previous example has time-varying magnitude and angle

$$x(t) = 2 \,.\, [1 + b \,.\, \sin(\omega_m t)] \,.\, \cos\,[\omega_0 t + g \,.\, \cos(\omega_a t)]$$

where ω_m and $\omega_a \ll \omega_0$.

Then

$$y_I(t) = [1 + b \,.\, \sin(\omega_m t)] \,.\, \cos\,[\omega_0 t + g \,.\, \cos(\omega_a t)]$$
$$y_Q(t) = [1 + b \,.\, \sin(\omega_m t)] \,.\, \sin\,[\omega_0 t + g \,.\, \cos(\omega_a t)]$$

So

$$\text{mag}\,[\mathbf{y}(t)] = 1 + b \,.\, \sin(\omega_m t)$$
$$\text{angle}\,[\mathbf{y}(t)] = \omega_0 t + g \,.\, \cos(\omega_a t)$$

Consequently the analytic signal may also be expressed as

$$\mathbf{y}(t) = [1 + b \,.\, \sin(\omega_m t)] \,.\, \exp\, j\,[\omega_0 t + g \,.\, \cos(\omega_a t)] \qquad \bullet$$

A generalised method of extracting the magnitude and phase information from a modulated signal is therefore to form the analytic signal equivalent and then determine its magnitude and angle. We shall demonstrate this operation shortly. Note that all the operations described here are easily accomplished using discrete-time processing, and in principle are also possible for continuous-time signals.

12.2   Operations using complex signals

12.2.1   Phase rotation

There are numerous cases where we wish to impart a phase-shift to a single-frequency signal. The most common is where the phase-shift represents information. Thus, suppose that we are transmitting information and can allow the *carrier* frequency to take on one of four different phases, $0°$, $90°$, $180°$, $270°$. Each time we send this signal, there is a choice of one out of four possibilities, and the Introduction points out that this set of states therefore represents 2 bits of potential information. The number of such signalling intervals that we are able to send per second depends upon the bandwidth of the transmission system. The analytic signal concept provides one method of generating such a signal, by rotating the corresponding complex phasor.

Consider a single-frequency signal which is represented by the analytic signal with complex magnitude $\mathbf{A}$:

$$\mathbf{x}(t) = \mathbf{A} \cdot \exp \mathrm{j}(\omega t) \tag{12.13}$$

Suppose now that we wish to add a phase-shift θ to the signal, which means rotating the analytic signal phasor by θ rad. This is simply done by multiplying by a complex operator $\mathbf{w}(t)$ to form the new analytic signal $\mathbf{y}(t)$:

$$\mathbf{w}(t) = \exp \mathrm{j}(\theta) \tag{12.14}$$

$$\begin{aligned} \mathbf{y}(t) &= \mathbf{x}(t) \cdot \mathbf{w}(t) \\ &= \mathbf{A} \cdot \exp \mathrm{j}(\omega t + \theta) \end{aligned} \tag{12.15}$$

In terms of analytic signals then, the operation is simply stated, but expanding into I and Q components, the arithmetic can also be deduced. Thus, we re-state the two equations 12.14 and 12.15, in terms of I and Q components:

$$\mathbf{w}(t) = \cos(\theta) + \mathrm{j}\,\sin(\theta) \tag{12.16}$$

$$\begin{aligned} \mathbf{y}(t) &= [x_I(t) + \mathrm{j}x_Q(t)] \cdot [\cos(\theta) + \mathrm{j}\,\sin(\theta)] \\ &= y_I(t) + \mathrm{j}y_Q(t) \end{aligned} \tag{12.17}$$

where
$$y_I(t) = x_I(t) \cdot \cos(\theta) - x_Q(t) \cdot \sin(\theta)$$
$$y_Q(t) = x_I(t) \cdot \sin(\theta) + x_Q(t) \cdot \cos(\theta)$$

Phase-shifting has therefore been carried out by the normal arithmetic processes of multiplication and addition, which is convenient in both

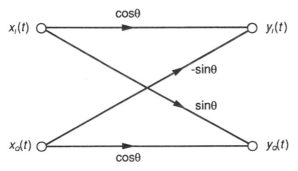

Figure 12.4 Signal flow-graph of phase rotation

continuous-time and in discrete-time signal-processing systems. However, note that the phase-shift is limited inherently to $\pm 180°$ or $\pm\pi$ rad, by the normal properties of the circular functions *sin* and *cos*. Figure 12.4 shows the signal flow-graph for this operation.

Notice that we have defined the full operation where an analytic signal is rotated and becomes a second analytic signal. If the real equivalent output signal only is required, for radio transmission for instance, then just the I-component need be taken which reduces the arithmetic complexity by half.

● **Example 12.5**   The signal of example 12.1 is now to be phase-shifted by $+60°$. Calculate the individual multiplications for the phase shifter when $t = 0\cdot03$ sec, and show that the desired phase-shift has been accomplished.

$$x(t) = 10\,.\,\sin(200t + 30°)$$

Now at $t = 0\cdot03$ sec, the argument $(200t + 30°)$ is $6\cdot524$ rad. Consequently the corresponding analytic signal is

$$x_I(t) = 5\,.\,\sin(200t + 30°) = 1\cdot191$$
$$x_Q(t) = -5\,.\,\cos(200t + 30°) = -4\cdot856$$

This yields a phasor angle of

$$\arctan[x_Q(t)/x_I(t)] = -1\cdot330 \text{ rad}$$

For the desired phase-shift, $\theta = 60°$, hence

$$\cos(\theta) = 0\cdot5 \quad \text{and} \quad \sin(\theta) = 0\cdot866$$

Hence, using equation 12.17,

$$y_I(t) = x_I(t)\,.\,\cos(\theta) - x_Q(t)\,.\,\sin(\theta)$$
$$= 4\cdot801$$
$$y_Q(t) = x_I(t)\,.\,\sin(\theta) + x_Q(t)\,.\,\cos(\theta)$$
$$= -1\cdot397$$

The phasor angle represented by these two components is therefore:

$$\varphi = \arctan[y_Q(t)/y_I(t)] = -0\cdot283 \text{ rad}$$

The difference in the two phasor angles is $1\cdot047$ rad or $60°$.   ●

12.2.2   Spectral rotation

In chapter 4 we introduced the idea of multiplying a signal by a sinusoidal component in order to change its frequency. Since frequency is the rate-of-change of phase, a complex phasor whose phase is increasing at a steady rate is rotating at constant frequency. Thus we may extend the ideas explored in the previous section to include this condition, and discover that the frequency of a signal can be shifted at will.

Quite simply, in place of θ in the equations above, we write $\omega_0 t$. Thus

$$\mathbf{y}(t) = \mathbf{x}(t) \cdot \exp \mathrm{j}(\omega_0 t)$$
$$= \mathbf{A} \cdot \exp \mathrm{j}[(\omega + \omega_0)t] \tag{12.18}$$

The frequency of the signal has been increased by ω_0 rad/sec, and the scheme is similar to figure 12.4, except that the multiplying coefficients are now time-varying. Figure 12.5 shows a schematic illustrating the complete process, which can be carried out in either continuous-time or in discrete-time regimes.

Also by this means, the phase of a signal can be measured relative to a reference phasor. To take one application, suppose that we generate a sinusoidal signal $x(t)$, and then pass it through a system which imparts a magnitude scaling of α and a phase-shift of β rad. The output signal from the system is $y(t)$, and a final processed signal is $z(t)$. Then modelling the signals by their equivalent phasors, the following relationships apply:

$$\mathbf{x}(t) = \exp \mathrm{j}(\omega_0 t) \tag{12.19}$$

$$\mathbf{y}(t) = \alpha \cdot \exp \mathrm{j}(\omega_0 t + \beta) \tag{12.20}$$

$$\mathbf{z}(t) = \mathbf{y}(t) \cdot \exp \mathrm{j}(-\omega_0 t) \tag{12.21}$$

$$z_I(t) = \alpha \cdot \cos(\beta) \quad \text{and} \quad z_Q(t) = \alpha \cdot \sin(\beta) \tag{12.22}$$

Figure 12.6 shows how this works out in practice, giving the real signals, the phasors and the processing required. The signal $y(t)$ has been compared

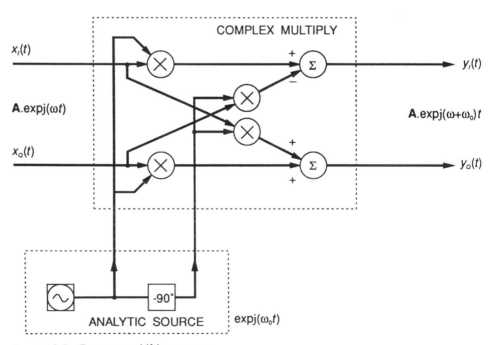

Figure 12.5 Frequency-shifting process

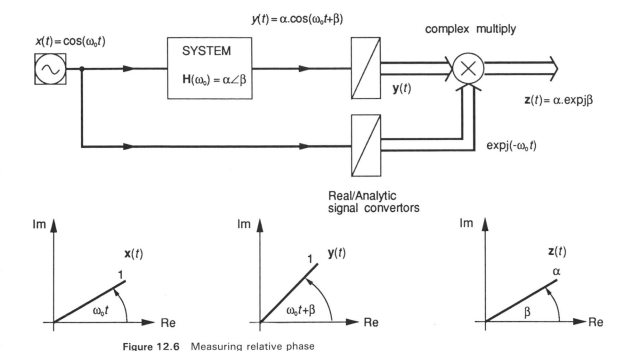

Figure 12.6 Measuring relative phase

with a reference signal at the same frequency, and the relative phase measured by resolving the unknown signal along the rotating axes defined by the reference. This is an extremely important operation in signal detection and measurement, although there are of course several ways of presenting the concept, and this explanation is but one of those.

Note that in the figure, the real–analytic signal converters ignore the factor of $\frac{1}{2}$, which is only introduced as a convenience for the analysis.

This technique is frequently invoked in instruments which measure system frequency response. A known frequency source is connected to the system, and the output signal is resolved into I and Q channels relative to the input signal phasor. Magnitude and phase are then measured using equations 12.11 and 12.12. Actual implementations may vary, but the principle remains the same.

A more general application of the technique allows us to extract information carried by a modulated carrier signal. Consider a transmitted signal at frequency ω_0 rad/sec, whose magnitude is modulated by an information signal $a(t)$, and whose angle is modulated by an information signal $\varphi(t)$. Then the real signal as received is

$$x(t) = a(t) \cdot \cos\left[\omega_0 t + \varphi(t)\right] \tag{12.23}$$

An elegant way of recovering the information is to frequency-shift the signal down to zero-frequency, having first turned it into an analytic signal $\mathbf{x}(t)$. This process is best illustrated by the schematic diagram of figure 12.7.

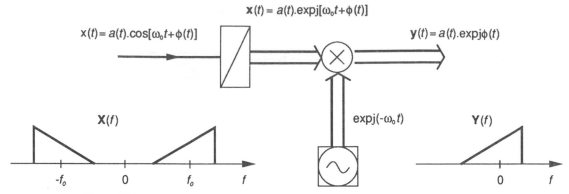

Figure 12.7 Demodulating a modulated signal

Frequency-shifting produces a complex signal $\mathbf{y}(t)$, which is not analytic since in general it will have a two-sided spectrum. Thus

$$y_I(t) = a(t) \cdot \cos[\varphi(t)]$$
$$y_Q(t) = a(t) \cdot \sin[\varphi(t)] \tag{12.24}$$

Although apparently complicated, the spectral rotation procedure is nevertheless ideal for discrete-time systems. However for high-frequency signals, it is often more convenient to simplify the process by frequency-shifting the two-sided spectrum of the original signal. In this case

$$y_I(t) = a(t) \cdot \cos[\varphi(t)] + a(t) \cdot \cos[2\omega_0 t + \varphi(t)]$$
$$y_Q(t) = a(t) \cdot \sin[\varphi(t)] - a(t) \cdot \sin[2\omega_0 t + \varphi(t)] \tag{12.25}$$

The $2\omega_0$ terms are removed by a simple lowpass filter in each of the two channels. Figure 12.8 shows this case.

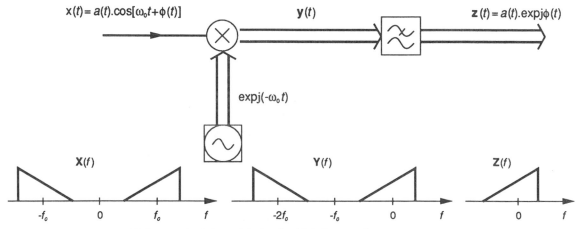

Figure 12.8 Practical demodulation for high frequencies

12.2.3 Phase-shifting—the Hilbert transformer

In the discussion so far, we have taken the $-90°$ phase-shifting operation for granted. Now if the signal is known to be a single frequency, then providing a $90°$ phase-shift is not difficult, particularly if the frequency is known and constant. However, for many cases of real interest, the signal occupies a finite bandwidth, and it is necessary to provide this phase-shift over the whole band. The ideal frequency response of this phase-shifter or *Hilbert transformer* is shown in figure 12.9a, together with an idealised discrete-time approximation in 12.9b.

In the continuous-time or analog world, there are known networks which approximate this performance over a limited bandwidth, and the design procedure is also well known. The networks are quite complicated, and of high order. We shall not study them here.

In the discrete-time world, we have already shown that a phase-shift of $-90°$ can be achieved at all positive frequencies up to the Nyquist frequency, by using a moving-average or all-zero digital filter whose unit-pulse response has odd symmetry about the centre (section 11.1.2). In that section we shaped the magnitude response to provide a differentiation operation, and here we use the same principle but contrive to make the magnitude response uniform over the frequency range.

We commence again with a *non-causal* filter, and the set of filter coefficients $\{a_r\}$ are arranged to have odd symmetry about $r = 0$. $a_0 = 0$ since the ideal frequency response also has a zero at zero frequency. The

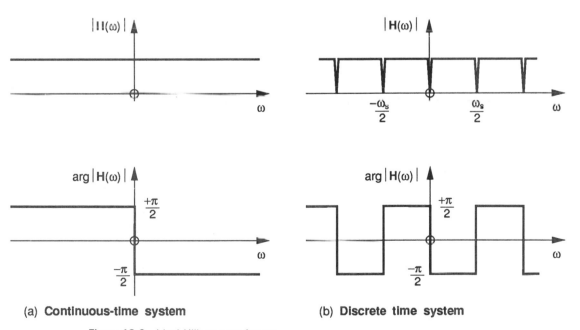

(a) **Continuous-time system**   (b) **Discrete time system**

Figure 12.9   Ideal Hilbert transformer

Table 12.1 *Coefficients for Hilbert transformer*

M	a_1	a_2	a_3	a_4	a_5	a_6	a_7	a_8	a_9	a_{10}	a_{11}
11	0·58069	0	0·08443	0	0·01019						
15	0·60762	0	0·13631	0	0·03224	0	0·00728				
19	0·61896	0	0·16340	0	0·05858	0	0·01706	0	0·00566		
23	0·62476	0	0·17852	0	0·07709	0	0·03173	0	0·01082	0	0·00463

frequency response of this type of filter $\mathbf{H}(\omega)$, and its difference equations are given as

$$\mathbf{H}(\omega) = -\mathrm{j}2 \cdot \sum_{r=1}^{M} a_r \cdot \sin(\omega T_s r) \tag{12.26}$$

$$y(n) = \sum_{r=1}^{M} a_r \cdot [x(n-r) - x(n+r)] \tag{12.27}$$

Notice that $\mathbf{H}(\omega) = 0$ at $\omega T_s = 0$ and π.

So we have to choose the set $\{a_r\}$ so that $|\mathbf{H}(\omega)|$ approximates to the square-wave function shown in figure 12.9b. This is not the place to enter into this design procedure in depth, so we just point out that the coefficients may be found by expanding the function $\mathbf{H}(\omega)$ as a Fourier series (appendix A2). This result is not ideal, and needs to be smoothed by using a *window* such as discussed in section 13.3.2 to reduce the size of magnitude ripples.

For the record, table 12.1 gives some coefficient values using a *Hamming Window*, and figure 12.10 shows the magnitude of the corresponding frequency responses. Notice that

- good approximations can be made
- approximations improve as higher-order filters are taken
- the phase response is identically $-90°$ for positive frequencies.

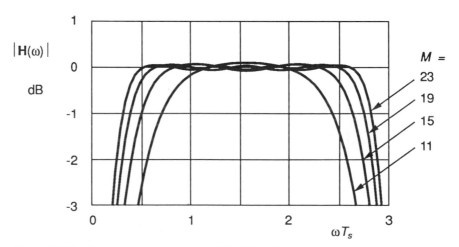

Figure 12.10 Frequency responses for Hilbert transformers

12.3 Some applications

12.3.1 Doppler radar

A classic application of complex signal processing lies in the field of using a radar system to measure the velocity of a remote object.

Figure 12.11 shows the idea of an extremely simplified radar system. A radio antenna projects a narrow beam of electromagnetic radiation at frequency f_0 Hz. At some distance d m, a target object (an aircraft for instance) has a velocity component towards the antenna of v m/s. A small amount of the incident radiation on the target is reflected, and ultimately arrives back at a receiving antenna where it is routed to a radio receiver. In practice it is common for there to be only one antenna, which alternates between transmit and receive duties.

Measurement of target velocity relies upon the *Doppler Frequency* effect. In the absence of any relative motion between target and antenna, the round trip from the antenna to the target and back encompasses a distance of $2d$ m, which is equivalent to a delay of $2d/c$ sec (where c m/s is the velocity of light, $\simeq 3 \times 10^8$ m/s). The incoming radio signal at the antenna has suffered a phase-shift of $2\pi . 2df_0/c$ rad, relative to the transmitted signal.

When the target is moving towards the antenna at a velocity v m/s, the round-trip distance is changing at a rate of $2v$ m/s, and so the relative phase is changing at a rate of $2\pi . 2vf_0/c$ rad/sec. Since frequency is rate-of-change of phase, the return signal frequency is apparently different from that transmitted, by an amount known as the *Doppler frequency f_D* Hz:

$$f_D = 2f_0 \frac{v}{c} \tag{12.28}$$

Measurement of this frequency change can therefore give an estimate of the target velocity, and frequency can be measured by several means; for example by a tunable bandpass filter, a zero-crossing counter, or a spectrum analysis algorithm as discussed in chapter 13. However, while these methods will give the magnitude of the frequency change, and hence the velocity which causes it, velocity is directional and can be either positive or negative. So we require a method of frequency measurement which takes account

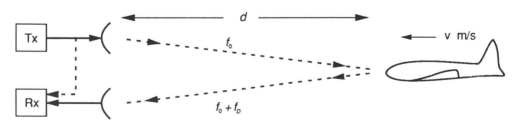

Figure 12.11 A simple radar velocity-measuring system

245

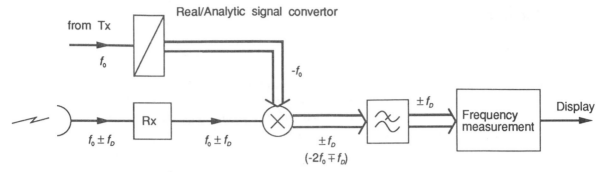

Figure 12.12 Doppler radar signal processing

of positive and negative frequencies, and that requires complex signal processing.

Velocity-measuring radars routinely resolve the incoming signal into I and Q channels, after mixing with the transmitted signal. Figure 12.12 shows a broad schematic of this process.

Many technological difficulties are encountered when carrying out these apparently simple operations, particularly as most radars are pulsed; and one of the most important is that of maintaining a stable reference at high radio frequencies. Processing is done using the schematic of figure 12.8 at radio frequency, and the signal is then digitised at the I and Q channel level.

Consider the typical case of an airport surveillance radar. The transmission is at a microwave frequency of 3 GHz with a wavelength of 10 cm. An aircraft velocity of 300 mph corresponds to 134 m/s. The Doppler frequency is therefore 2·68 kHz, which is in the audio range and easily handled by digital processing.

A road traffic velocity measurement system may transmit at 10 GHz (3 cm), and for a vehicle speed of 60 mph would give a Doppler frequency of 1·79 kHz. Such frequencies are easily measured.

● **Example 12.6**   Consider the case of a 9 GHz Doppler radar, when the range of vehicle speeds is 10 to 120 mph.

Using equation 12.28, the range of Doppler frequencies is 0·27 to 3·22 kHz.

For processing, a $90°$ phase-shift network is required at 9 GHz, to form the analytic equivalent of the transmitter frequency.

After using this analytic signal to multiply the incoming radio frequency signal, frequency components occur in the range of $\pm 0·27$ to $\pm 3·22$ kHz, and similar deviations around 18 GHz. The lowpass filters therefore have an easy task in separating off the baseband signals.   ●

12.3.2   Single-sideband (SSB)

A further classic application of the principle of complex signals lies in the generation of single-sideband (SSB) signals. We pointed out in section 5.1

that modulated signals develop two sidebands, one above the carrier frequency and the other below it. Since one is the conjugate of the other, this structure is redundant, and either sideband could be transmitted on its own in order to convey the information, while reducing the necessary transmitted signal power.

The principle of SSB generation is simple to explain using the analytic signal. Suppose first that we have a sinusoidal signal $x(t)$, to modulate on to a carrier frequency of f_0 Hz. Then using conventional double-sideband modulation, the modulating signal $x(t)$ forms the *envelope* of the carrier signal, forming the modulated signal $y(t)$:

$$x(t) = 2a \cdot \cos(\omega_m t) \tag{12.29}$$

$$\begin{aligned} y(t) &= x(t) \cdot \cos(\omega_0 t) \\ &= a \cdot \cos[(\omega_0 - \omega_m)t] + a \cdot \cos[(\omega_0 + \omega_m)t] \end{aligned} \tag{12.30}$$

Single-sideband operation is automatically achieved, if we use the analytic equivalent of the modulating signal to form a *complex envelope* on an analytic carrier signal:

$$\mathbf{x}(t) = a \cdot \exp \mathrm{j}(\omega_m t) \tag{12.31}$$

$$\begin{aligned} y(t) &= \mathrm{real}[\mathbf{x}(t) \cdot \exp \mathrm{j}(\omega_0 t)] \\ &= a \cdot \cos(\omega_m t) \cdot \cos(\omega_0 t) - a \cdot \sin(\omega_m t) \cdot \sin(\omega_0 t) \\ &= a \cdot \cos[(\omega_0 + \omega_m)t] \end{aligned} \tag{12.32}$$

Notice that in this case we require the real part of the result, since we wish to transmit this signal in the real world, via a radio transmitter for instance. The initial signal processing is carried out in a combination of continuous-time and discrete-time processing as convenient. Figure 12.13 shows the outline of the SSB generator in terms of real signals and processing.

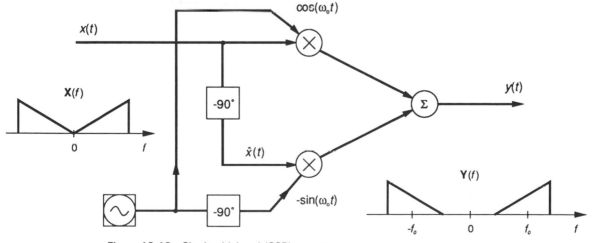

Figure 12.13 Single-sideband (SSB) generator

247

This style of SSB generation is known as the *phasing method*, and although it can work very well in discrete-time processing because the balance between the I and Q channels is maintained, it is a weak method to use for radio frequency applications. Other techniques are available for these cases, and SSB is widely used in radio transmission because it is economical in bandwidth.

A typical SSB modulator for a radio system is required to modulate an audio signal (0·3 to 3 kHz) on to an intermediate carrier frequency of, say, 10·7 MHz.

● **Example 12.7** A Hilbert transformer network must provide a 90° phase shift over the frequency range 0·3 to 3 kHz. For many types of phase-shifter, this is only possible within some tolerance.

Suppose that a phase-shifter has an error of 2° at 1 kHz. What effect does this have on the final signal spectrum?

Since frequency rotation does not alter the modulating signal, we need consider only the formation of this complex modulating signal. So, if the Hilbert transformer gives a phase-shift of $(\pi/2 + \delta)$ rad instead of $\pi/2$ rad, then compared with equation 12.31 the complex modulating signal becomes

$$\mathbf{x}(t) = \cos(\omega t) + \mathrm{j}\,\sin(\omega t + \delta)$$

This signal is no longer analytic, and so must have a negative-frequency phasor as well as a positive-frequency phasor. We discover the magnitudes of these phasors by first expanding the cos and sin terms as sums of exponentials. Thus

$$2\,.\,\mathbf{x}(t) = [1 + \exp \mathrm{j}(\delta)]\,.\,\exp \mathrm{j}(\omega t) + [1 - \exp \mathrm{j}(-\delta)]\,.\,\exp \mathrm{j}(-\omega t)$$

The positive-frequency phasor, which will give rise to the upper sideband frequency $(\omega_0 + \omega)$, has a complex magnitude of $[1 + \exp \mathrm{j}(\delta)]/2$. Since we are considering a small angle error δ, the actual magnitude of this component is approximately 1.

The negative-frequency phasor, which will give rise to a spurious lower-sideband frequency $(\omega_0 - \omega)$, has a complex magnitude of $[1 - \exp \mathrm{j}(-\delta)]/2$, which gives an approximate actual magnitude of $\delta/2$.

Consequently, we see that the output signal spectrum is approximately correct, but that there is an additional *lower-sideband* at $(\omega_0 - \omega)$ as well as one at $(\omega_0 + \omega)$. The ratio of these two for $\delta = 2\pi/180$ is approx 57·3 or 35 dB. This then is the penalty for not having the Hilbert transformer exact. ●

12.3.3 Telecontrol example

In our statement of the telecontrol example in section I.4.1, we included the requirement that the value of a current at the outstation should be transmitted back to the control centre. The range of current values is 0 to 100 A, resolved in steps of 0·5 A.

In section 5.1.4 during discussion of the benefits of frequency shifting and modulation and demodulation, it was suggested that this signal might be carried on a carrier frequency of 10 kHz. We can now investigate some techniques for accomplishing this task, using the analytic signal directly to modulate the carrier signal. These proposals might not actually be used in practice, but they do allow an elegant solution to the problem, particularly if discrete-time signals are employed.

The first solution to try uses phase-modulation of the carrier, after the style of figure 12.4. However, a total of 200 increments of current are required to be transmitted, and since this technique is limited to a phase excursion of less than $\pm 180°$, each increment can correspond to no more than $1·8°$. Such a scheme places heavy demands upon the stability of the frequency source, and on the phase measuring system. This is not a practical proposition.

Frequency-shifting of the carrier is more practical, since the shift can be quite large. Suppose for the sake of example that the varying current signal can be made to alter the frequency of a simple sinusoidal oscillator over the range 0 to 400 Hz, as the current varies from 0 to 100 A. In discrete-time processing, a simple oscillator may be made by varying the rate at which a set of sine wave values is read out from a fixed memory.

So we now have a signal frequency which varies from 0 to 400 Hz, and this must be frequency-shifted to the neighbourhood of 10 kHz, to give an output signal which varies from 10·0 to 10·4 kHz. This is easily accomplished by the SSB principle of figure 12.13, and for these frequencies can readily be carried out by discrete-time processing.

Reception of the varying frequency signal can be a mirror operation to the transmitting process. The incoming signal, which occupies the frequency range 10·0 to 10·4 kHz, will be shifted down to the range 0 to 0·4 kHz, and then measured by counting the number of periods that occur in a certain time. Notice that the resolution is required to be within 2 Hz of the true value, and that this means the counter must accumulate zero-crossing counts for a period of 0·5 sec, whether the signal is shifted down from 10 kHz or not. In practice, it is a little easier to do the counting at baseband.

One particular advantage of using complex signal rotation lies in the choice of sample frequency for a discrete-time process. Suppose that the incoming frequency is f_0 Hz, and we require to frequency change it down to 0 Hz by conventional means. Multiplying by $\cos(\omega_0 t)$, we generate frequency components at 0 Hz and at $2f_0$ Hz. If aliasing is to be avoided, then

$$f_s > 4f_0 \tag{12.33}$$

If this limit is not obeyed and aliasing occurs, it might be difficult to filter out the $2f_0$ component.

Now with complex spectral rotation, the result is a component at 0 Hz alone, and no filtering is necessary. At intervening stages in the multiplication (see figure 12.5), frequency components at $2f_0$ do occur, but they cancel out. So even if these components are aliased, cancellation will still take place and the result stands. Consequently, the sampling frequency limit need cater only for the original frequency f_0. Hence

$$f_s > 2f_0 \qquad\qquad (12.34)$$

12.3.4 System simulation

One prime example of complex signal processing in use is when we wish to simulate a modulated-carrier system. The carrier frequency conveys no intrinsic information but is introduced for practical reasons, allowing the signal to be propagated over a radio path, or multiplexed with other similar channels over a cable system. If we wish to simulate the performance of the system using a digital computer, the carrier frequency can therefore be ignored.

In order to retain full information about the modulated signal, including phase modulation and distortion, a complex model must be used. This approach is common.

STUDY QUESTIONS

1 Explain the terms *complex phasor*, *Hilbert transform*, *analytic signal*.

2 Give the complex phasor equivalent to the signal $x(t)$:

   $x(t) = 5 . \cos(300t + 2)$

3 Calculate the magnitude and phase of the phasor in question 2.

4 A discrete-time version of the signal in question 2 is

   $x(n) = 5 . \cos(0 \cdot 5n + 2)$

   Determine the corresponding phasor, and its magnitude and angle.

5 Determine the magnitude of the analytic signal which represents the real signal:

   $x(t) = 10 . \cos(300t) + 8 . \cos(500t)$

6 What distinguishes an analytic signal from a general complex signal?

7   The discrete-time signal in question 4 is to be phase-shifted by $-45°$ using a complex phase-shifter. Calculate the coefficients involved in the calculation at $n = 8$, and verify that the operation has been done correctly.

8   A certain phase-modulated signal can take on the four angles of $\pm 45°$ and $\pm 135°$. It is to be detected by a spectral rotation to zero frequency. Determine the ideal I and Q channel outputs for an input signal magnitude of 1 V.

9   A 10 cm Doppler radar is to measure the velocity of targets in the range 100 to 1000 mph. What range of detected frequencies occurs? Sketch a suitable signal processing scheme for detection.

10   An upper-sideband SSB signal is formed by the phasing method. Determine the maximum phase error which can occur, so that the spurious lower-sideband component is at least -30 dB relative to the upper-sideband component.

13 Signal Processes: Spectral Analysis and the DFT

OBJECTIVES

To introduce the practical problems involved in estimating the *spectrum* of a discrete-time signal. In particular to

a) Investigate the limitations of analysing a finite-length signal record
b) Study the effects of a time-window.

COVERAGE

The concept of a signal spectrum is introduced by reviewing the models for continuous-time signals.

The Discrete-Fourier-Transform (DFT) was previously introduced in chapter 7, and now we discuss its practical evaluation, which is constrained by the finite record-length of the signal, and by the finite time available for calculation. These lead to an approximate representation of the original signal by a finite number of phasors, which summarise the signal in terms of a set of complex coefficients.

The algorithm finds practical embodiment in the Fast-Fourier-Transform (FFT), which we discuss in terms of its principle and its advantages.

Finite record-length generates spectral leakage, which is a form of spectral distortion, and we discover its importance before introducing the shaped time-windows which reduce its effect. Discrete-frequency evaluation of the signal spectrum is easier to handle.

A few applications are mentioned, to give an inkling of how widespread is the DFT in signal processing.

13.1 The nature of a signal spectrum

In order to compare several signals, or to calculate the effect of a signal process on a given signal, it is necessary to construct a *model* which describes that signal on a mathematical basis. The most prevalent signal model uses the *complex phasor* as a basis function, leading to the *signal spectrum* which describes the signal in terms of a frequency variable. As well as enabling signals to be compared, in order to classify sounds for instance, knowledge of the signal spectrum also enables specific features of a system to be identified, like a frequency of vibration or modulation of a carrier.

Further, since the signal spectrum can be calculated from the time description of the signal, knowledge of the impulse response of a system enables the corresponding frequency response to be found.

13.1.1 Continuous-time signals

Under this modelling regime, which was introduced in chapter 4, a bandlimited signal $x(t)$ is represented by a set of N phasors, each of frequency ω_k and complex magnitude $\mathbf{C}_k$:

$$x(t) = \sum_{k=1}^{N} \mathbf{C}_k . \exp j(\omega_k t) \tag{13.1}$$

Such a model describes a class of signals where $x(t)$ is periodic, because the basis functions, the complex phasors, are themselves periodic. The periodic frequency of the signal is given by the *Highest-Common-Factor (HCF)* shared by the set of frequencies $\{\omega_k\}$:

$$\omega_p = \text{HCF}[\{\omega_k\}] \tag{13.2}$$

● **Example 13.1** A modulated signal is given by

$$x(t) = [1 + 0 \cdot 3 . \cos(100t)] . \cos(2000t)$$

This gives rise to phasors at 1900, 2000, 2100 rad/sec. Their factors are respectively:

$$\{19, 100\}, \ \{2, 2, 5, 100\}, \ \{3, 7, 100\}$$

The HCF is 100, showing that the periodic frequency of the whole signal is 100 rad/sec.

(The periodic frequency can be calculated either in Hz or in rad/sec.)   ●

A special case of this model is the *Fourier series*, where each phasor frequency is a multiple of the periodic frequency, or *fundamental frequency*

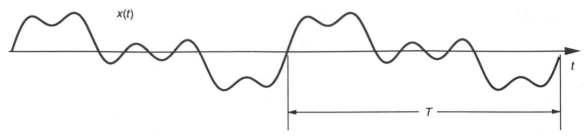

Figure 13.1 Signal representable by Fourier series

Table 13.1 *Fourier coefficients*

k	-5	-2	-1	1	2	5
$\mathbf{C}_k$	j0·25	j0·4	j0·5	$-$j0·5	$-$j0·4	$-$j0·25

ω_p (appendix A2):

$$\omega_k = k \cdot \omega_p \tag{13.3}$$

An important property of all these phasors, is that they must occur in conjugate pairs if the original signal $x(t)$ is real. Thus

$$\mathbf{C}_{-k} = \mathbf{C}_k^* \tag{13.4}$$

An example of a signal which is described by a Fourier Series is given in figure 13.1.

Having described such a signal by the set of coefficients $\{\mathbf{C}_k\}$, we may now compare two similar signals, or calculate the result of applying this signal to a linear filter having a frequency response $\mathbf{H}(\omega)$ (see section 4.2).

Most real signals are not periodic, and so we must modify the model. Consider first a set of phasors whose frequencies are given by equation 13.3. The fact that $x(t)$ is not periodic may be represented by the statement that $\omega_p \to 0$. The number of phasors N consequently tends to infinity and the modified model is thus:

$$x(t) = \frac{1}{2\pi} \int_{-\infty}^{\infty} \mathbf{X}(\omega) \cdot \exp j(\omega t) \, d\omega \tag{13.5}$$

The discrete set of phasors has been replaced by a continuum of phasors, whose frequency ω varies smoothly and whose magnitudes are described by $\mathbf{X}(\omega)$. Leaving aside the problems involved with measuring this function of frequency $\mathbf{X}(\omega)$, we note that it uniquely describes the signal $x(t)$.

$$\mathbf{X}(\omega) = \int_{-\infty}^{\infty} x(t) \cdot \exp j(-\omega t) \, dt \tag{13.6}$$

The pair of equations 13.5 and 13.6 are the *Fourier Transform* pair, and relate signal descriptions in time and in frequency domains, for a non-periodic energy signal (Appendix A3).

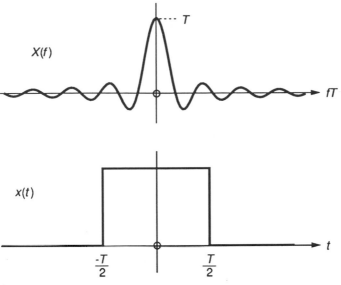

Figure 13.2 Example of Fourier Transform pair

● **Example 13.2** The simplest form of signal to consider is the rectangular pulse, which is of magnitude 1 over the range $-T/2 \leqslant t \leqslant T/2$.

This is sometimes referred to in shorthand form as the *rect* function:

$$x(t) = \text{rect}\,[t/T]$$

Integration of equation 13.6 is particularly simple in this case, and yields

$$\mathbf{X}(\omega) - \frac{2}{\omega}\sin(\omega T/2)$$

or $\mathbf{X}(f) = T.\text{sinc}(fT)$

where $\text{sinc}(u) = \sin(\pi u)/(\pi u)$.

These two functions are illustrated in figure 13.2. ●

The *spectrum* of a signal is its description in terms of the frequency variable, and is $\{C_k\}$ or $\mathbf{X}(\omega)$ depending upon which type of signal is being considered. *Spectral analysis* is the process whereby the signal spectrum is determined or estimated. Whereas a finite number of measurements can determine the details of the set $\{C_k\}$, an infinite number of measurements is necessary in order to discover the function $\mathbf{X}(\omega)$, unless its form is known. Consequently, an approximate measurement is normally made, with selectable degrees of precision.

Figure 13.3 shows fragments of two voice signals. The two signals are evidently similar, and yet they represent different audio sounds and are therefore distinct. In order to compare them, their frequency spectra should be determined, but they are not periodic signals and hence the model of

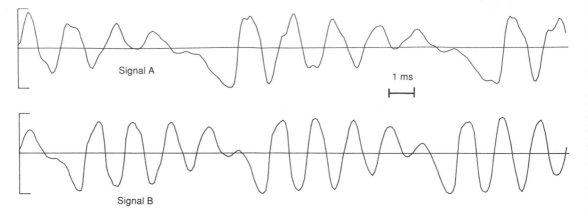

Signal A

1 ms

Signal B

Figure 13.3 Example of two voice signal waveforms

equation 13.5 must be used. We shall show later that using a discrete-time signal analysis and restricting the time of observation for each signal, we are able to arrive at an adequate estimate of the true spectrum. Given this information, the sounds that these two signals represent can be synthesised.

13.1.2 Discrete-time signals — the DFT

A discrete-time signal is described by the set of sample values $\{x(n)\}$, which are taken at intervals of T_S sec. Realistic signals are not periodic, so that the frequency description of this signal will follow equation 13.5 but with

a) the continuous variable t replaced by nT_S
b) the independent variable of integration normalised to ωT_S—for convenience
c) the integration limits replaced by $\pm\pi$ because of the sampling theorem:

$$x(n) = \frac{1}{2\pi} \int_{-\pi}^{\pi} \mathbf{X}(\omega) . \exp \mathrm{j}(\omega T_S n) \, \mathrm{d}(\omega T_S) \qquad (13.7)$$

Since the signal exists only at discrete times, the spectrum $\mathbf{X}(\omega)$ is given by (section 7.2.3):

$$\mathbf{X}(\omega) = \sum_{n=-\infty}^{\infty} x(n) . \exp \mathrm{j}(-\omega T_S n) \qquad (13.8)$$

This pair of equations for discrete-time signals corresponds to the pair 13.5 and 13.6 for continuous-time signals. They are known variously as the *Fourier Transform* of a discrete-time signal, as the *Discrete-time Fourier Transform* or simply as the *Discrete Fourier Transform (DFT)*. We shall refer to them in these notes as the DFT. A little more detail is given in appendix A5.

● **Example 13.3**   Consider the discrete-time version of the signal in example 13.2. Then

$$x(n) = 1 \qquad 0 \leqslant n \leqslant M - 1$$
$$ = 0 \qquad \text{elsewhere}$$

Applying equation 13.8, we see that

$$|\mathbf{X}(\omega)| = \frac{\sin(M\omega T_S/2)}{\sin(\omega T_S/2)}$$

Notice that we have considered only the magnitude of the spectrum $\mathbf{X}(\omega)$.

Notice too that the original signal $x(n)$ starts from $n = 0$. This condition has been chosen because an even number of non-zero points cannot be obtained so simply from a function that straddles $n = 0$.

Figure 13.4 illustrates this pair of functions for $M = 16$, and should be compared with figure 13.2. They are similar, and become more so as M increases.   ●

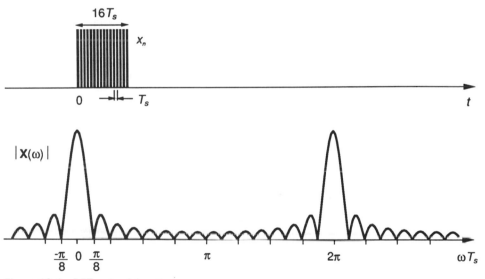

Figure 13.4   DFT pair of functions

13.1.3   Some properties of DFT spectra

The DFT is the basis of all discrete-time spectral analysis, and we shall consider in section 13.2 how it may be evaluated for realistic signals. Some basic properties of signal spectra are illustrated below for the discrete-time signal case, using the DFT relationships of equations 13.7 and 13.8.

● **Periodicity**   The DFT signal spectrum is periodic in ω_S, which follows from equation 13.8 since $\mathbf{X}(\omega)$ is the weighted sum of equally spaced exponentials.

- **Symmetry**   For a real signal, the phasors in equation 13.7 must occur in conjugate pairs. Hence the spectrum must have real-even and imaginary-odd symmetry.
- **Delay**   Consider the spectrum of a signal which has been delayed by mT_S sec. By substituting into equation 13.8, we note that

$$\text{If} \quad x(n) \Leftrightarrow \mathbf{X}(\omega)$$

$$\text{then} \quad x(n-m) \Leftrightarrow \mathbf{X}(\omega).\exp \mathrm{j}(-m\omega T_S) \tag{13.9}$$

This property has also been discussed in section 7.2.4.

- **Example 13.4**   A certain signal is described thus:

$$x(-1) = -1$$
$$x(1) = 1$$
$$x(n) = 0 \qquad |n| \neq 1$$

Applying this information to the calculation of the spectrum from equation 13.8:

$$\mathbf{X}(\omega) = -2\mathrm{j}.\sin(\omega T_S)$$

This function is plotted in figure 13.5. Notice that the spectrum is periodic, and that it possesses real-even, imaginary-odd symmetry.   ●

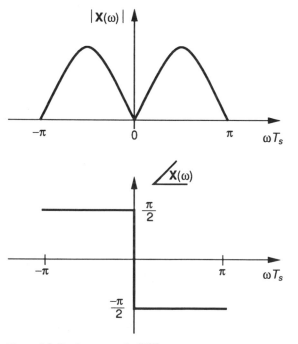

Figure 13.5   An example DFT

● **Example 13.5** Consider the effect of delaying the signal in example 13.4 by $2T_S$ sec. Then

$$x(1) = -1$$
$$x(3) = 1$$
$$x(n) = 0 \qquad n \neq 1 \text{ and } n \neq 3$$

Application of equation 13.8 gives

$$\mathbf{X}(\omega) = -\exp \mathrm{j}(-\omega T_S) + \exp \mathrm{j}(-3\omega T_S)$$
$$= \exp \mathrm{j}(-2\omega T_S) \cdot [-\exp \mathrm{j}(\omega T_S) + \exp \mathrm{j}(-\omega T_S)]$$
$$= -\exp \mathrm{j}(-2\omega T_S) \cdot [2\mathrm{j} \cdot \sin(\omega T_S)]$$

The magnitude of the spectrum is unchanged from the previous example, but an additional phase term reflects the delay of the signal by $2T_S$ (equation 13.9). ●

13.2 Practical difficulties in evaluating a spectrum

Although radio-frequency spectral analysis is carried out with analog filters, a great deal of spectral analysis is today carried out with digital hardware, leading to a discrete-time version of the spectrum. We now consider the problems latent in this procedure.

The signal that we wish to analyse is most likely to be analog or continuous-time, but that can easily be converted to discrete-time form (chapter 6), having first been bandlimited. Unless that signal is also time-limited, then we are unable to obtain a full description of its frequency spectrum, and must be content with an estimate.

Referring back to equation 13.8, we note two practical limitations:

a) An infinite summation is not physically possible.
b) Computation time of the spectrum is limited, so that it can only be evaluated at a finite number of frequencies ω.

These difficulties are crucial, and are now to be considered in some detail, because without a proper appreciation of their effects, the resulting spectrum 'estimate' is valueless.

13.2.1 Finite summation

When evaluating equation 13.8 in order to find a signal spectrum, there is no difficulty in truncating the summation to a finite number of signal samples N; the only query is what effect this has. Reducing the number of samples used in the calculation reduces the information input to the process, and hence the signal spectrum will be defined incompletely.

Consider the original signal $x(n)$, which is unlimited in its extent over the 'time' axis. Now we are to take only N samples, and the section which we are to use for analysis lies in the region $0 \leqslant n \leqslant (N-1)$. Effectively the original signal $x(n)$ is being viewed through a restrictive *window* $w(n)$, and

$$w(n) = 1 \qquad 0 \leqslant n \leqslant (N-1)$$
$$ = 0 \qquad \text{elsewhere} \tag{13.10}$$

The signal presented for analysis is $x'(n)$ and

$$x'(n) = x(n) \cdot w(n) \tag{13.11}$$

This effect is illustrated in figure 13.6.

Now, when it comes to evaluating the signal spectrum, we take the DFT of the modified signal in equation 13.11. The transform of a product is the convolution of the individual transforms (appendices A3, A4, A5), so denoting the observed spectrum calculated over N points as $\mathbf{X}_N(\omega)$, we write

$$\mathbf{X}_N(\omega) = \mathbf{X}(\omega) * \mathbf{W}(\omega) \tag{13.12}$$

$\mathbf{W}(\omega)$ is the spectrum corresponding to the window function $w(n)$, and distorts the observed spectrum by being convolved with the true spectrum. The true spectrum is therefore viewed through a window which alters its shape, and can *never be determined exactly*. The ideal form for $\mathbf{W}(\omega)$ is a delta function $\delta(\omega)$, but this is not possible, as section 13.3 will show.

It is not easy to visualise the effect of a convolution, but in section 13.3 we shall explore this effect in more detail, trying to put some quantitative measure on it. For the moment however, we shall take a few simple examples to illustrate what can happen.

For the simple window defined in equation 13.10, a *rectangular* window, the DFT equation simplifies to

$$\mathbf{X}_N(\omega) = \sum_{n=0}^{N-1} x(n) \cdot \exp \mathrm{j}(-\omega T_S n) \tag{13.13}$$

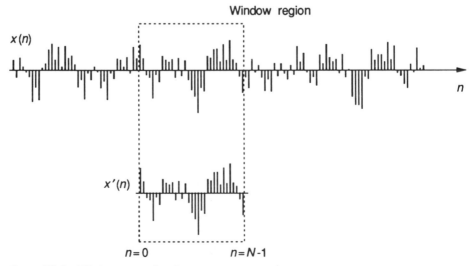

Figure 13.6 Window on a signal

● **Example 13.6**   Consider a phasor signal:

$$x(n) = \exp\, j(\omega_1 T_S n)$$

This extends beyond the window extent of $n = N$, and so we may expect the windowed spectrum to be distorted. Applying the convolution equation 13.12 and since $\mathbf{X}(\omega) = \delta(\omega - \omega_1)$:

$$\mathbf{X}_N(\omega) = \mathbf{W}(\omega - \omega_1)$$

The spectrum corresponding to the window is therefore shifted to a new frequency, ω_1. The significance of this will be seen in section 13.3.1.   ●

● **Example 13.7**   Consider the impulse response of an ideal lowpass filter. A common measurement technique is to observe the impulse response of a system and then use the DFT to infer the frequency response (section 7.2.3).

For this example, we take

$$x(n) = [\sin(n\pi/3)]/n$$

This is the impulse response of the ideal lowpass filter having a uniform frequency response between $-\omega_S/6$ and $\omega_S/6$.

Without detailing the algebra, we calculate the windowed response $\mathbf{X}_N(\omega)$ for several values of N, which are shown in figure 13.7. Notice how that the approximation gets better as N gets larger.   ●

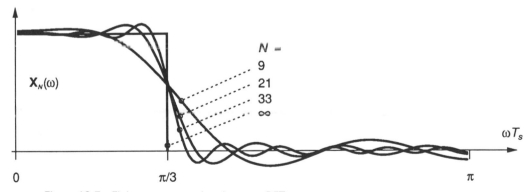

Figure 13.7   Finite-sum approximation to a DFT

13.2.2   Discrete-frequency

Although a calculation of the DFT is distorted by operating only on a finite set of signal values, there is the further restriction that calculation can take place only for a finite set of frequencies. The phasor model therefore changes from that of equations 13.7 and 13.8 to something like the discrete set of phasors of equation 13.1. The key question now is: how many frequency points should be calculated?

Several arguments can be advanced, but perhaps the simplest is to regard the finite-frequency DFT as a Fourier series. Figure 13.8 shows the original signal, and the windowed section which is to be analysed. If we regard this section of N points as the basis of a periodic repeat, then we may analyse it as a Fourier series. The period is then NT_S, and the periodic frequency is ω_S/N. The maximum frequency allowed by the sampling theorem is $\omega_S/2$. Broadly then, $N/2$ positive-frequency phasors are generated, and $N/2$ negative-frequency phasors, giving N in total, although for a real signal these two sets will be conjugates.

An astute reader may well be wondering where the zero-frequency phasor fits into the picture. Actually the phasors at $\omega_S/2$ and $-\omega_S/2$ are identical (equation 13.13), so only one of these two is an independent parameter. It is usual therefore to evaluate phasors at $(N/2 + 1)$ frequencies in the range 0 to $\omega_S/2$ rad/sec, and at $(N/2 - 1)$ negative frequencies, making a total of N.

Reviewing this information, we denote the interval along the frequency

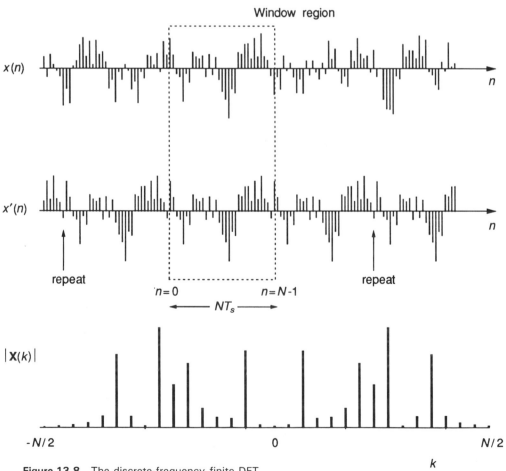

Figure 13.8 The discrete-frequency finite DFT

axis as Ω_S rad/sec or δ_S Hz. This then is the *fundamental frequency*, or the frequency difference between any adjacent pair of phasors. Two vital expressions can now be established, which relate the prime parameters of the discrete-frequency DFT:

$$N.\Omega_S = \omega_S \qquad (13.14)$$

$$N.\delta_S = f_S \qquad (13.15)$$

● **Example 13.8** A certain digital system is sampled at 5 kHz. Calculate the spacing of the frequency samples when 128 and 1024 points respectively are taken in the DFT.

From equation 13.15 $\quad \delta_S = f_S/N$
For $N = 128$ $\quad \delta_S = 39 \cdot 0625$ Hz or $\Omega_S = 245 \cdot 437$ rad/sec
For $N = 1024$ $\quad \delta_S = 4 \cdot 8828$ Hz or $\Omega_S = 30 \cdot 6796$ rad/sec

Note that the number of points N is a power of 2, which is very common, for reasons which will become apparent. ●

Now that the frequency scale has been discretised, the DFT equations can be recast in a new form. Substituting the new conditions into equation 13.8,

$$\mathbf{X}(k\Omega_S) = \sum_{n=0}^{N-1} x(n).\exp \mathrm{j}(-k\Omega_S T_S n) \qquad (13.16)$$

However, the exponent reduces to $-2\pi kn/N$, and it is also usual to refer to the frequency variable only by its index k. So the pair of equations is normally written:

$$\mathbf{X}(k) = \sum_{n=0}^{N-1} x(n).\mathbf{W}^{-kn} \qquad (13.17)$$

$$x(n) = \frac{1}{N} \sum_{n=0}^{N-1} \mathbf{X}(k).\mathbf{W}^{kn} \qquad (13.18)$$

where $\mathbf{W} = \exp \mathrm{j}(2\pi/N)$.

The frequency spectrum is now a *vector* containing N complex elements, and the region around each discrete-frequency is known as a *bin* or frequency slot.

Notice that the DFT and its inverse have now become simple numerical operations, and that they are very similar. The $\mathbf{W}$ *twiddle factor* holds the key to the evaluation of the DFT, and forms the basis for the transformation. The spectrum so produced has certain properties, which will be illustrated by a trivial example. The DFT is most usually computed on digital hardware for a large number of points, but we use a numerical calculation merely to show the principle.

● **Example 13.9** Calculate the discrete-frequency DFT over 6 points, of the following signal.

$$x(n) = 1 \qquad n = 0, 1$$
$$ = 0 \qquad \text{elsewhere}$$

Now $\mathbf{W} = \exp j(\pi/3)$ then

$$\mathbf{X}(0) = \Sigma x(n) \quad = 2 \cdot 0$$
$$\mathbf{X}(1) = 1 + \mathbf{W}^{-1} = 1 \cdot 73 \ \angle -0 \cdot 52$$
$$\mathbf{X}(2) = 1 + \mathbf{W}^{-2} = 1 \cdot 0 \ \angle -1 \cdot 05$$
$$\mathbf{X}(3) = 1 + \mathbf{W}^{-3} = 0$$
$$\mathbf{X}(4) = 1 + \mathbf{W}^{-4} = 1 \cdot 0 \ \angle 1 \cdot 05$$
$$\mathbf{X}(5) = 1 + \mathbf{W}^{-5} = 1 \cdot 73 \ \angle 0 \cdot 52$$

Notice that the zero-frequency value $\mathbf{X}(0)$ is always real if the signal is real.

$\mathbf{X}(3)$ corresponds to $k = N/2$, and is at the *folding* or *Nyquist* frequency. Its value is always real for a real signal, but not necessarily always zero as here.

$\{\mathbf{X}(1), \mathbf{X}(5)\}$ and $\{\mathbf{X}(2), \mathbf{X}(4)\}$ form conjugate pairs, since they are reflected about the folding frequency.

These particular values are illustrated in figure 13.9.   ●

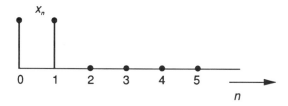

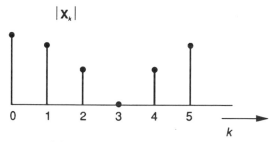

Figure 13.9 DFT numerical example

More interesting results occur when an increased number of signal points is taken, and figure 13.10 shows the result of taking a 512-point DFT of the signals in figure 13.3.

Although both spectra are non-trivial, it is now possible to distinguish between these two signals, and perhaps to classify them according to some pre-arranged knowledge of the spectra which correspond to various sounds.

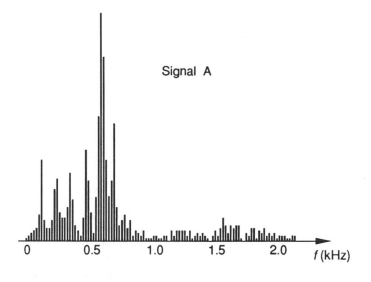

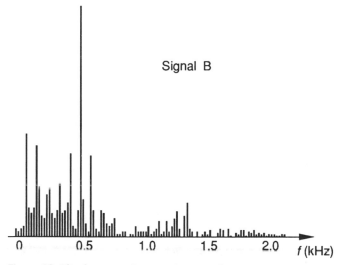

Figure 13.10 Spectra of two voice waveforms

13.2.3 The Fast-Fourier-Transform (FFT)

We have shown that the DFT represents the frequency spectrum of a given discrete-time signal, and have derived a compact numerical form for its calculation. However, the real genius and value of the DFT lies in the speed with which this numerical expression can be calculated in its so-called *Fast* form. The FFT is one of the most efficient computing algorithms ever devised, and many of the current applications of digital signal processing would not be possible without it.

It is not our purpose to investigate the structure of the FFT algorithm, since we are concerned with understanding and applying the concepts behind signal processing. The FFT exists in many different forms, and is usually available as a standard routine on a general-purpose computer system. However for completeness, we shall outline the principle behind it, and briefly mention the order of magnitude of its advantage.

First, in the direct mode of equation 13.17, each of the N points in the spectrum is evaluated by the sum of N products, giving a total of N^2 *operations*. An *operation* in this sense is a complex multiply-and-add. Broadly, multiplication is a more intensive task than addition, and so the number of these operations should be minimised for best computational advantage.

Refer back again to equation 13.17. As the two indices k and n progress through their ranges of 0 to $(N-1)$, the twiddle factor raised to the power (kn) passes through many rotations, and hence the same value is repeated many times in the overall calculation. Decomposition of the algorithm to the fast form takes advantage of this redundancy.

Consider the input signal $x(n)$, and for convenience take it over a total of $2N$ points where $0 \leqslant n \leqslant (2N-1)$.

Divide it into two halves, $u(n)$ taking the even-numbered samples and $v(n)$ the odd-numbered samples. Then

$$\left. \begin{array}{l} u(n) = x(2n) \\ v(n) = x(2n+1) \end{array} \right\} \quad 0 \leqslant n \leqslant (N-1) \tag{13.19}$$

Now we may form an expression for the $2N$-point transform of the input signal, and express it in terms of the N-point transforms of the two half-signals $u(n)$ and $v(n)$:

$$\mathbf{X}_{2N}(k) = \sum_{n=0}^{2N-1} x(n) . \mathbf{W}_{2N}^{-kn}$$

$$= \sum_{n=0}^{N-1} x(2n) . \mathbf{W}_{2N}^{-2kn} + \sum_{n=0}^{N-1} x(2n+1) . \mathbf{W}_{2N}^{-k(2n+1)}$$

$$= \mathbf{U}_N(k) + \mathbf{W}_{2N}^{-k} . \mathbf{V}_N(k) \tag{13.20}$$

Thus, one $2N$-point transform, which requires $4N^2$ operations, is replaced by the combination of two N-point transforms, requiring $N(2N+1)$ operations. The saving in computational operations is almost 50%.

The decomposition can be pressed further, since

$$\mathbf{X}_{2N}(k+N) = \mathbf{U}_N(k) - \mathbf{W}_{2N}^{-k} . \mathbf{V}_N(k) \tag{13.21}$$

This pair of equations forms a *butterfly*, as the signal-flow graph of figure 13.11 shows, involving only one complex multiplication.

Such an operation is in effect a 2-point transform, and is used to decompose an N-point transform, provided that N is a power of 2, ie. $N = 2^m$. An N-point transform can be calculated from two $N/2$-point transforms, by $N/2$ butterfly operations.

One such strategy is illustrated in figure 13.12, and is called *Decimation-*

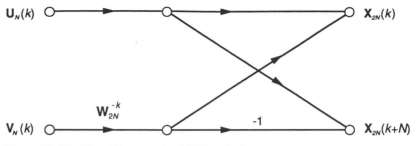

Figure 13.11 Signal-flow graph of FFT butterfly

in-Time (DIT). $N/2$ such 2-point transforms are calculated at the first pass through the algorithm. The second pass calculates $N/4$ 4-point transforms, and so on through all m passes, until the final stage is 2 $N/2$-point transforms. At each pass, a new N-point signal vector is calculated, and since each pair of calculated points depends on a unique pair of input points, the calculated vector stays *in-place*, requiring only N elements of storage at all times during the calculation. Figure 13.12 gives the broad outline of this procedure.

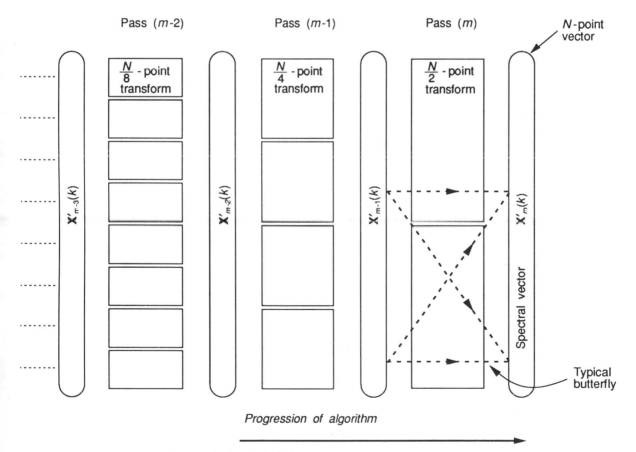

Figure 13.12 An FFT calculation

Table 13.2 *FFT computational effort*

N	32	1024	32768
N^2	10^3	10^6	10^9
$(N/2) . \log_2 N$	80	5120	$2 \cdot 45 \times 10^5$
Ratio	13	205	4369

Selection of the $\mathbf{U}(k)$ and $\mathbf{V}(k)$ inputs to each butterfly from the previous signal vector, and choosing the correct coefficient $\mathbf{W}_{2N}^{-k}$ at each stage, are complicated issues. These are the technicalities of the FFT and various strategies are adopted in practice, but such details are not our concern here, we are merely pointing out that the FFT is desirable. Most often it exists as a program routine, and the user can apply it very quickly.

Since each pass always incorporates $N/2$ butterflies, the overall computation effort is

$$\frac{N}{2} . \log_2 N \quad \text{butterflies} \tag{13.22}$$

This figure may be compared with the N^2 operations required of the direct method. In practice there are many additional factors which govern the actual computational effort required to evaluate the FFT, but the comparison in Table 13.2 gives a first-order indication of the likely gains in time.

13.3 Errors in the result

The greatest difference between the true signal spectrum and the observed signal spectrum is due to the finite number of points which are considered, that is the *windowing* problem. It can be shown that calculating the spectrum only at discrete frequencies presents no loss of information, since the N spectral points contain the same information as the N signal values. The sampling theorem can be shown to operate in both the time and the frequency domains with equal effect.

In 13.2.1 we introduced the idea of windowing, while here we calculate the *spectral leakage* produced by it, and then go on to discover how the effect can be reduced by shaping the window. A further error is sometimes produced because the DFT calculating engine appears to the signal like a parallel bank of bandpass filters, which is called the *picket fence* effect.

13.3.1 Spectral leakage

Consider a rectangular window, as introduced in section 13.2.1:

$$w(n) = 1 \qquad 0 \leqslant n \leqslant (N-1)$$
$$= 0 \qquad \text{elsewhere}$$

The spectrum corresponding to this window is calculated by the method shown in example 13.3, and is

$$\mathbf{W}(\omega) = \exp j[-(N-1)\omega T_S/2] \cdot \frac{\sin(N\omega T_S/2)}{\sin(\omega T_S/2)} \qquad (13.23)$$

The phase term expresses the fact that the centre of the window is at $(N-1)/2$, and will be ignored for the time being. When calculating the spectrum on a discrete-frequency scale, we replace ω by $k\omega_S/N$ and denote the new spectral function as $\mathbf{W}'(k)$. Then

$$|\mathbf{W}'(k)| = \frac{\sin(\pi k)}{\sin(\pi k/N)} \qquad (13.24)$$

The window spectral function is shown in figure 13.13.

Now in order to see what effect this window has, consider a signal which consists of two phasors, one of greater magnitude than the other:

$$x(n) = A_1 \cdot \exp j(2\pi q_1 n/N) + A_2 \cdot \exp j(2\pi q_2 n/N) \qquad (13.25)$$

where $A_1 \gg A_2$.

The object of analysing this signal with the DFT operation is to measure the magnitudes A_1 and A_2. The true signal spectrum is

$$\mathbf{X}(\omega) = A_1 \cdot \delta(\omega - \omega_1) + A_2 \cdot \delta(\omega - \omega_2) \qquad (13.26)$$

When observed through the window and in discrete-frequency:

$$\mathbf{X}'_N(k) = A_1 \cdot \mathbf{W}'(k - q_1) + A_2 \cdot \mathbf{W}'(k - q_2) \qquad (13.27)$$

The observed spectrum is therefore spread out by the convolution with $\mathbf{W}(\omega)$. The first component has the larger magnitude, and it generates a spectrum of the shape shown in figure 13.13, spreading out from $k = q_1$

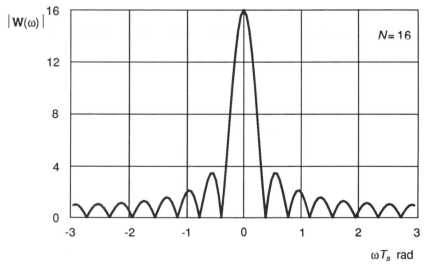

Figure 13.13 Spectrum of rectangular window

This effect is known as *leakage*, and can mask a smaller component by spreading signal energy over all the other frequency bins.

● **Example 13.10** Consider the case where there is a single frequency component in the signal.

$$f_S = 4 \text{ kHz}, \ N = 8, \text{ then } \delta_S = 500 \text{ Hz}$$

Now let the single input frequency be at $1 \cdot 125$ kHz, which corresponds to $q = 2 \cdot 25$. Then

$$|\mathbf{X}_N'(k)| = \frac{|\sin[\pi(k - q)]|}{\sin[\pi(k - q)/8]}$$

Figure 13.14 shows this observed spectrum, and overlaid on it is the envelope corresponding to the window spectrum of figure 13.13.

Notice that if q was an integer, there would be no spectral leakage. Notice too that since q is not an integer in this case, the main spectral lobe is spread between two adjacent spectral points.   ●

We may now consider a general expression for the 'goodness' of the spectral analysis when there are two frequency components in the signal, as in equation 13.25.

Since the magnitude of the component at q_2 is the smaller of the two, it is the one which is most likely to be obscured by leakage. Assume that q_2 is an integer, concentrate on this frequency, and form the signal-to-interference ratio:

$$\frac{\langle \text{wanted signal} \rangle}{\langle \text{unwanted signal} \rangle} = \frac{A_2 \cdot \mathbf{W}'(0)}{A_1 \cdot \mathbf{W}'(q_2 - q_1)} \tag{13.28}$$

Notice that this ratio depends on the relative magnitude of the two signal

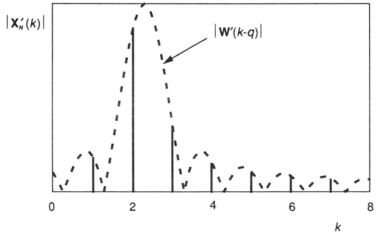

Figure 13.14 Spectral leakage in operation

components, and also on the leakage characteristic defined by the window spectral function of figure 13.13. In particular, if $(q_2 - q_1)$ is an *integer* then the ratio is infinite and the measurement of the small magnitude is unencumbered, since the small component falls at a zero of the window spectral function.

Further, the manner in which $\mathbf{W}(k)$ falls away with increasing k determines the *selectivity* of the spectral analysis.

● **Example 13.11** A certain signal consists of two frequency components:

1) Frequency $1 \cdot 2109$ kHz, magnitude $2 \cdot 0$ V
2) Frequency $2 \cdot 500$ kHz, magnitude $0 \cdot 04$ V

$f_S = 10$ kHz and $N = 128$.
Calculate the effect of the leakage from component 1 on component 2. Now the frequency scale spacing is $\delta_S = 78 \cdot 125$ Hz.

Hence $\quad q_1 = 15 \cdot 5 \quad$ and $\quad q_2 = 32$

Component 1 predominates, causing leakage on to component 2, which would otherwise be clear.

Evaluating the ratio ⟨wanted signal⟩/⟨unwanted signal⟩ as in equation 13.28, we see that the leakage component at $2 \cdot 500$ kHz is

$\quad (0 \cdot 04 . 128)/(2 \cdot 0 . 2 \cdot 54) = 1 \cdot 01$ or $0 \cdot 07$ dB

The leakage contribution is therefore about the same magnitude as the component that we are trying to measure. ●

The examples taken above have been simple ones in order to illustrate the magnitude of this effect, and to allow calculation by hand. In practice, each component in the input signal is modelled by both positive and negative frequency components, and the net leakage is the vector sum of the individual leakage spectra. Realistic signals have many more frequency components anyway. Figure 13.15 shows the case for $N = 16$ which is dominated by a conjugate pair of components at $q = 10 \cdot 5$ and $5 \cdot 5$. The two graphs show the spectrum with and without two small components at $q = 3$ and 13. The general effect of leakage is to hide the small-magnitude frequency-components in other bins or frequency slots. The signals being analysed in this case are

$x_1(n) = \cos(4 \cdot 12n)$
$x_2(n) = \cos(4 \cdot 12n) + 0 \cdot 1 . \cos(1 \cdot 18n)$

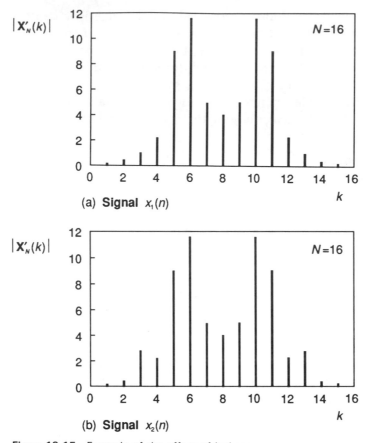

Figure 13.15   Example of the effect of leakage

13.3.2   Windows reduce leakage

The rectangular window is the worst shape of all, and by its leakage renders a DFT spectral analyser almost unusable. Other forms of window can be used to reduce leakage, although at the expense of selectivity. We shall now outline the development of a simple set of windows, and then summarise the characteristics of some of the windows in common use.

One clue to leakage reduction lies in the fact that the skirt of the window spectral function $\mathbf{W}(\omega)$ is oscillatory, and a certain class of windows exploit this effect. Recapping, a rectangular window has a spectral function as follows (equation 13.23):

$$\mathbf{W}(\omega) = \exp j\,[-(N-1)\omega T_S/2] \cdot \frac{\sin(N\omega T_S/2)}{\sin(\omega T_S/2)} \tag{13.29}$$

When calculated at discrete-frequencies, the spectral function appears like this:

$$\mathbf{W}'(k) = \exp j\,[-(N-1)\pi k/N] \cdot \frac{\sin(\pi k)}{\sin(\pi k/N)} \tag{13.30}$$

This function is identically zero when k is an integer ($k \neq 0$), but has local maxima when k is of the form [⟨integer⟩ $+ 0 \cdot 5$]. At this worst case therefore, the $\sin(\pi k)$ term takes values of ± 1, but the phase term advances by approximately π radians each time k is incremented. The net effect is that adjacent sample values of $\mathbf{W}'(k)$ are of similar magnitude and remain of similar sign. (Note that a window which straddles $k = 0$ does not have the phase term, and the values alternate in sign.)

In order to smooth the skirt of the window spectral function, we can therefore form a new, averaged spectrum $\mathbf{Y}_N(\omega)$, from three sample values of the original windowed spectrum $\mathbf{X}_N(\omega)$. When the discrete-frequency form of the DFT is used, the odd number of samples included in the average will enable the averaged value to appear at one of the discrete-frequencies and not between two of them.

$$\mathbf{Y}_N(\omega) = \alpha \cdot \mathbf{X}_N(\omega) - \frac{(1 - \alpha)}{2} \cdot \mathbf{X}_N(\omega - \omega_S/N) - \frac{(1 - \alpha)}{2} \cdot \mathbf{X}_N(\omega + \omega_S/N)$$

(13.31)

In terms of the discrete-frequency spectrum:

$$\mathbf{Y}_N'(k) = \alpha \cdot \mathbf{X}_N'(k) - \frac{(1 - \alpha)}{2} \cdot \mathbf{X}_N'(k - 1) - \frac{(1 - \alpha)}{2} \cdot \mathbf{X}_N'(k + 1)$$

(13.32)

The new effective window spectrum can be calculated by substituting the form of rectangular window spectrum from equation 13.30. This is a tedious calculation, and does not emerge in a closed form, so we work out first a numerical example, and then summarise the window spectral functions in figure 13.17. However, following on from equation 13.32:

$$\mathbf{W}_2'(k) - \alpha \cdot \mathbf{W}'(k) - \frac{(1 - \alpha)}{2} \cdot \mathbf{W}'(k - 1) - \frac{(1 - \alpha)}{2} \cdot \mathbf{W}'(k + 1)$$

(13.33)

● **Example 13.12** Consider example 13.11 and re-work it using this averaged window, to see how much improvement is effected.

In summary, we have two frequency components in the original signal spectrum:

1) Magnitude $2 \cdot 0$ at a discrete-frequency such that $q_1 = 15 \cdot 5$.
2) Magnitude $0 \cdot 04$ at a discrete-frequency such that $q_2 = 32$.

$N = 128$

Component 1 produces considerable leakage at the frequency of component 2, and its magnitude is determined by $|\mathbf{W}'(16 \cdot 5)|$.

Now we average, using equation 13.33 and setting $\alpha = 0 \cdot 5$, which defines a *Hanning* window.

First, calculate the appropriate values of the unmodified window

273

spectral function:

k	mag$[\mathbf{W}'(k)]$	arg$[\mathbf{W}'(k)]$
15·5	2·693	1·191 rad
16·5	2·538	1·166
17·5	2·401	1·141

Now we add these values after weighting them with the coefficients as in equation 13.33. The revised window spectral coefficient is then

$$\mathbf{W}'_2(16·5) = 0·5 . \mathbf{W}'(16·5) - 0·25 . \mathbf{W}'(15·5) - 0·25 . \mathbf{W}'(17·5)$$

and $\quad |\mathbf{W}'_2(16·5)| = 0·00454$

We are now in the position to calculate the 'goodness' ratio, remembering that $\mathbf{W}'_2(0) = \alpha . N$.

$$\frac{\langle \text{wanted signal} \rangle}{\langle \text{unwanted signal} \rangle} = \frac{0·04 \times 64}{2·0 \times 0·00454}$$

$$= 282 \quad \text{or} \quad 49 \text{ dB}$$

In example 13.11, using the rectangular window, this ratio was about 0 dB, so averaging the spectrum has improved the discrimination of the DFT spectral analyser by 49 dB. This is a worthwhile improvement. ●

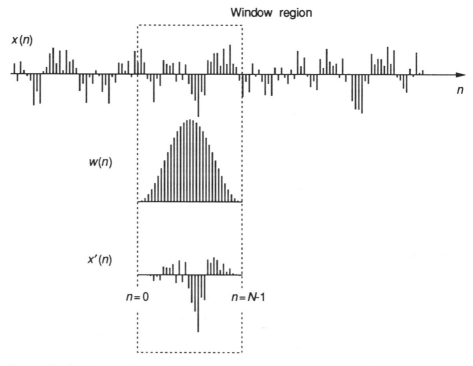

Figure 13.16 A shaped time-window

Table 13.3 *Details of some useful windows*

Name	Time-window	Max. sidelobe dB	3 dB bandwidth	Sidelobe roll-off dB/oct
Rectangular	$\mathrm{rect}\,[n/N - \tfrac{1}{2}]$	-13	$0 \cdot 89$	-6
Triangular	$2n/N: \quad 0 \leqslant n \leqslant N/2$ $2(1 - n/N): \quad N/2 \leqslant n \leqslant N-1$	-27	$1 \cdot 28$	-12
Hanning	$0 \cdot 5\,[1 - \cos(2\pi n/N)]$	-32	$1 \cdot 44$	-18
Hamming	$0 \cdot 54 - 0 \cdot 46\,.\,\cos(2\pi n/N)$	-43	$1 \cdot 30$	-6
Blackman	$0 \cdot 42 - 0 \cdot 5\,.\,\cos(2\pi n/N) + 0 \cdot 08\,.\,\cos(4\pi n/N)$	-58	$1 \cdot 68$	-18
Kaiser $(\alpha = 2 \cdot 5)$	$\dfrac{I_0[\pi\alpha\sqrt{\{1 - (2n/N - 1)^2\}}\,]}{I_0(\pi\alpha)}$	-57	$1 \cdot 57$	-6

($I_0[\ ]$ *is the modified Bessel function*)

The calculation outlined in example 13.12 is fairly tedious, although systematic and amenable to computer evaluation, but spectral smoothing can be applied simply and mechanically by using the *time-window* approach. Referring back to equation 13.31, we take the inverse DFT to form a time domain description of the operation.

Since $\quad \mathrm{IDFT}\,[\mathbf{X}_N(\omega - \omega_S/N)] = x(n)\,.\,w(n)\,.\,\exp j(-2\pi n/N)$ \hfill (13.34)

$$y(n) = x(n)\,.\,w(n)\,.\,[\alpha - (1 - \alpha)\,.\,\cos(2\pi n/N)] \hfill (13.35)$$

where $w(n)$ represents the rectangular window.

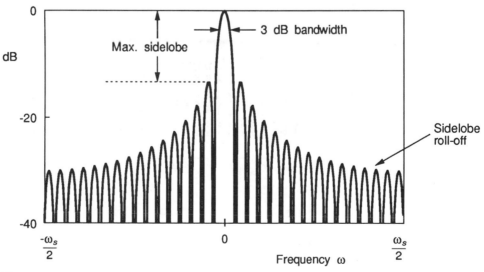

Figure 13.17 Window parameters

275

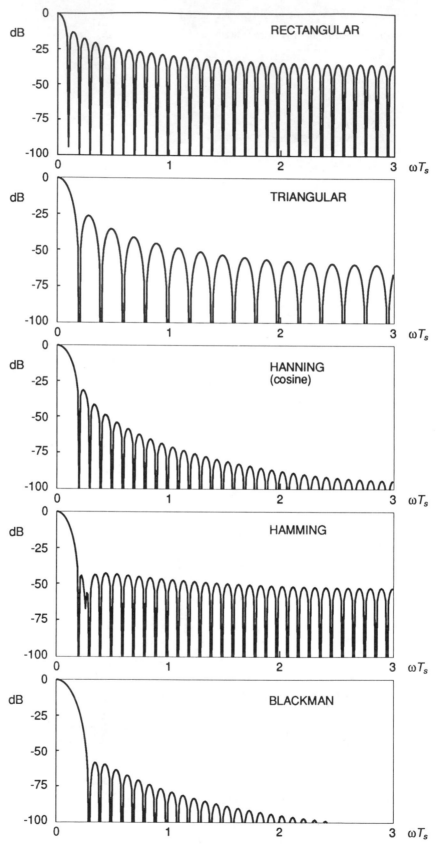

Figure 13.18 Spectral shapes of windows in Table 13.3 ($N = 64$)

Thus, multiplying the input signal by the composite window

$$w_2(n) = w(n) \cdot [\alpha - (1 - \alpha) \cdot \cos(2\pi n / N)] \tag{13.35}$$

enables the averaged spectrum to be produced, with its attendant reduction of leakage. This is the manner in which most windows are applied, although sometimes it is convenient to average spectral values after having obtained them via the DFT. Frequently, the sample values of the time-window are calculated in advance, and then read out from a table as required for use.

Figure 13.16 shows the shaped time-window being applied to the signal illustrated in figure 13.6.

Two distinct values of α are used. A Hanning (or raised-cosine) window uses $\alpha = 0 \cdot 5$ and a Hamming window uses $\alpha = 0 \cdot 54$. The characteristics of these and other common windows are given in Table 13.3. Figure 13.17 shows the parameters that are quoted, and figure 13.18 illustrates the spectral functions produced by these windows. Notice that the *Reciprocal Spreading* rule applies here, in that good sidelobe suppression implies a wider central lobe to the response; reduced leakage away from the dominant frequency can only be achieved by compromising the ability to detect two signal frequencies close together.

13.3.3 The picket fence

In our discussion of leakage, we have shown that truncating the signal record to N points distorts the perceived spectrum by convolving the true signal spectrum $\mathbf{X}(\omega)$ with the window spectral function $\mathbf{W}(\omega)$. While this effect may be reduced by making $\mathbf{W}(\omega)$ narrower, it can never be eliminated.

From this view, the measured spectrum is seen as a whole; that is the complete function $\mathbf{X}_N(\omega)$, or the spectral vector $\{\mathbf{X}_N'(k)\}$. However, we may also see the DFT process from the point of view of a single output, and discuss its response to a varying frequency input.

Consider then, a time-window $w_2(n)$, whose spectral function is $\mathbf{W}_2(\omega)$, and let the input signal be a phasor $x(n) = \exp \mathrm{j}(\omega_1 T_S n)$. The observed spectrum is therefore $\mathbf{W}_2(\omega - \omega_1)$.

Evaluating a discrete-frequency DFT, we concentrate attention on a single element in the spectral vector, at frequency

$$\omega = k \cdot \omega_S / N \tag{13.36}$$

As the input frequency ω_1 is varied, the output of the DFT process, at this frequency, is $\mathbf{W}_2(k\omega_S / N - \omega_1)$, which is illustrated in figure 13.19.

The graph informs us that the output from a single DFT *bin* is influenced by this weighting. A composite input signal generates the following output:

$$x(n) = \sum_r a_r \cdot \exp \mathrm{j}(\omega_r T_S n) \tag{13.37}$$

277

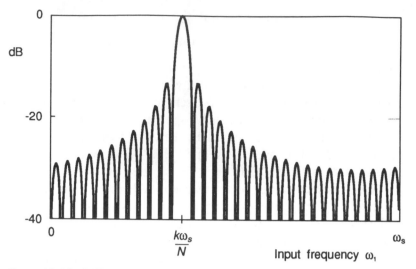

Figure 13.19 DFT response at single frequency bin

$$\mathbf{X}_N^k(k) = \sum_r a_r \cdot \mathbf{W}_2(k\omega_S/N - \omega_r) \tag{13.38}$$

The window spectral function $\mathbf{W}_2(\omega)$ therefore describes the frequency weighting applied to all signals which contribute to the spectral output $\mathbf{X}_N^k(k)$, and hence this may be used to assess the effect of broadband noise on the DFT measurement, but more of that at some other time.

The *picket fence* effect is a description of this frequency-selective process. A signal in the centre of the frequency bin is multiplied with a weighting factor of $\mathbf{W}_2(0)$, whereas a signal with a frequency on the boundary between

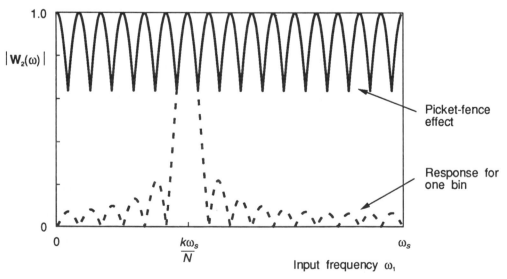

Figure 13.20 Picket fence effect ($N = 16$)

two bins has a weighting factor of $\mathbf{W}_2(\omega_S/2N)$. The net effect for all bins is that the DFT exhibits a ripple across its frequency range, as shown in figure 13.20.

● **Example 13.13** Calculate the magnitude of the picket fence effect for a rectangular window, and for a Hanning (cosine) window.

For the rectangular window,

$$|\mathbf{W}_2(\omega)| = \frac{\sin(N\omega T_S/2)}{\sin(\omega T_S/2)}$$

So $|\mathbf{W}_2(0)| = N$

$|\mathbf{W}_2(\omega_S/2N)| = 1/\sin(\pi/2N) \simeq 2N/\pi$

Hence the ratio is $0 \cdot 637$ or $-3 \cdot 9$ dB.

In terms of power, the ratio is $0 \cdot 41$ which is quite severe.

For the Hanning window, by averaging together three adjacent values from the rectangular window, we discover that

$$|\mathbf{W}_2(0)| = N/2$$
$$|\mathbf{W}_2(\omega_S/2N)| \simeq 4N/3\pi$$

Hence the ratio is $0 \cdot 848$ or $-1 \cdot 4$ dB.

In terms of power, the ratio is $0 \cdot 72$ which is a real improvement. ●

The picket fence effect is of greatest importance when the signal consists of many frequency components, or has a continuous frequency spectrum, particularly for random signals. Notice that although a frequency component at the edge of the bin is attenuated relative to its true value, it does register in *both* the adjacent bins through the leakage effect.

13.4 Applications

Spectral analysis via the FFT is a common technique in a variety of signal processing applications. Those that follow have been selected to demonstrate further points than the basic principles developed previously in the chapter. Although the principles of spectral analysis are straightforward enough and amenable to calculation, a user requires considerable experience before valid and reliable results can be obtained from the spectral analysis of unknown signals.

13.4.1 Voice analysis

Although many signals have constant spectra, other spectra are time-varying, and speech or voice signals are typical of this class. The content

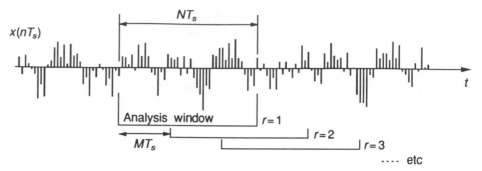

Figure 13.21 Analysis of a time-varying signal

and shape of the spectrum follows the changing sound of the utterance as a function of time. So the signal is analysed by taking a succession of spectral estimates, whose spacing is carefully selected, and then its meaning may be extracted.

Consider a signal sequence $\{x(n)\}$, which is analysed by taking N-point FFTs at intervals of MT_S sec. Thus, N point-values are taken from the signal and analysed, and then the reference point is moved forward M points while the process is repeated, as indicated in figure 13.21.

Formally, when using a window weighting sequence $\{w(n)\}$, the rth discrete-frequency finite-DFT is

$$\mathbf{X}_N(k,r) = \sum_{n=0}^{N-1} x(n+rM) \, . \, w(n) \, . \, \mathbf{W}^{-nk} \tag{13.39}$$

where $\mathbf{W} = \exp \, \mathrm{j}(2\pi/N)$.

The spectrum is now a function of the frequency index k, and the time index r. Samples occur along the frequency axis at intervals $1/NT_S$ Hz, and along the time axis at intervals of MT_S sec. Phase is not important in this application, so a power spectrum is plotted, $|\mathbf{X}(k,r)|^2$. Figure 13.22 shows a common way of displaying the result.

Choice of the advancing parameter M depends upon the effective bandwidth of the spectral analyser, and is constrained again by the *Reciprocal Spreading* principle (section 7.2.4.3).

Now the effective bandwidth of the analyser depends upon the precise window function chosen. If a *Hanning window* is selected (Table 13.3), then the 3 dB bandwidth is $1 \cdot 44/NT_S$ Hz. There are $N/2$ bands in the frequency range 0 to $f_S/2$ Hz, but because of overlaps they yield a smaller number of independent estimates of the spectrum.

The sampling theorem (section 6.2) dictates that the time-varying spectrum must be sampled at a rate which is at least twice the bandwidth of each analyser channel, if no significant information is to be lost. However, the 3 dB bandwidth is not a good parameter to use, since for a *random signal* such as we have here, it is the *autocorrelation function* of the

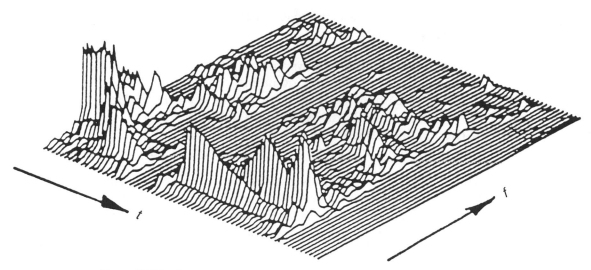

Figure 13.22 Spectrograph, for a time-varying voice spectrum [Courtesy: Boll, *IEEE Transactions ASSP-27*, 1979, pp. 113–120]

bandlimited signal which determines the sampling rate (chapters 15 and 16), and this is not directly obtainable from the 3 dB bandwidth.

The rate at which spectrum estimates are being made is $1/MT_S$ per second, and taking a pessimistic view by using the 3 dB bandwidth since at present we have nothing else, we conclude that this rate must be $> 2 \cdot 88/NT_S$ Hz for a Hanning window. Hence, loosely:

$$M < \frac{N}{2 \cdot 88} \tag{13.40}$$

Taking a value of M near this upper limit gives fairly coarse resolution along the time axis, although all significant information is included. Taking a much smaller value of M means that successive spectral estimates are not independent, but on the other hand do show the trends in spectral development.

● **Example 13.14** (Oppenheim, *IEEE Spectrum*, vol. 7, pp. 57–62, August 1970) A certain voice signal has a bandwidth of 5 kHz and is sampled at 10 kHz. A Hanning window is used with the FFT, and two types of analysis are carried out, 'narrowband' and 'wideband'.

Wideband analysis shows clearly the spectral variation with time.

$N = 128$, hence the spectral resolution is

$\delta = 1/NT_S = 78 \cdot 125$ Hz

The effective analyser 3 dB bandwidth is

$1 \cdot 44 \, . \, \delta = 112 \cdot 5$ Hz

From equation 13.40, $M < 44 \cdot 4$, so that spectra are to be evaluated at intervals of $< 4 \cdot 4$ ms.

In the practical case referenced here, spectra were evaluated every 2·4 ms, giving $M = 24$.

Narrowband analysis shows fine detail in the frequency domain, but only slow trends in the time domain.

$N = 512$, so that the spectral resolution is $\delta = 19·53$ Hz.

The effective 3 dB bandwidth is 28·1 Hz.

The limiting value of M is 177·8, or an interval of 17·8 ms.

In practice an interval of 9·6 ms was used, or $M = 96$. ●

This technique is commonly used to analyse voice waveforms, comparing synthesised voice with a recorded signal, and for speech coding to reduce signal information content. Other examples of time-varying spectra include the analysis of an internal-combustion-engine cylinder pressure.

13.4.2 Transient analysis

The frequency response of a system is often difficult and time-consuming to measure, particularly if the time constants are of the order of seconds, minutes or even hours. On the other hand, the impulse response is fairly easy to determine, by exciting the system input with a 'short sharp shock' like a hammer blow for a mechanical structure or a short pulse for an electrical system.

Consider such a system, having an impulse response $h(t)$ and excited by a short pulse of magnitude A and width τ sec. Provided that τ is small compared with any significant features on $h(t)$, then the input pulse approximates to an impulse of weight $A\tau$, and the output signal $y(t)$ is

$$y(t) \simeq A\tau . h(t) \tag{13.41}$$

Measuring this signal at discrete intervals T_S sec yields the set of samples $\{y(n)\}$, which is a scaled and sampled version of the impulse response. Taking the DFT of this sequence yields the following signal spectrum:

$$\mathbf{Y}(\omega) = A\tau \sum_{n=-\infty}^{\infty} h(n) . \exp \mathrm{j}(-\omega T_S n) \tag{13.42}$$

Provided that the sampling frequency $1/T_S$ Hz is chosen to be greater than twice the expected bandwidth of the system so that aliasing is excluded, this signal spectrum approximates to the frequency response of the system, $A\tau . \mathbf{H}(\omega)$. A numerical form of DFT is of course used, so that an N-point discrete-frequency DFT evaluates the frequency response at frequencies of k/NT_S Hz.

Now the impulse response of a system is a transient which decays steadily, and let us say that its values become insignificant after a time T sec. Then the greatest number of independent points that can be taken by sampling is

$$N = T/T_S \tag{13.43}$$

Using the FFT on $\{y(n)\}$, then, yields a frequency response estimate at intervals of $1/NT_S$ Hz, or $1/T$ Hz. No more information can be gleaned from this transient by changing the sampling frequency or by taking more points. Figure 13.23 shows an example of an impulse response and its derived frequency response.

Certain practical points are often important. First, the choice of sampling frequency is crucial, and in a trial measurement it is wise to examine the

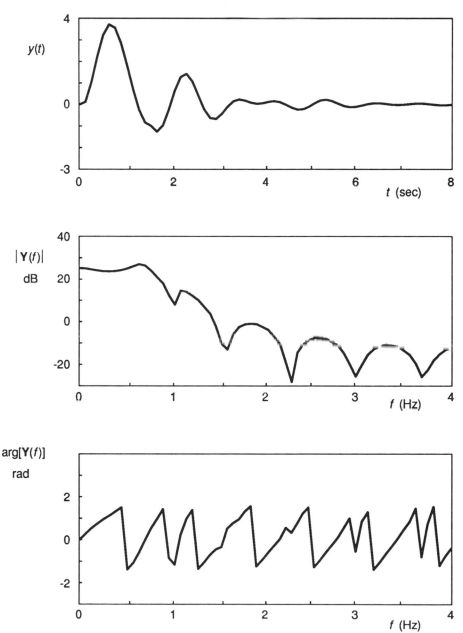

Figure 13.23 Determination of frequency response from impulse response

derived 'response' in the region of the Nyquist frequency to see if aliasing might have occurred (section 6.2).

Second, if the transient must be truncated for any practical reason, then some tapering window will be necessary in order to suppress unwanted ripples in the perceived response.

Third, if an anti-alias filter is used ahead of the analog-to-digital converter, then this should have linear phase characteristics to avoid unnecessary distortion of the measurement, or alternatively have a known response so that the measured system response may be corrected.

13.4.3 Digital filtering

In our discussion of the Z-transform (chapter 8) we made the point that a signal processing transform is intended to replace a convolution operation by a multiplication of transforms. Such a technique yields important simplications for analytical calculations, but it also holds for real processing, using the numerical form of the DFT.

Suppose that we have a signal sequence $\{x(n)\}$, which is to be processed by a digital filter having a unit-pulse response $\{g(n)\}$. Instead of convolving these two sequences, we may first transform them, and then multiply their transforms. Expressed in symbolic form, the output signal sequence $\{y(n)\}$ is in principle given by

$$y(n) = \text{IDFT}[\ \text{DFT}[x(n)] \cdot \text{DFT}[g(n)]\ ] \tag{13.44}$$

The computational advantage of this strategy is not obvious until we examine it in detail. Let the unit-pulse sequence be FIR, and have a length of L points, while the signal sequence contains N points. A direct convolution of the two sequences requires $L \cdot N$ multiply-add operations.

The transformation route requires the following number of complex-multiply-add operations:

FFT on $\{x(n)\}$	$(N/2) \cdot \log_2(N)$
FFT on $\{g(n)\}$	$(N/2) \cdot \log_2(N)$
Product of transforms	N
Inverse FFT to give $\{y(n)\}$	$(N/2) \cdot \log_2(N)$

Each complex-multiply-add operation is broadly equivalent to 4 real-multiply-add operations, so we may say that if the transformation sequence is to be beneficial then

$$6 \cdot \log_2(N) < (L - 4) \tag{13.45}$$

For $L = 128$ for instance, any value of N in the range $128 \leqslant N \leqslant 1 \cdot 66 \times 10^6$ offers an improvement in processing load. In practice other benefits accrue since the DFT of $\{g(n)\}$ needs only to be calculated once, and also there are compact FFT routines for real-data which almost halve the computational effort required for the general complex-data version.

However, as in all practical enterprises, the trade-offs are not all in one direction. Multiplication of transforms generates a *circular convolution* which distorts the result in an unacceptable fashion.

Without labouring the mathematical principle, we point out first that the result of an inverse-DFT is a *periodic* time sequence. Inverse transformation of a finite-DFT $\mathbf{X}_N(k)$ gives

$$x(n) = \frac{1}{N} \sum_{k=0}^{N-1} \mathbf{X}_N(k) . \exp \mathrm{j}(2\pi kn/N)$$

$$= x(n + rN) \tag{13.46}$$

where r is an integer.

The index n is therefore effectively modulo-N, and the function $x(n)$ is periodic in N. Alternatively, since the spectral function is sampled, it follows that the corresponding time sequence is periodic. Convolution of

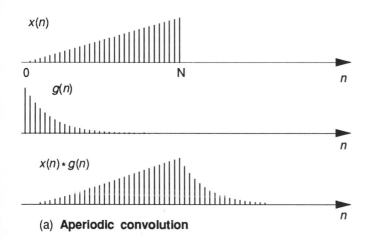

(a) **Aperiodic convolution**

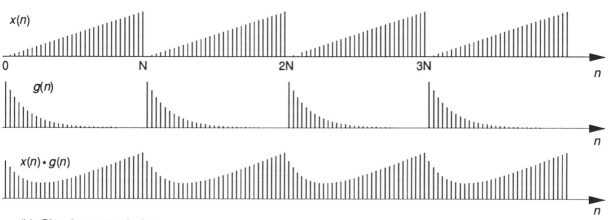

(b) **Circular convolution**

Figure 13.24 Convolution

285

such sequences is written as in equation 13.47, and since all indices are modulo-N, the resulting sequences are *periodic* or *circular*:

$$y(n) = \sum_{m=0}^{N-1} x(m) \cdot g(n - m)$$

$$= y(n + rN) \tag{13.47}$$

Figure 13.24 illustrates this point with two simple functions, $x(n)$ and $g(n)$. Notice how circular convolution causes the function to wrap around and add extra products to the convolution expression of equation 13.47. Clearly this effect must be avoided.

There are several strategies for avoiding the effects of circular convolution, and we describe briefly the *Overlap-save* method in order to illustrate the principle. Following the discussion above, we consider a filter which has a unit-pulse response $\{g(n)\}$ of length L points. A $2L$-point frequency response is first formed, by padding $\{g(n)\}$ out with L zeros:

$$\mathbf{H}_{2L}(k) = \sum_{n=0}^{L-1} g(n) \cdot \mathbf{W}_{2L}^{-kn} \tag{13.48}$$

where $\mathbf{W}_{2L} = \exp \mathrm{j}(2\pi/2L)$.

Using the transform to convolve the expanded unit-pulse sequence with a $2L$-point signal sequence, yields a sequence of $2L$ points. Since we have forced the second half of the $2L$ points in the unit-pulse response to zero, the first L points of the result suffer from circular overlap, while the second set of L points is a true aperiodic convolution of the first half of the input signal.

Assuming that the signal to be filtered is a long or infinite sequence, then the procedure outlined in Table 13.4 may be followed, which is illustrated in figure 13.25.

Table 13.4 *Overlap-save procedure*

Repeat:

    Take a block of $2L$ points from the input signal, $\{x(n)\}$

    Take a $2L$-point DFT to form the spectrum $\mathbf{X}_{2L}(k)$

    Multiply transforms: $\mathbf{V}_{2L}(k) = \mathbf{X}_{2L}(k) \cdot \mathbf{H}_{2L}(k)$

    Take the inverse transform of $\mathbf{V}_{2L}(k)$ to form the $2L$-point set $\{v(n)\}$

    Save the second set of L points from $\{v(n)\}$ to form $\{y(n)\}$

        $y(n) = v(n + L) \qquad 0 \leqslant n \leqslant L - 1$

    Shift the origin of the input signal by L points, overlapping the original set by L points

End

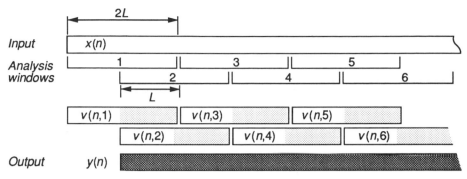

Figure 13.25 Overlap-save convolution

STUDY QUESTIONS

1 $x(t) = 2\{5 . \cos(100t) + 2 . \sin(150t)\}\cos(1000t)$

List the complex magnitudes and angular frequencies of the eight phasors which represent this signal. What is its periodic frequency?

2 Calculate the frequency spectrum of the following signal, by the Fourier Transform:

$$x(t) = \cos(\omega_0 t) \qquad |t| < T/2$$
$$- 0 \qquad \text{elsewhere}$$

3 Calculate the DFT for this signal:

$$x(n) = \exp j(\omega_U T_S n) \qquad 0 \leqslant n \leqslant M - 1$$
$$= 0 \qquad \text{elsewhere}$$

Note that the Geometric Progression sum is $\displaystyle\sum_{r=0}^{N-1} a^r - \dfrac{1 - a^N}{1 - a}$

4 Find the discrete-time signal whose DFT is $X(\omega) = \cos(\omega T_S)$. What property of this signal guarantees that the DFT will be real?

5 Using example 13.6 as a basis, write down the symbolic form of the DFT for a signal record of N points, when

    a) $x(n) = \cos(\omega_1 T_S n)$
    b) $x(n) = \sin(\omega_1 T_S n)$

6 Calculate the discrete-frequency DFT over 8 points, for the signal:

$$\{x(n)\} = \{0, 1, 2, 0, 0, 0, 0, 0\}$$

7 Explain why the FFT calculates the finite-frequency finite-DFT more rapidly than the direct form of calculation. What order of improvement is gained when $N = 512$?

8 A certain signal has a periodic frequency of 5·125 kHz, and is of the form:

$$x(t) = 5 \cdot \exp j(\omega_1 t)$$

A digital spectrum analyser samples the signal at 64 kHz and calculates the discrete-frequency DFT on the basis of 256 sample points. Calculate the leakage from this signal, at a frequency of 8 kHz.

9 The spectrum calculated in question 8 is now smoothed to give a revised spectrum $Y'_N(k)$:

$$Y'_N(k) = -0 \cdot 24 X'_N(k-1) + 0 \cdot 52 X'_N(k) - 0 \cdot 24 X'_N(k+1)$$

Calculate the new value of leakage at 8 kHz, and the improvement over the unsmoothed spectrum.

10 Suppose that the signal in question 8 now has an additional component of magnitude 0·2 at 8 kHz. Calculate the 'goodness' ratio for the two cases in questions 8 and 9.

COMPUTATIONAL EXERCISES

Signal processing is a 'doing' subject, and the limit of hand calculation is quickly reached when considering a topic such as spectral analysis. The best way to discover more is to find a computer system with an FFT routine on it, and then start to experiment with realistic numbers of points. Some suitable exercises might be:

1 Generate a single frequency signal, $x(n) = \cos(\omega T_s n)$, and find its spectrum. Look for the symmetry and for spectral leakage. Vary the frequency ω and the number of points N to gain a good appreciation of how they interact. Choose $N = 32, 64, 128 \ldots 2048$.

2 Apply one of the time-windows in Table 13.3 to the signal. Note the improvement in spectral leakage. Experiment with Hanning, Hamming, Blackman and Triangular windows.

3 Now generate a multi-component signal of your choice and attempt to analyse it by the FFT. Investigate the effect of improved time-windows.

4 Try deducing the spectrum of

$$x(n) = \cos[\theta_0 n + a \cdot \sin(\theta_m n)]$$
$$\theta_0 = 2\pi q_0 / N \quad \text{and} \quad \theta_m = 2\pi q_m / N$$

Choose for instance, $q_0 = N/4$ and $q_m = 1, 2, 3, 4, 5 \ldots$. Let $a = \pi/4$.

14 Random Signals: Amplitude Measures

OBJECTIVES

To show how random discrete-time signals can be described by definite parameters, and to apply this information. In particular to

a) Use the statistical properties of signal samples to describe the signal
b) Show how these may be used to predict the performance of such a signal when it is processed

COVERAGE

We first explore the meaning of randomness, and mention typical situations where random signals occur. Arising from this discussion, we assert that only *average* parameters are significant, and develop the Cumulative-Distribution Function (cdf) and the Probability-Density Function (pdf) as valid descriptors of a random signal. Applying moments to the distributions yields the concept of *variance*. Typical cases include the Gaussian density function and the uniform density function.

Non-linear amplitude operations on signals, like a square-law detector for instance, are used to demonstrate further application of these concepts.

System applications include quantisation noise, data transmission and radar signals.

14.1 Randomness introduced

Practical signals always have an element of randomness or uncertainty about them. If a signal is *deterministic*, that is described by a known mathematical function with constant and known coefficients, then it can convey no information. For a signal to convey information, it must retain an element of surprise.

For instance, any measurement of a physical quantity such as position, velocity, voltage, power, etc. is a random quantity since the value is not predictable before the measurement takes place. Where the measured variable changes slowly with time and measurements are made at regular intervals, then each measured value may be estimated beforehand and so the amount of information generated by the measurement process, the uncertainty in the measured value, is reduced.

So when we speak of a *random signal*, we are referring to a waveform or series of signal sample values which cannot accurately be predicted. Any kind of communication signal, such as voice or video information, falls into this category, as well as numerous signals in the field of measurement and control. The error signal in a closed-loop control system, for instance, is intrinsically random, since the loop is designed to compensate for unknown disturbances within it.

● **Example 14.1**   A certain signal $x(t) = a \cdot \cos(\omega t + \varphi)$ has a known frequency but unknown phase. It is integrated over a time T sec. Thus

$$y(t) = \int_0^T x(t) \, dt = a[\sin(\omega T + \varphi) - \sin(\varphi)]$$

The integral, which we normally use to obtain the mean value of a signal, yields a random result if the phase is unknown.   ●

An important class of random signals is that due to *noise*. Electron motion in circuit resistors and transistors contributes a random noise signal which is added to the signal being received, and can overwhelm very small signals.

● **Example 14.2**   A resistance R Ω at an absolute temperature of T °K is at the input of an amplifier with bandwidth B Hz. It generates a random noise voltage with rms value V:

$$V = 2\sqrt{(kTBR)} \qquad k \text{ is Boltzmann's constant}$$

For $R = 1$ kΩ, $B = 5$ MHz, $T = 290$ °K, then $V = 9 \, \mu V$.   ●

In many situations, measurement accuracy is often compromised by extraneous electrical noise and interference from electrical machinery or broadcast radio signals. When planning instrumentation for measuring parameters like temperature, flow-rate in a manufacturing plant for instance, these influences must be anticipated. In addition to any uncertainties associated with the signal being measured, random *noise* confuses the measurement and adds destructive uncertainty. The general effect of noise is to reduce the amount of true information which can be received or measured.

14.2 Randomness described

Since the waveform description of a random signal has little meaning, we must look for other ways of classifying and describing random signals. The first concept is that of the *ensemble* or *set* of sample functions, and we follow that with a presentation of *average* properties sufficient to describe samples of a random signal and make it amenable to analysis.

14.2.1 Ensembles and random variables

Since we are unable to describe each random signal in precise terms, we settle for the description of a class of similar signals, an *ensemble* or set of

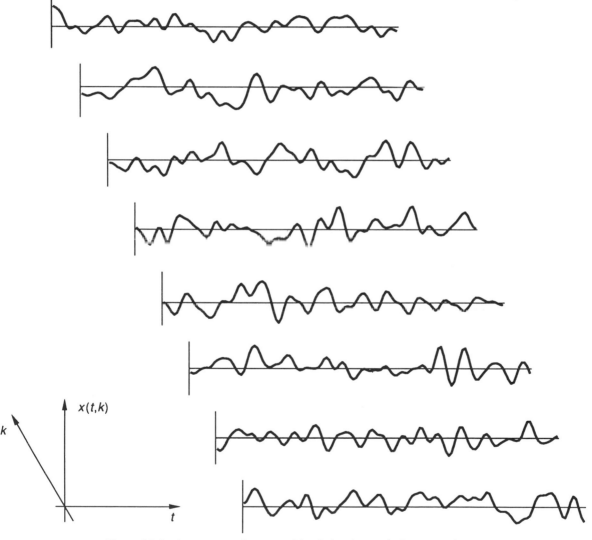

Figure 14.1 A representative ensemble of signal records from a random source

sample functions. Each signal that belongs to this *ensemble* is generated by the same *source* or *process*, and shares certain common properties with all other members of the *ensemble*. Each member will have the same average parameters such as mean value, rms value, bandwidth and so on, but each member of the ensemble will be distinct from all the others and be unique. Consequently, even if we know these average parameters for a given process, we still cannot construct any *particular* member of the set in the general case, we may only describe it by its average parameters.

Figure 14.1 shows eight sample functions out of a particular ensemble $\{x(t, k)\}$, from a process which is capable of generating an infinite number of similar waveforms. Each waveform differs from the others, and so there is no point in analysing the fine detail of any particular function of time, since it is unique and will never occur again. Average properties form the only valid description, and they may be obtained along the time axis as a *time average*, or across the different members of the ensemble, as the *ensemble average* or *statistical average*.

● **Example 14.3**   A simpler example, of a finite-ensemble, is the set of sinusoidal waveforms with a finite number of different phases:

$$x(t, \varphi) = \cos(\omega t + \varphi)$$

Such a waveform is frequently used for transmitting digital data.

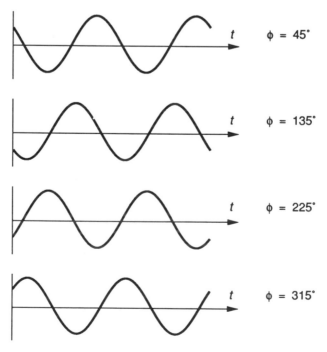

Figure 14.2   4-phase random signal ensemble

Figure 14.3   A random signal sequence

Suppose that the phase φ can take on 4 possible values:

$$\{\varphi\} = \{45°, 135°, 225°, 315°\}$$

Let the extent of the signalling element be 2 cycles of the carrier frequency, then the four members of the ensemble are shown in figure 14.2.

When transmitted in a stream, the resulting signal would look as in figure 14.3, where the various phases follow in random sequence. Each signalling element has 4 possible forms, and so carries 2 bits of potential information. The receiver has now to decode this information, by measuring the phase of the sinusoidal signal over 2-cycle intervals.   ●

Even with this superficial introduction, we can see that there will be problems ahead when trying to measure the properties of a random process. For each measurement, we have available only one member of the ensemble of waveforms which together make up the random process, and so any deductions we might make from a single waveform (or signal record) will be subject to considerable uncertainty.

On the other hand, once we know sufficient about the random process which is generating the signal ensemble, the ensemble average properties may be developed analytically, and then the precision of a particular measurement may be defined. In the discussion which follows, we shall assume a knowledge of the characteristics of the random processes that we encounter, and thence work out a number of useful descriptors of the signals themselves. Determining such parameters by practical measurement only is an altogether more difficult exercise, which we shall leave well alone!

Before defining some more precise descriptors for random signals, we must first introduce some broad definitions, and comment upon certain terms which are used in all treatments of random signals. Discrete-time signals provide the easiest definitions, and we shall deal mostly with these during the remainder of the discussion.

Thus, the nth sample of a discrete-time random signal may be written as $x(n)$. Now this is an unknown value, and the set of the nth samples from each member of the ensemble will cover a wide range of values. So we use the term *random variable* to describe the whole gamut of possible values, and give it the symbol X. The random variable X conforms to the properties of the random source and is defined by them.

● **Example 14.4**   Consider the ensemble of signals introduced in example

14.2:
$$x(t, \varphi) = \cos(\omega t + \varphi)$$

Now let these functions be described in discrete-time, sampling interval T_S. Then

$$x(n, \varphi) = \cos(2\pi f T_S n + \varphi) \qquad \text{where } f T_S = 0 \cdot 2, \text{ say}$$

Now consider the random variable at $n = 3$, which is defined by the equation above.

The random variable is therefore: $X(3) = \cos(6\pi f T_S + \varphi)$

The values of the random variable as φ changes are

$$\{X(3)\} = \{-0 \cdot 16, 0 \cdot 99, 0 \cdot 16, -0 \cdot 99\} \qquad \bullet$$

The random variable X describes a single sample point in the random signal, but the *random process* or *stochastic process* is described by a *time series* of random variables $X(n)$. The properties of a single sample or measurement can be obtained by observing a single random variable, and are known as *first-order statistics*, so mean and mean-square-value are examples of such properties. More information must be taken into account when analysing a random process, so we should consider the relationships between a sample value and its neighbours, which means preserving the sequence order of random variables. For this we need statistics of two or more random variables, of which *second-order* statistics are the most useful and determine the bandwidth and correlation properties of the random process.

This chapter considers first-order statistics only, while chapter 15 deals with some of the second-order statistics parameters. However, before concentrating only upon first-order statistics, we present a few more general definitions which may help to retain the broad overview which we are trying to achieve.

A *stationary* process is one whose statistics are not a function of time; from which it follows that all random variables in the sequence are identical and any sample time can be used for measurement of the process. An example of a non-stationary process is that of a radio receiver with automatic-gain-control (agc) (figure 10.2). When a signal is suddenly applied to the input of the receiver, the gain is at a maximum, but decreases gradually as the presence of the input signal is sensed. Since the first stage of the amplifier generates electronic noise, the noise power at the amplifier output is also reduced with time, leading to a change in the statistics of this random process. In practice, true stationarity is a theoretical concept, implying that all high-order statistics are stationary, so processes are often accepted as *wide-sense stationary* if a few of the simpler properties are shown to be stationary.

An *ergodic process* is one where the time average and the statistical average are equal. Referring to figure 14.1 we note that this process ensemble has two dimensions, time and ensemble-member number. In the real world it is most convenient to measure mean values of whatever sort along the time axis, whereas for calculation purposes it is most convenient to assess mean values by ensemble averaging using the properties of the random variable (as we shall see in later sections). It is therefore vital to ensure that the averages obtained by these two methods are the same, and an *ergodic process* is one where this is so. Proof of ergodicity is not easy, but we shall assume it to be true unless there are obvious reasons in the form of the signals to assume otherwise. Ergodicity requires that the process is stationary.

Two random variables $X(n)$ and $X(n-r)$ are said to be *independent* when $X(n)$ is completely unaffected by $X(n-r)$, or vice versa. Independence affects the information content of random variables. For instance, sampling a bandlimited signal at the Nyquist rate ensures that successive samples are truly independent, and this is one of the valuable spin-offs of the sampling theorem. Increasing the sampling rate generates more samples, but no further information about the signal waveform, since adjacent samples are now dependent in some way. In the realm of mechanical systems, successive sample values are often dependent since the system is inertia-limited rather than being bandlimited. Thus, in a radar system measuring the range to a moving target, the likely position of the target at a small interval of time in the future can be predicted because of the dependence between successive estimates of target range.

So the idea of dependence is very important, but we have many other factors about random processes to discover, before we can pursue this idea with benefit. It will surface again in chapter 15.

14.2.2 Cumulative-distribution-function (cdf)

We now commence to define amplitude distribution functions, which describe the behaviour of a random variable. These functions are couched in probability terms, since the next value taken by the random variable cannot be predicted.

The cumulative-distribution-function (cdf) has the symbol $F_X(x)$, and has the disarmingly simple definition:

$$F_X(x) = \text{Prob}[X \leqslant x] \tag{14.1}$$

So $[X \leqslant x]$ defines the event where the random variable X lies below some value x, and the cdf is the probability that this event will occur.

Figure 14.4 illustrates a practical interpretation of this situation. The sequence $x(n)$ is one discrete-time signal generated by a random process. We set an arbitrary threshold level x, and then record the number of samples for which the sample value is less than x. In the limit, this number, expressed as a fraction of the total number of samples considered, is the cdf.

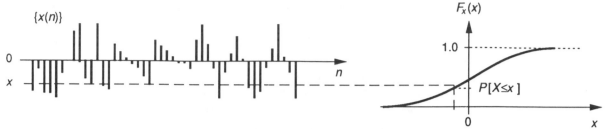

Figure 14.4 The cumulative-distribution-function illustrated

The cdf is hence an eminently practical concept, but to determine it accurately requires a very large number of samples to be measured.

If we set the threshold at a very low value, then there is a vanishing probability that any samples fall below it. Correspondingly, if we set the threshold at an extremely large value, it becomes certain that all samples will lie below it. These properties are formalised by equations:

$$F_X(x) \rightarrow 0 \qquad \text{as } x \rightarrow -\infty \tag{14.2}$$

$$F_X(x) \rightarrow 1 \qquad \text{as } x \rightarrow \infty \tag{14.3}$$

Further, it is certain that the function $F_X(x)$ will increase monotonically from zero to unity over the range of values of X, since probability is always a positive quantity.

14.2.3 Probability-density-function (pdf)

Although the cdf is easy to define and to measure, the probability-density-function (pdf) is far more useful in calculations. The pdf is given the symbol $f_X(x)$, and is defined in terms of the cdf:

$$f_X(x) = \frac{\mathrm{d}F_X(x)}{\mathrm{d}x} \tag{14.4}$$

Since $F_X(x)$ is monotonic, $f_X(x)$ must always be positive. In order to see what the pdf actually represents, consider the probability that the random variable X lies in the range $a \leqslant X \leqslant b$. Then

$$\text{Prob}[a \leqslant X \leqslant b] = F_X(b) - F_X(a) \tag{14.5}$$

$$= \int_a^b f_X(x)\,\mathrm{d}x \tag{14.6}$$

The pdf therefore represents the *density* of the probability distribution, and to obtain an actual probability it is necessary to take an *area* under the pdf. If the random variable X can take a continuous range of values, then the probability that $[X = b]$ is zero, but the probability that $[b \leqslant X \leqslant (b + \delta b)]$ is finite.

Combining these points we may state a simple but vital property of the pdf:

$$\int_{-\infty}^{\infty} f_X(x) \; \mathrm{d}x = 1 \tag{14.7}$$

It is now time to consider some simple numerical examples, in order to find our bearings in this subject. We assume that the pdf is known, and is of a simple form. Notice that the pdf and cdf are deterministic functions, although they characterise a random variable; but they can only be inferred from observation of the random variable to within some statistical error.

● **Example 14.5**   The random variable of a certain signal can take any value in the range -1 V to $+3$ V. Sample values occur within this range with equal probability.

The pdf is therefore uniform or constant between the values $x = -1$ and $x = 3$.

Applying equation 14.7, the magnitude of the pdf is $0\cdot25$ V^{-1}. Therefore

$$f_X(x) = 0\cdot25 \qquad -1 \leqslant x \leqslant 3 \text{ V}$$

Applying equation 14.4,

$$F_X(x) = 0\cdot25(x+1) \qquad -1 \leqslant x \leqslant 3 \text{ V}$$

Calculate now the probability that the signal is negative.

$$\mathrm{Prob}\,[X < 0] = F_X(0) = 0\cdot25$$

The probability that the random variable exceeds $1\cdot5$ is

$$\mathrm{Prob}\,[X > 1\cdot5] = 1 - F_X(1\cdot5) = 0\cdot375$$

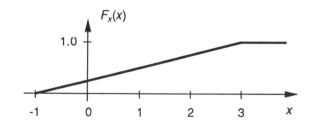

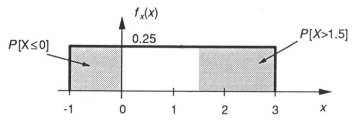

Figure 14.5   Uniform pdf example

297

These probabilities are marked as areas on the pdf diagram (figure 14.5). In simple cases like this, the area can often be assessed more quickly than doing the algebra! ●

● **Example 14.6**   The signal in example 14.5 is now quantised into 4 distinct levels such that the random variable takes the values:

$$\{-0\cdot5, 0\cdot5, 1\cdot5, 2\cdot5\} \text{ V}$$

Each level has an equal probability of occurring

The pdf consists now of a series of impulse functions, since the random variable is allowed to have a value only at one of these points. The probability of any one of these levels occurring is 1/4 or $0\cdot25$.

Thus:

$$f_X(x) = 0\cdot25\delta(x+0\cdot5) + 0\cdot25\delta(x-0\cdot5)$$
$$+ 0\cdot25\delta(x-1\cdot5) + 0\cdot25\delta(x-2\cdot5)$$

The corresponding cdf is a staircase function, as shown in figure 14.6. For this signal:

$$\text{Prob}[X < 0] = F_X(0) = 0\cdot25$$
$$\text{Prob}[X > 1\cdot5] = 1 - F_X(2) = 0\cdot25$$
$$\text{Prob}[X \geqslant 1\cdot5] = 1 - F_X(1) = 0\cdot5$$

Notice the ambiguity that emerges at values like $1\cdot5$, and that the results are subtly different from those when the random variable values are on a continuous scale. ●

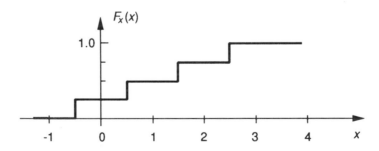

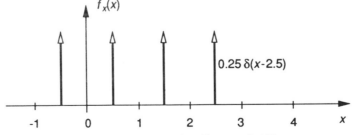

Figure 14.6   Discrete-valued signal, uniform probability

14.2.4 Moments and variance

The amplitude distribution functions are interesting, and can be used directly, as we have done, to calculate the probability that a certain signal level is exceeded or not. This is of extreme practical value, as we shall show in the applications of section 14.4.

The distribution functions have a further important role to play in the analysis and understanding of random signals. By the unlikely operation of taking moments of the pdf, we are able to calculate the mean value and the mean-square-value of the random variable. The link between taking moments and probability seems sufficiently obscure as to justify a small detour.

Consider a set of N sample values $\{x(n)\}$, characterised by a random variable X whose pdf $f_X(x)$ we know. Now the mean or arithmetic average of the set of samples is given by η_X, and

$$\eta_X = \lim_{N \to \infty} \frac{1}{N} \sum_{n=1}^{N} x(n) \tag{14.8}$$

Notice that with a finite number of samples N, this operation could be carried out by a practical accumulator or integrator, but that the result would have a statistical variation and itself be a random variable. The ideal case where $N \to \infty$ can never be determined exactly in practice. However, this is a perfectly respectable argument to use for a theoretical calculation!

Now consider the impact of the pdf on this calculation. Figure 14.7 shows the general form of pdf, and includes some nomenclature.

The equation 14.8 adds together the sample values irrespective of sequence and without classifying them in any way. However, from the pdf we do know how many of them fall into a given range of values. Take the small range of values $[x \leqslant X \leqslant (x + \delta x)]$, and suppose that N_X of the N values fall within this range. The contribution of this sub-class of values to the overall summation in equation 14.8 is $x \cdot N_X$, since these N_X samples all have essentially the same value x.

Now the probability that a given sample falls within this range is $f_X(x) \cdot \delta x$. If we take N samples from the signal, and let $N \to \infty$, then the

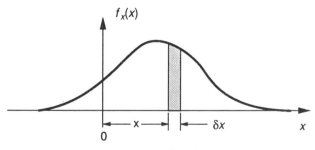

Figure 14.7 Probability density function

299

actual number of samples falling in this range is

$$N_X = N \cdot f_X(x) \cdot \delta x \tag{14.9}$$

Thus, the calculation of the average in equation 14.8 can be re-written

$$\eta_X = \frac{1}{N} \sum_x N \cdot x \cdot f_X(x) \cdot \delta x \tag{14.10}$$

Making $\delta x \to 0$ and pressing to the limit in the usual mathematical manner changes the summation into an integral. Assuming that the process is ergodic, we then equate the mean value η_X with the *expected value* or *expectation* $E[X]$, which is the value we would expect all estimates of the mean to fall around:

$$E[X] = \eta_X = \int_{-\infty}^{\infty} x f_X(x) \, dx \tag{14.11}$$

Notice that this equation defines the *first moment* of the pdf diagram, and locates its centre-of-gravity; hence $E[X]$ is often called the first moment of the distribution.

A similar argument, but based on the mean of the squares, yields the second moment of area:

$$E[X^2] = \int_{-\infty}^{\infty} x^2 f_X(x) \, dx \tag{14.12}$$

The mean-square-value (msv) of a signal waveform is normally regarded as the signal power, and if the random process is ergodic then $E[X^2]$ is synonomous with the power P as determined by a time average:

$$P = \lim_{N \to \infty} \frac{1}{N} \sum_{n=1}^{N} x^2(n) \tag{14.13}$$

Higher-order moments $E[X^r]$ may also be defined, but they are not so generally useful as the first two. Notice that the first moment $E[X]$ locates the centroid of the pdf, while the second moment $E[X^2]$ measures its spread about the origin; x^2 giving greater weight to occurrences away from the origin, as well as not distinguishing between positive and negative values.

Although the mean-square-value of a waveform is important, it is usually more helpful to calculate the spread of the pdf about its mean value, which we call its *variance*. In particular, if the mean value is large and the spread of signal values about it is small, then $E[X^2] \simeq (E[X])^2$, so we need the variance in order to separate similar signals. Following the conventions which we have used already, the variance (which is the square of *standard deviation*) is given by

$$V[X] = \sigma_X^2 = E[(X - \eta_X)^2] \tag{14.14}$$

Now, expanding the expectation term:

$$V[X] = E[X^2] + E[\eta_X^2] - E[2\eta_X X]$$

Since η_X is a constant, $E[\eta_X^2] = \eta_X^2$; and $E[2\eta_X X] = 2\eta_X^2$; hence

$$V[X] = E[X^2] - \eta_X^2$$

or $\qquad V[X] = E[X^2] - (E[X])^2 \qquad\qquad\qquad\qquad\qquad (14.15)$

The *variance* of a random variable is therefore the *mean-square-value* minus the *square of the mean*. In common parlance we might say that

$$E[X^2] \quad \text{represents the total signal power}$$

$(E[X])^2$ or η_X^2 represents the 'dc power' in the signal

$V[X]$ represents the 'ac power', or that component of signal power which causes fluctuations about the mean. Sometimes called the *fluctuation power*.

● **Example 14.7** Let us calculate the moments for the signal whose random variable is given in example 14.5:

$$E[X] = \int_{-1}^{3} 0 \cdot 25 x \, \mathrm{d}x = 1 \cdot 0$$

$$E[X^2] = \int_{-1}^{3} 0 \cdot 25 x^2 \, \mathrm{d}x = 2 \cdot 333$$

Hence $\qquad V[X] = 1 \cdot 333$

If these signal samples are in volts, the mean value is $1 \cdot 0$ V and the total power is $2 \cdot 333$ V^2. The variance is $1 \cdot 333$ V^2 so the standard deviation is $\sqrt{1 \cdot 333}$ or $1 \cdot 155$ V.

If the mean signal value is the parameter we wish to measure, then in this case the fluctuation noise is overwhelming the desired signal value. We normally express the quality of the measurement in terms of *Signal-to-Noise Ratio (SNR)*. Thus

$$\text{SNR} = 1 \cdot 0 / 1 \cdot 333 \text{ or } -1 \cdot 25 \text{ dB} \qquad ●$$

● **Example 14.8** Similar calculations can be done for the discrete-valued signal in example 14.6, except that summations are used instead of integrals.

$$E[X] = \Sigma \, x \cdot P(x) = 0 \cdot 25(-0 \cdot 5 + 0 \cdot 5 + 1 \cdot 5 + 2 \cdot 5) = 1 \cdot 0$$
$$E[X^2] = \Sigma \, x^2 P(x) = 0 \cdot 25(0 \cdot 25 + 0 \cdot 25 + 2 \cdot 25 + 6 \cdot 25) = 2 \cdot 25$$
$$V[X] = E[X^2] - (E[X])^2 = 1 \cdot 25 \qquad ●$$

14.2.5 The Gaussian process

Most error and noise mechanisms encountered in the real world have a probability density function which tails off smoothly at high and low deviations from the mean. The Gaussian function describes such pdfs, and the Central Limit Theorem establishes the mathematical credibility for

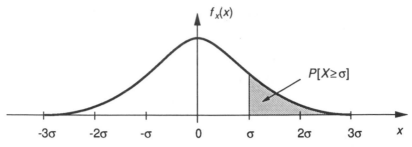

Figure 14.8 Gaussian pdf

assuming that this function applies to many random variables in the real world.

The Gaussian pdf, for a signal whose mean is zero and whose variance is σ^2, is defined by

$$f_X(x) = \frac{1}{\sigma\sqrt{(2\pi)}} \exp - \left(\frac{x^2}{2\sigma^2}\right) \tag{14.16}$$

Figure 14.8 shows the general form of this function, and the shaded area represents the probability that $[X \geqslant \sigma]$.

The important fact about the Gaussian pdf is that it is completely defined from a knowledge of the signal variance σ^2. The variance is an easy parameter to measure, and if we know that the signal we are dealing with is Gaussian, then we can immediately calculate the probability that a certain signal level is exceeded or not. So

$$P[X \geqslant a] = \int_a^\infty f_X(x)\,\mathrm{d}x$$

$$= \frac{1}{\sigma\sqrt{(2\pi)}} \int_a^\infty \exp - \left(\frac{x^2}{2\sigma^2}\right)\,\mathrm{d}x \tag{14.17}$$

Unfortunately, this integral cannot be solved analytically. However, the solution to this equation is of such general importance that numerical tables have been calculated for unity standard deviation, to enable all useful cases to be solved. There is a further problem, in that there are several different types of table, and the user must be sure that the right one is selected! We list below several common alternatives, but this list is not exhaustive.

Area table

$$P[X \geqslant a] = 1 - \Phi[a/\sigma]$$

where $\quad \Phi[y] = \frac{1}{\sqrt{(2\pi)}} \int_{-\infty}^y \exp - \left(\frac{x^2}{2}\right)\,\mathrm{d}x \tag{14.18}$

Q function (often used for data transmission calculations)

$$P[X \geqslant a] = Q[a/\sigma]$$

where
$$Q[y] = \frac{1}{\sqrt{(2\pi)}} \int_y^\infty \exp - \left(\frac{x^2}{2}\right) dx \qquad (14.19)$$

This function is tabulated in appendix A8.

Thus
$$Q[y] = 1 - \Phi[y] \qquad (14.20)$$

Also
$$Q[y] \simeq \frac{1}{y\sqrt{(2\pi)}} \exp - \left(\frac{y^2}{2}\right) \quad \text{if } y > 4 \qquad (14.21)$$

Error function

$$P[X \geqslant a] = \tfrac{1}{2}(1 - \text{erf}[a/(\sigma\sqrt{2})])$$

where
$$\text{erf}[y] = \frac{2}{\sqrt{\pi}} \int_0^y \exp - (x^2) \, dx \qquad (14.22)$$

The *complementary error function* is sometimes used:

$$\text{erfc}[y] = 1 - \text{erf}[y]$$

Also
$$\text{erfc}[y] = 2Q[y\sqrt{2}] \qquad (14.23)$$

● **Example 14.9**   A certain signal has a mean value of 2 V, and has a fluctuating component with Gaussian pdf and variance $0\cdot333$ V^2. What is the probability that a particular sample value is below 0 V or above 3 V?

The standard deviation is $\sigma = \sqrt{0\cdot333} = 0\cdot577$ V.
The mean value x_m is $2\cdot0$ V (see figure 14.9).
Then $P[X \geqslant a] = Q[(a - x_m)/\sigma]$
Now, for $a = 3$ $(a - x_m)/\sigma = 1\cdot73$
From the table of $Q[\ ]$ functions in appendix A8:

$$P[X \geqslant 3] = Q[1\cdot73] = 0\cdot0419$$

Also $P[X \leqslant b] = Q[(x_m - b)/\sigma]$
For $b = 0$ $(x_m - b)/\sigma = 3\cdot47$

$$P[X \leqslant 0] = Q[3\cdot47] = 0\cdot0003$$

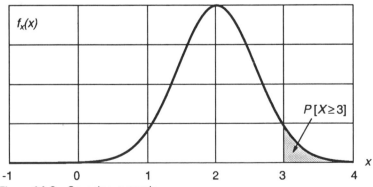

Figure 14.9   Gaussian example

303

The desired probability is therefore $0 \cdot 0422$ or, say, $4 \cdot 2 \times 10^{-2}$.

Notice that, in probability calculations, the order of magnitude of the result is usually more important than the number of significant figures. •

14.3 Transformations on random signals

Whilst we have developed some descriptions of random signals, and introduced measures like the amplitude-probability functions and mean values of various sorts, we have given no hint as to how the *processing* of such signals might be described. So now we make a move into the region of random-signal processing; it is a tentative movement but does make a start on the general problem. Handling random signals is going to be a more abstruse operation than that for deterministic signals, but fortunately there are certain operations which can be described easily.

A discrete-time random signal has sample values $\{x(n)\}$ described by a random variable X, and is processed to become a signal with sample values $\{y(n)\}$ described by random variable Y. At this point in our discussions, we are considering the instantaneous transformation of one random variable into another, and not considering operations like linear filtering which we leave for chapter 15.

The simplest kind of signal amplitude transformation is linear scaling. The random variable properties follow immediately, by considering equations 14.11, 14.12, 14.7, where

$$Y = k \cdot X$$

Then
$$E[Y] = k \cdot E[X] \tag{14.24}$$

$$E[Y^2] = k^2 \cdot E[X^2] \tag{14.25}$$

and
$$f_Y(y) = (1/k) \cdot f_X(kx) \tag{14.26}$$

• **Example 14.10** Suppose that the signal described in example 14.5 is amplified by 5 times, i.e. $k = 5$.

Then the mean value becomes $5 \times 1 \cdot 0 = 5 \cdot 0$ V
the mean-square-value becomes $25 \times 2 \cdot 333 = 58 \cdot 333$ V^2
the variance becomes $33 \cdot 333$ V^2

The pdf becomes

$$f_Y(y) = 0 \cdot 25/5 = 0 \cdot 05 \text{ V}^{-1} \qquad -5 \leqslant Y \leqslant 15 \text{ V} \qquad •$$

There are however many practical cases where the relationship between Y and X is not so simple, and follows some non-linear function. Let

$$Y = g[X] \tag{14.27}$$

where $g[\ ]$ represents an arbitrary function.

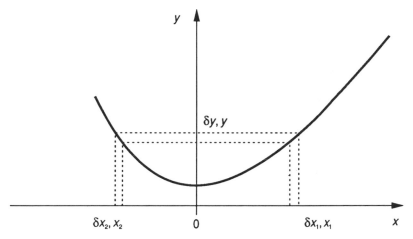

Figure 14.10 A non-linear functional relationship: $y = g[x]$

Figure 14.10 shows a typical case relating x and y, where a certain output value y may be generated by an input at either x_1 or x_2. In practical terms, this is similar to a square-law detector.

Now the probability that output random variable Y is within the region $[y \leqslant Y \leqslant (y + \delta y)]$ is $f_Y(y) . \delta y$. Similar expressions apply to the two alternative conditions of the input signal, so that

$$f_Y(y) . \delta y = f_X(x_1) . \delta x_1 + f_X(x_2) . \delta x_2 \tag{14.28}$$

Taking the limit:

$$f_Y(y) = f_X(x_1) \left| \frac{\mathrm{d}x_1}{\mathrm{d}y} \right| + f_X(x_2) \left| \frac{\mathrm{d}x_2}{\mathrm{d}y} \right| \tag{14.29}$$

Using equation 14.29, we can therefore find out the probability-density-function for the processed-signal random variable, and hence calculate the mean values and various probabilities. There is a more straightforward method of obtaining the means however, by substituting equation 14.29 and $y = g[x]$ directly into an equation like 14.12:

$$E[Y^r] = \int (g[x])^r . f_X(x) . \mathrm{d}x \tag{14.30}$$

where $r = 1$ or 2.

Several points should be noted:

a) The *magnitude* of $(\mathrm{d}x/\mathrm{d}y)$ is taken, since probability contributions must always be positive.

b) This case is not limited to a two-valued curve, and equation 14.29 can be generalised to any number of terms.

c) If $(\mathrm{d}x/\mathrm{d}y)$ is discontinuous at any point, then equation 14.29 cannot be used, and an *ad hoc* technique must be employed, which is illustrated in example 14.12.

305

d) Linear scaling is a special case where $(dx/dy) = 1/k$, as discussed above.

● **Example 14.11** A square-law function is sometimes used for signal detection, and is described by

$$y = x^2$$

Such a detector has an input-signal whose random variable X has the following pdf:

$$f_X(x) = \exp - (x) \qquad 0 \leqslant x < \infty$$
$$= 0 \qquad\qquad\quad x < 0$$

Find the properties of the output-signal random variable Y.

Now $dx/dy = 1/(2\sqrt{y})$ and

$$f_Y(y) = \frac{\exp - (\sqrt{y})}{2\sqrt{y}} \qquad \text{for } 0 \leqslant y < \infty$$

Notice that $f_Y(y)$ approaches infinity as $y \to 0$, which indicates that the processed signal goes tangential to the axis at certain points.

Now, calculating the mean value:

$$E[Y] = \frac{1}{2} \int_0^\infty \sqrt{y} \cdot \exp - (\sqrt{y}) \cdot dy = 2$$

Also $\quad E[Y] = \int_0^\infty x^2 \cdot \exp - (x) \cdot dx = 2$

The second-moment of Y can also be calculated in similar fashion. ●

● **Example 14.12** A signal rectifier has the characteristics:

$$y = x \qquad 0 \leqslant x < \infty$$
$$= 0 \qquad x < 0$$

Find the pdf of the output-signal random variable if the input-signal is Gaussian.

$$f_X(x) = \frac{1}{\sigma\sqrt{(2\pi)}} \cdot \exp - \left(\frac{x^2}{2\sigma^2}\right)$$

For $x \geqslant 0$, $(dx/dy) = 1$ and $f_Y(y) = f_X(x)$.

For $x < 0$, $y \equiv 0$ so an impulse function occurs at the boundary of this region, at $y = 0$. The probability of the random variable entering this region is $\frac{1}{2}$. Hence

$$f_Y(y) = \tfrac{1}{2}\delta(y) + \frac{1}{\sigma\sqrt{(2\pi)}} \cdot \exp - \left(\frac{y^2}{2\sigma^2}\right) \qquad y \geqslant 0$$

$$= 0 \qquad\qquad\qquad\qquad\qquad\qquad\qquad y < 0$$

These functions are illustrated in figure 14.11. ●

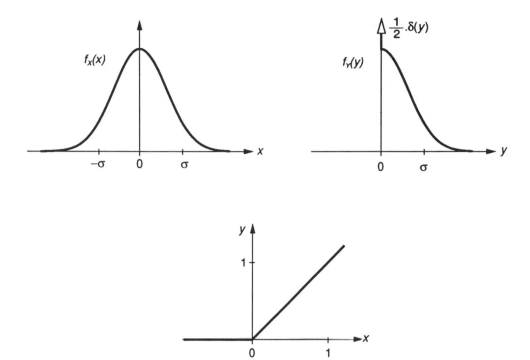

Figure 14.11 Rectified Gaussian signal

14.4 Applications

The definitions of randomness and the parameters with which we measure randomness have been explained above, but may seem far removed from practical engineering. Such concepts however are keystones in certain applications, and we now outline a few of these.

14.4.1 Quantisation noise and bias

The process of converting a continuous-valued signal into one which has only discrete values of amplitude is known as *quantising*, and inevitably generates errors in its representation of the original signal. Analog-to-digital conversion is one example of quantisation which has been discussed in chapter 6, but there are a number of other cases where a signal is quantised into only a few levels, for instance when converting a receiver signal into binary or ternary form.

Consider a random signal with sample values $\{x(n)\}$ which can vary continuously in magnitude. They are described by the random variable X, and pdf $f_X(x)$. Suppose that quantisation levels are equally spaced with an interval q between each pair.

When a sample is quantised, a value x is represented by a quantised variable y_p, which corresponds to the nearest allowed value, the pth. The difference between x and the nearest quantised level is the error γ:

$$\gamma = x - p \cdot q \qquad (14.30)$$

There are two interpretations of the term *nearest level*:

    a) Rounding arithmetic

$$(p - 0 \cdot 5) \cdot q < x < (p + 0 \cdot 5) \cdot q \qquad (14.31)$$

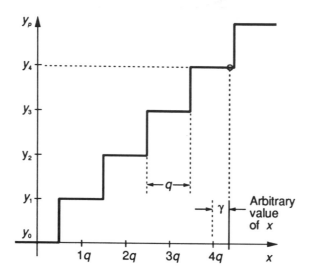

(a) **Rounding arithmetic**

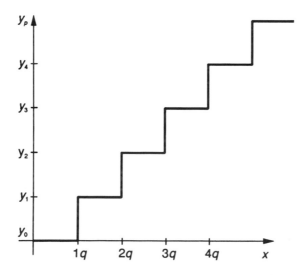

(b) **Truncating arithmetic**

Figure 14.12 Quantisation of continuous signal

b) Truncating arithmetic

$$p.q < x < (p+1).q \qquad (14.32)$$

Rounding is more normal, and both processes are described in figure 14.12.
The probability that the input signal quantises into the level y_p is given by

$$P[Y = y_p] = \int_{x1}^{x2} f_X(x) \, dx \qquad (14.33)$$

where the limits x_1, x_2 are defined by equations 14.31 or 14.32.

● **Example 14.13**   A certain random signal has a pdf $f_X(x)$ which is given by

$$f_X(x) = 2(1 - x/3)/3 \qquad 0 \leqslant x \leqslant 3$$

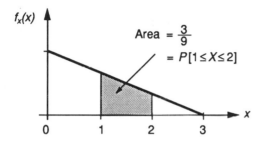

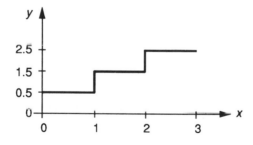

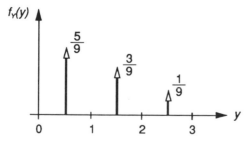

Figure 14.13   Quantisation example

309

It is quantised by a device whose characteristic is shown in figure 14.13, using truncating arithmetic and providing outputs only at 0·5, 1·5, 2·5.

Calculate the pdf for the quantised signal.

Carry out the integrations of equation 14.33. If the output signal is y, and has a random variable Y and pdf $f_Y(y)$:

$$f_Y(y) = [5\delta(y - 0\cdot5) + 3\delta(y - 1\cdot5) + \delta(y - 2\cdot5)]/9$$

This too is shown in figure 14.13.　　●

Quantisation noise arises in analog-to-digital conversion, because the discrete-valued digital signal must always differ from the continuous-valued analog signal. Several hundred quantising intervals is normal for analog-to-digital conversion, so we may assume that $f_X(x)$ is constant within each quantising interval q. There will of course be a varying probability that each interval is selected, which is reflected in the pdf of the output signal as shown in the above example, but the quantisation noise is the aggregate of the discrepancy at all available levels. We can now define the quantisation noise for the two cases:

(1) **Rounding**: where the error γ is bounded by

$$-q/2 \leqslant \gamma \leqslant q/2$$

Since the error is equally likely to appear anywhere within this range, its random variable is described by the uniform pdf, of height $1/q$. Thus

$$f_\Gamma(\gamma) = 1/q \qquad -q/2 \leqslant \gamma \leqslant q/2 \tag{14.34}$$

Applying equations 14.11 and 14.12,

$$E[\Gamma] = 0 \qquad E[\Gamma^2] = q^2/12 \qquad V[\Gamma] = q^2/12$$

(2) **Truncating**: where the error γ is bounded by

$$0 \leqslant \gamma \leqslant q$$

The random variable is therefore described by a uniform pdf such that

$$f_\Gamma(\gamma) = 1/q \qquad 0 \leqslant \gamma \leqslant q \tag{14.35}$$

Hence

$$E[\Gamma] = q/2 \qquad E[\Gamma^2] = q^2/3 \qquad V[\Gamma] = q^2/12$$

So the variance of the quantisation error is the same for both rounding and truncation, but the latter introduces a constant bias. Remember that this result assumes a constant pdf $f_X(x)$ within each quantising interval, which excludes cases like that in example 14.13.

14.4.2   Data transmission and errors

A typical data transmission link is shown schematically in figure 14.14. Digital data is converted into analog form for transmission over the medium, which might be a cable, a radio path or an optical fibre. When arriving at the receiver, the original signal is distorted, attenuated and disturbed by noise. Noise in transmission arises from crosstalk from adjacent circuits, interference from other electrical sources, and possibly natural noise signals.

In amplifying the received signal, the receiver may add additional noise to the minute signal (see example 14.2). The detector has now to separate the signal from its accompanying noise, and the simplest strategy is to sample the incoming signal and then compare the result with a threshold.

Many practical systems have to modulate the data on to an audio or radio-frequency carrier signal, and elaborate schemes are used. However, we are concerned at the moment just to demonstrate the application of random signal properties and measures, so we take the simplest form of data transmission, baseband signalling, where a pulse is transmitted for a 1 and no pulse for a 0. This type of signalling is often called *polar* or ON/OFF signalling, and is illustrated in figure 14.15.

The figure shows that a wrong decision is made occasionally, due to presence of the noise, and we set out to calculate the probability of a wrong decision, the error probability or error rate.

Let us first examine what happens at the sampler. The input signal is $s(t)$ and the noise is $n(t)$, but the signal presented to the sampler is $x(t)$:

$$x(t) = s(t) + n(t) \tag{14.36}$$

We have no means of knowing how much of the combined signal is due to noise and how much due to the data signal. The sampler converts the incoming signal into discrete samples, whose positions are chosen to be at the centre of each signalling interval, where the bandlimited data pulse (if it exists in the interval) will be at a maximum. Making sure that sampling occurs at exactly the right time is a complete story in itself; suffice to say that this operation of *symbol synchronisation* is readily achieved and presents few problems in practice.

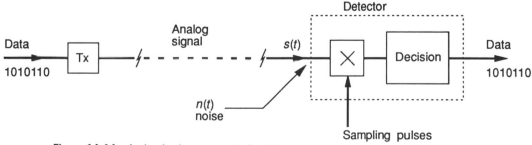

Figure 14.14   A simple data transmission link

311

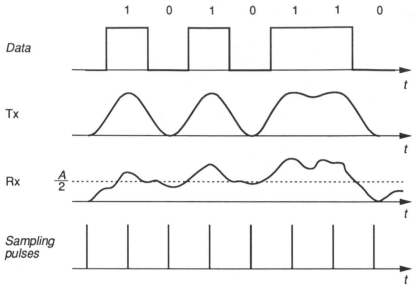

Figure 14.15 Polar signalling waveforms

After sampling therefore, we have a sample whose value in the absence of noise would be 0 or A, but which is disturbed by Gaussian noise of variance N. We must distinguish between two cases:

CASE 1: pulse transmitted $E[X] = A$ $V[X] = N$
CASE 2: no pulse transmitted $E[X] = 0$ $V[X] = N$

For this simple detector, we attempt to distinguish between these two cases by setting a threshold value, say B, and quantising the signal at this boundary. Let $B = \alpha A$, where $\alpha < 1$. Figure 14.16 then shows the probability-density-functions for the two cases, and we can define the conditions for error:

CASE 1: pulse transmitted $\text{Prob}[\text{error}] = \text{Prob}[X_P < B]$
$$= Q[(A - B)/\sqrt{N}]$$
$$= Q[(1 - \alpha)A/\sqrt{N}] \qquad (14.37)$$

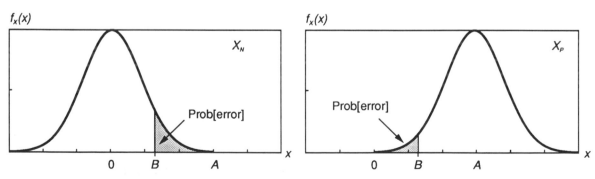

Figure 14.16 Pdfs for pulse detection

CASE 2: no pulse transmitted

$$\text{Prob [error]} = \text{Prob}[X_N > B]$$
$$= Q[B/\sqrt{N}]$$
$$= Q[\alpha A/\sqrt{N}] \qquad (14.38)$$

Notice that these two probabilities are equal if $\alpha = 0 \cdot 5$, which is obvious from the symmetry of the problem anyway. This is the condition normally chosen, although due to unforeseen changes in signal level, operation away from the optimum does occur.

In order to make the results more general, it is usual to normalise the ratio $A/\sqrt{N}$ in terms of the *signal-to-noise power ratio (SNR)*; so we need to calculate the signal power. For this exercise we take the 'signal power' of the sample values, ignoring what happens to the analog signal in between the samples. Thus the signal power when a pulse is transmitted is A^2; and the signal power when no pulse is transmitted, is zero.

Suppose that the probability of transmitting a pulse is P_1, then the probability of not transmitting a pulse is $(1 - P_1)$. The signal power is S, and

$$S = P_1 . A^2 \qquad (14.39)$$

$$\text{SNR} = P_1 . A^2/N \qquad (14.40)$$

Now pulse and no-pulse symbols usually occur with equal probability, so

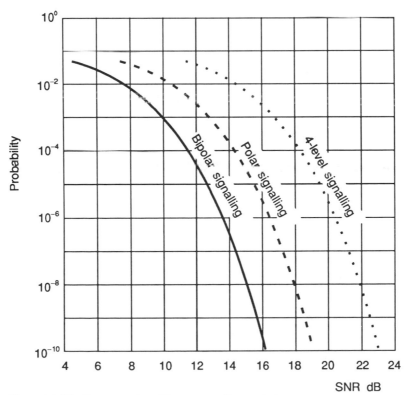

Figure 14.17 Error rate for different signalling techniques

313

that normally $P_1 = 0 \cdot 5$. Combining the effects of equations 14.37 and 14.38, the overall probability of a symbol error is P_e:

$$P_e = \tfrac{1}{2} . Q[\sqrt{(SNR/2)}] + \tfrac{1}{2} . Q[\sqrt{(SNR/2)}]$$
$$= Q[\sqrt{(SNR/2)}] \tag{14.41}$$

This is a standard result, when the threshold is spaced equally between the two signalling levels ($\alpha = 0 \cdot 5$), and when the two symbols occur with equal probability ($P_1 = 0 \cdot 5$). It is plotted in figure 14.17, together with other similar curves which we shall discuss in a moment. Notice that the probability of error approaches $0 \cdot 5$ as the SNR is decreased, and approaches zero as the SNR is increased.

● **Example 14.14** *Bipolar signalling* is a variant of the above scheme, where signals of alternate polarity are transmitted. Thus, to send a 1 for instance the signal level is $+ A$, and to send a 0 the level is $- A$. Calculate the probability of error for optimum decision levels.

Maintaining the equal-error criterion for both signalling states, the threshold $B = 0$.
The probability of any one symbol being in error is then $Q[A/\sqrt{N}]$.
The signal power is $S = \tfrac{1}{2} . A^2 + \tfrac{1}{2} . A^2$ for equal probability of symbols. Hence

$$P_e = \tfrac{1}{2} . Q[\sqrt{(SNR)}] + \tfrac{1}{2} . Q[\sqrt{(SNR)}]$$
$$= Q[\sqrt{(SNR)}]$$

This also is plotted in figure 14.17. Notice the doubling or 3 dB improvement in performance; the curve is the same as for polar signalling, but shifted along by 3 dB.   ●

● **Example 14.15** A certain optical receiver operates on a polar binary signal. It delivers a signal of peak magnitude 50 mV to the decision device, in the presence of noise having an rms value of 20 mV. The threshold is set erroneously at 30 mV. Estimate the probability of error for each signalling state.

In terms of the equations above: amplitude $A = 50$ mV; noise $\sqrt{N} = 20$ mV; threshold $B = 30$ mV; so that $\alpha = 0 \cdot 6$.
When a pulse is transmitted, probability of error (from appendix A8) is

$$P_1 = Q[(1 - \alpha)A/\sqrt{N}]$$
$$= Q[1 \cdot 0] = 0 \cdot 159$$

When no pulse is transmitted, the probability of error is

$$P_2 = Q[\alpha A/\sqrt{N}]$$
$$= Q[1 \cdot 5] = 0 \cdot 067$$

Assuming that both states are equally probable, the net probability of symbol error is

$$P_e = \tfrac{1}{2}(0 \cdot 159 + 0 \cdot 067) = 0 \cdot 113 \qquad \bullet$$

These ideas may be extended to multi-level signalling. A four-level system offers the choice of 1 out of 4 possible symbols at each sampling interval, and hence conveys 2 bits of information per symbol. Table 14.1 shows the signalling levels and thresholds for such a scheme, and figure 14.18 indicates a typical non-bandlimited noise-free signalling waveform.

The problem is replete with symmetry, so only one probability calculation has to be done. The distance between any signalling level and the nearest threshold is $A/3$, and the probability that this distance is exceeded is given by $Q[A/(3\sqrt{N})]$. Symbols β and γ have twice the likelihood of error when compared with the other two, since each is bounded by two thresholds. We assume that a noise magnitude which exceeds two thresholds in tandem is unlikely.

We have now to calculate the signal power (the mean-square-value), assuming that all symbols are equi-probable. Hence

$$S = 0 \cdot 25(A^2 + A^2/9 + A^2 + A^2/9)$$
$$= 5A^2/9$$

Hence   Prob[one error] $= P_1 = Q[\sqrt{(SNR/5)}]$

Table 14.1   *4-level signalling scheme*

Signal	Data (dibit)	Sig. level	Threshold
α	00	A	
	....	....	$2A/3$
β	01	$A/3$	
	....	....	0
γ	11	$-A/3$	
	....	....	$-2A/3$
δ	10	$-A$	

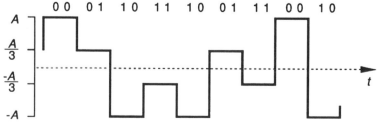

Figure 14.18 Typical 4-level signal

315

and averaging this over all symbols:

$$P_e = 0 \cdot 25(1 + 2 + 2 + 1) \cdot P_1$$
$$= 1 \cdot 5 \cdot Q[\sqrt{(SNR/5)}] \tag{14.42}$$

Notice that this is in the same form as equation 14.41 for the polar signalling case, except that a penalty of $2 \cdot 5$ times or 4 dB is added to the SNR for a given error performance. This curve is also plotted in figure 14.17.

Each 4-level symbol transmits the equivalent of 2 binary symbols, so if the symbol rate is say R symbols/sec (or *bauds*), then the data rate is $2R$ bits/sec. However, in return for this advantage, each symbol is more susceptible to error, and has to operate at an SNR 7 dB greater than that for binary bipolar signalling to maintain the same symbol error rate. Such trade-offs are part and parcel of communication system design.

14.4.3   Radar and false-alarms

Radar is a descriptive word derived from *Radio Detection and Ranging*, and radar systems are used extensively for diverse applications such as aircraft detection and tracking, road vehicle speed measurement, intruder alarms, weather tracking and inter-planetary distance measurement. If the radar is moving over a surface, the one-dimensional distance measurements are extended to two-dimensions, and high-resolution visual images can now be constructed. An earth satellite, for instance, can fly a radar imaging system which is able to map details on the ground with the resolution of a few metres.

Radar systems depend upon transmission at high radio frequencies, often in the microwave region, or upon acoustic radiation, and their performance depends in the first place upon the design of the rf part of the system—transmitter, antennas and receiver. However, the key to successful system design lies in signal processing, both inside and after the receiver. The range and resolution of the radar can be improved greatly by appropriate signal processing.

We shall now describe broadly the radar principle, and then show how random signal properties are key parameters in the design. Figure 14.19 shows the principle.

Transmitter and receiver are coupled to an antenna through a *diplexer*, which ensures that transmitted power does not go into the receiver, and that all the received signal power passes to the receiver. A short pulse of electromagnetic energy is transmitted and reflected from any target within the radar beam.

A target at distance d from the antenna generates a round-trip delay of $2d/c$ sec, where c is the speed of light. For instance if $d = 1$ km then the delay is $6 \cdot 7$ μs, an easily measured time. Multiple targets in the same direction from the antenna generate separate pulses at different delays, and hence can be detected individually.

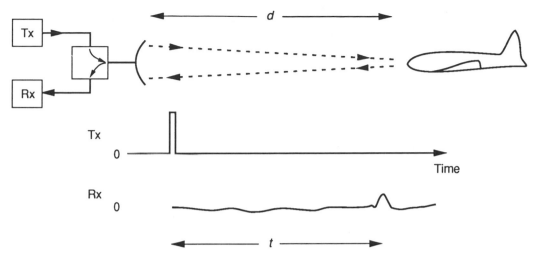

Figure 14.19 Radar principle

The received signal is extremely small, since the target does not reflect much of the incident energy, and also the propagation in both directions obeys the inverse-square law, so that overall propagation loss $\propto d^{-4}$. The limiting performance of the radar is therefore determined by its ability to measure extremely small signals, in the presence of noise which may be generated in the receiver or may be collected by the antenna from the sky.

Assume first that the noise is Gaussian. After the transmitted pulse has been sent, the receiver is open to receive reflected signals. Let us suppose that the receiver output signal is sampled at intervals of T_S sec. Each of these intervals defines a *range cell*, and the detection process is to decide whether or not there is a valid target within the distance defined by each range cell. Note that each range cell corresponds to a distance in space of $cT_S/2$ metres; this is the limiting resolution of the system and defines the bandwidth which is required. By the sampling theorem (section 6.2), the bandwidth W must be $> 1/(2T_S)$.

Between each transmitted pulse, there is time for a large number of range cells, each of which is looking at a point in space at a unique distance from the antenna. So we concentrate upon one particular range cell, and examine the signal to see the likelihood of detecting a pulse if it exists (the *detection probability* P_D), or of falsely detecting a pulse when none is present (the *false-alarm probability* P_N).

The pulse is detected by setting a threshold, and when the signal in the range cell exceeds this threshold it is assumed that a pulse is present. Suppose that the pulse amplitude is A, then let the threshold value be αA. The random variable when a pulse is received is X_P and when no pulse is received is X_O. Thus

$$P_D = \text{Prob}\,[X_P \geqslant \alpha A] \tag{14.43}$$

$$P_N = \text{Prob}\,[X_O \geqslant \alpha A] \tag{14.44}$$

317

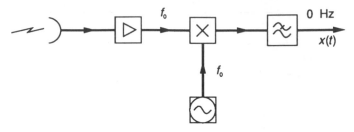

(a) **Coherent receiver structure**

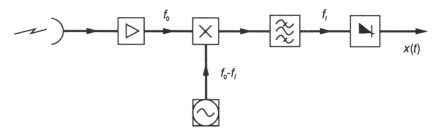

(b) **Non-coherent receiver structure**

Figure 14.20 Alternative receivers

The noise has variance N and is Gaussian, but the statistics of the received signal depend upon the type of receiver in use. For a *coherent* receiver, where the incoming signal is frequency-shifted to form a complex baseband signal (section 12.2.2), as indicated in figure 14.20a, the noise remains Gaussian. For a receiver which shifts the signal down to an *intermediate frequency (IF)*, as shown in figure 14.20b, an envelope detector is used and the noise-only envelope has Rayleigh statistics while the noise-plus-signal envelope statistics are now Rician (see equation 14.51). Taking first the coherent receiver, which yields the simpler Gaussian statistics, we write the SNR as γ, and

$$\gamma = A^2 / N \tag{14.45}$$

Thus

$$P_N = Q\left[\alpha\sqrt{\gamma}\right] \tag{14.46}$$

$$P_D = 1 - Q\left[(1-\alpha)\sqrt{\gamma}\right] \tag{14.47}$$

So, using equation 14.46 to establish the necessary threshold parameter α for a given false-alarm probability P_N, we can then find the probability of detection P_D for a given SNR γ. The results are summarised in the graphical chart of figure 14.21, which is fundamental to radar signal detection.

This design chart assumes a fixed and known signal amplitude A, but in practice the signal amplitude, or target strength, is also represented by a

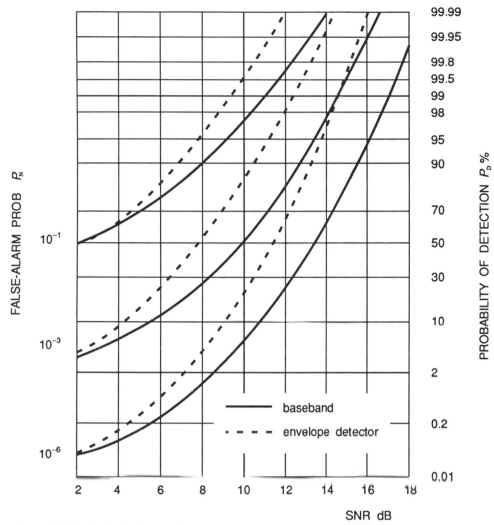

Figure 14.21 Radar design chart

random variable, so that radar system performance is influenced by far more parameters than we have disclosed here.

The false-alarm probabilities may seem low, but are seen in proper perspective when it is remembered that the range-cells are scanned at a rate of $1/T_S$ per sec, and each one of them contributes to a potential false-alarm. For example, suppose that T_S is $1 \mu s$, corresponding to a range-cell of 150 m, then a false-alarm probability of say 10^{-6} generates one false-alarm per second.

Now take the non-coherent receiver, using an envelope detector. If we assume that the carrier peak magnitude at the point of detection is A, and that the noise variance is N, then the SNR we write as γ, which is the ratio of average signal power during the pulse to average noise power:

$$\gamma = A^2/2N \tag{14.48}$$

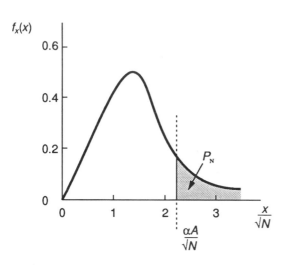

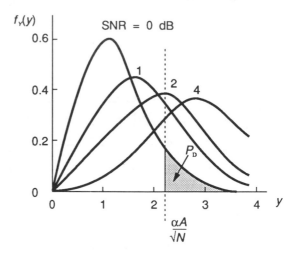

(a) **Rayleigh distribution - noise only**   (b) **Ricean distribution - signal and noise**

Figure 14.22 Pdfs for an envelope detector

In the absence of a pulse, the noise-only envelope statistics are Rayleigh, so the signal applied to the threshold device after the envelope detector has a random variable X, and

$$f_X(x) = \frac{x}{N} \cdot \exp - \left(\frac{x^2}{2N}\right) \tag{14.49}$$

Fortunately the Rayleigh pdf is integrable, so

$$P_N = \exp - (\alpha^2 . \gamma) \tag{14.50}$$

This equation enables the threshold parameter α to be determined for a given false-alarm probability P_N, which is then used in the Rician pdf of the signal-plus-noise envelope. For a random variable $Y = X/\sqrt{N}$, the Rician pdf is

$$f_Y(y) = (y/\sqrt{N}) \cdot \exp - (\gamma + y^2/2) . I_0[y\sqrt{(2\gamma)}] \tag{14.51}$$

where $I_0(a)$ is the modified Bessel function of the first kind of zero order.

These pdfs are illustrated in figure 14.22. Evaluating the probability of detection P_D using equation 14.43, we arrive at the broken-line set of curves in the design chart of figure 14.21.

Fortunately, the Rician pdf approaches a Gaussian form when the signal-to-noise ratio is moderate, say >4. Under this condition, which applies to the simple cases we are considering by way of example, the probability of detection P_D corresponds to Gaussian noise of variance N, superimposed on the peak envelope signal of magnitude A:

$$P_D = 1 - Q[(1 - \alpha)\sqrt{(2\gamma)}] \tag{14.52}$$

● **Example 14.16** A certain radar receiver uses an envelope detector. During the received pulse, the peak ac signal at the detector is $2 \cdot 0$ V, and the rms noise level is $0 \cdot 38$ V. Find the probability of detection if the false-alarm probability is set to 10^{-6}.

The signal-to-noise ratio from equation 14.48 is $\gamma = 13.85$.

For a false-alarm probability of 10^{-6}, the threshold level is $\alpha = 0 \cdot 999$ (equation 14.50), so the probability of detection $P_D \simeq 0 \cdot 5$.

Now we might ask, what false-alarm probability would result from adjusting the threshold so that the probability of detection is $0 \cdot 9$?

Since $P_D = 0 \cdot 9$, we can use equation 14.52 and work the tables backwards to discover that

$$(1 - \alpha)\sqrt{(2\gamma)} \simeq 1 \cdot 28$$

Hence $\alpha = 0 \cdot 757$ and the false-alarm probability is

$$P_N = 3 \cdot 6 \times 10^{-4} \qquad ●$$

STUDY QUESTIONS

1   What is the difference between a random variable and a random process?

2   Verify that the process described in example 14.4 is ergodic, by calculating the mean and mean-square-value at $n = 2$ and $n = 10$.

3   A certain random signal has a probability-density-function $f_X(x)$ which is described by

$$\begin{aligned} f_X(x) &= ax & 0 \leqslant x \leqslant 1 \\ &= a & 1 < x \leqslant 2 \\ &= 0 & \text{elsewhere} \end{aligned}$$

What readings would be obtained if this signal was presented to

  a) a mean-reading meter
  b) a true-rms meter
  c) a negative-reading peak meter.

4   A certain random signal is known to have a pdf of the form:

$$f_X(x) = b(1 - ax) \qquad 0 \leqslant x \leqslant 1/a$$

The measured mean power is $1 \cdot 5$ V^2.
  Determine for this signal:

  a) the cumulative-distribution-function (cdf)
  b) the mean value
  c) the probability that the signal will exceed 2 V.

5 The signal described in question 4 is now applied to a quantiser whose transfer characteristic is defined by figure 14.12b, with $q = 1$ and $y_p = p$. Determine the pdf, mean and variance of the output signal. Sketch a typical output signal sequence.

6 A certain signal whose pdf is $f_X(x)$ is processed by a square-law detector $y = x^2$. Sketch the pdf of the output signal and calculate its mean value and, its variance.

$$f_X(x) = (1/3) \cdot (1 - |x|/3) \qquad -3 \leqslant x \leqslant 3$$

7 The output signal mentioned in question 6 is now clipped at a level of $y = 7$. Determine the probability that clipping occurs.

8 A certain random signal is known to have Gaussian pdf, with zero mean. What is the probability that a given sample will exceed the rms value of the signal?

9 A certain random signal has a Gaussian pdf with a standard deviation of 2 V and a mean value of 1 V. Calculate the mean power of the signal, and the probability that a given sample will be positive.

10 A certain optical digit communication channel operates in the ON/OFF binary mode. The peak signal at the receiver is -40 dBm, while the effective Gaussian noise level is -50 dBm. (Note: dBm is the unit of *decibels relative to 1 mW*.) Determine the average bit error probability if the decision threshold is set at 40% of the peak signal.

11 The envelope detector of a certain radar receiver has Gaussian noise with an rms value 433 mV, and is to operate with a false-alarm probability of 10^{-4}. Determine the probability of detection when the pulse signal is $1 \cdot 5$ V rms.

15 Random Signals: Autocorrelation Function and the Averaging Process

OBJECTIVES

To show how random discrete-time signals can be averaged. In particular to

a) Introduce the Autocorrelation Function for discrete-time signals
b) Discuss the principles behind the averaging process for random signals
c) Show why such averages are required, and how they are implemented.

COVERAGE

The autocorrelation function is introduced for discrete-time signals, and its broad properties established via the joint and conditional probabilities of two random variables.

Addition of two random signals introduces the principles behind the moving-average processor. A discussion of signal averaging in radar shows the practical value of simple averaging.

A discussion of the general problem of measuring signals in noise leads on to a calculation of the performance of a moving-average processor, and yields results which may be applied to any measurement process with a Gaussian disturbance. A quick-design graph is displayed for this purpose.

A recursive averager is then investigated, enabling a direct comparison to be made between these two processes. A quick-design graph is given for the recursive averager, and the wide range of applications of this principle is indicated.

Our previous discussions of random signals in chapter 14 have been limited to amplitude statistics, defined by the properties of a *random variable*. Since each sample has been considered independently of those adjacent to it, first-order statistics have been adequate. We now move into a different regime and begin to consider the relationships between pairs of samples, which clearly demands the use of second-order statistics.

The type of operation that we are about to analyse is that of *averaging*, which will now be extended beyond the material of chapter 10 and into the realm of random signals. Most commonly, a signal of known form is obscured by a noisy disturbance as we saw in chapter 14, and the objective of averaging is to improve the signal-to-noise ratio. For a signal consisting of a constant value, this is accomplished by an *averaging filter* which reduces the noise power relative to the signal power, and we shall investigate the performance of both *non-recursive* and *recursive* forms of such a filter. For a pulse signal, the *matched-filter* is the optimum linear processor, which we shall meet in chapter 16.

First, however, we must investigate that property of a signal which is known as the *autocorrelation function*. Having understood this thoroughly, we can then apply its properties in order to understand linear filtering for random signals. A related signal measure is the *power spectrum*, which brings us back to the familiar ground of frequency response, and links the properties of filtered random signals with those of deterministic signals, but this we leave until chapter 16.

15.1 Autocorrelation function

We commence with a bald statement of what the autocorrelation function is. For a stationary random process consisting of the sequence of random variables $\{X(n)\}$, the autocorrelation function $R_X(k)$ is

$$R_X(k) = E[X(n) . X(n-k)] \tag{15.1}$$

The autocorrelation function is therefore the statistical average of the product of two random variables, which are a distance k sample times apart in the same signal stream. Remember that in section 2.1.3 we introduced the idea of correlation by expressing the correlation coefficient as a *time-average*. If the random process is ergodic, then these two operations are equivalent. The time-average operation also enables us to find the autocorrelation function for a deterministic signal, and although that is not our immediate aim, the following example shows how this is done.

● **Example 15.1** Find the autocorrelation function of an arbitrary sinusoidal signal.

$$x(n) = A . \cos(\omega T_S n + \varphi)$$

Hence

$$R_X(k) = \text{avg}\,[A \cdot \cos(\omega T_S n + \varphi) \cdot A \cdot \cos(\omega T_S (n - k) + \varphi)]$$
$$= (A^2/2) \cdot \cos(\omega T_S k)$$

The autocorrelation function for a sinusoidal signal is therefore a sinusoid of the same period in k as it was in n, but is independent of the phase of the signal. Thus, a given autocorrelation function satisfies an infinite number of different signals, and this property is to be compared with the ensemble properties for random signals that we introduced in Chapter 14. ●

The expectation process was defined in equation 14.11 for the single-variable case and is now expanded to the two-variable case. However, the nomenclature gets a bit confused with individual random variables $X(n)$ and $X(n - k)$, so for convenience we define new random variables as follows:

Let $\quad X \equiv X(n) \quad$ and $\quad Y \equiv X(n - k)$ $\qquad$ (15.2)

Then $\quad E[X \cdot Y] = \displaystyle\int_{-\infty}^{\infty} \int_{-\infty}^{\infty} x \cdot y f_{XY}(x, y)\ \mathrm{d}x\ \mathrm{d}y$ $\qquad$ (15.3)

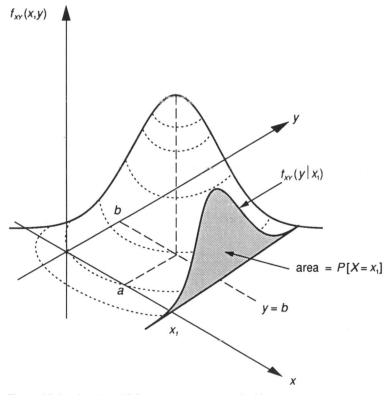

Figure 15.1 A joint pdf $f_{XY}(x, y)$, mean value (a, b)

The probability density function $f_{XY}(x, y)$ is known as the *joint pdf*, since it expresses the probability that both *X and Y* fulfil certain conditions; specifically that $[x \leqslant X \leqslant (x + \delta x)]$ and $[y \leqslant Y \leqslant (y + \delta y)]$. So $f_{XY}(x, y)$ is a function which is defined over the $\{x, y\}$ plane. Figure 15.1 attempts to give a pictorial view of such a function.

In order to appreciate the autocorrelation function, we must understand the nature and meaning of the joint pdf, so we now digress slightly in order to investigate this point. Appendix A9 summarises some of the major properties of the probabilities of two variables.

15.1.1 Joint and conditional probabilities of two variables

Although we have referred to a joint pdf in equation 15.3, the concepts come easier if we consider joint probabilities instead, which is what we will do for the moment. We are therefore referring implicitly to a random variable which can take only discrete amplitude values. For example, the values of a random variable X might be $\{-1, 0, +1\}$. So we consider two random variables X, Y, which are of this sort.

The joint probability that $[X = a, Y = b]$ is influenced first by the probability that $[X = a]$. The second influence is from the random variable Y, but very often this is constrained by the value that X has already assumed, so a *conditional probability* is required. Putting these ideas into a mathematical statement we express Bayes' Theorem:

$$P[X = a, Y = b] = P[X = a] \, . \, P[Y = b \,|\, X = a] \tag{15.4}$$

Also

$$P[X = a, Y = b] = P[Y = b] \, . \, P[X = a \,|\, Y = b] \tag{15.4a}$$

$P[Y = b \,|\, X = a]$ represents the *probability that* $[Y = b]$, *given that* $[X = a]$, and is a *conditional* probability.

The simplest case to consider is where the random variable Y is not influenced at all by random variable X, but is *independent* of it. Whatever the value of X, Y is not altered by it. In this case the conditional probability becomes

$$P[Y = b \,|\, X = a] = P[Y = b] \tag{15.5}$$

$$P[X = a, Y = b] = P[X = a] \, . \, P[Y = b] \tag{15.6}$$

or, more generally,

$$P[X, Y] = P[X] \, . \, P[Y] \tag{15.7}$$

Before applying this information to the autocorrelation function it will be helpful to amplify these ideas by a simple example. The concept of *independence* is most important for random signals.

● **Example 15.2** A certain ternary signalling system uses signal levels of $-A, 0, +A$, and all symbols occur with equal probability. Assume that successive symbols are independent. Calculate the joint probability for each possible pair of adjacent symbols.

Let the first symbol of each pair be x_i, and the second symbol of each pair be y_j.

Since all symbols are equally probable, $P[x_i] = 1/3$ for all i.

Now since the symbols are independent, each symbol y_j is not influenced by the previous symbol x_i, and the conditional probability is

$$P[y_j \mid x_i] = P[y_j] = 1/3$$

The joint probability $P[x_i, y_j] = P[x_i] . P[y_j] = 1/9$ for all $\{i, j\}$.   ●

● **Example 15.3** The signalling system of example 15.2 is now constrained so that no two adjacent symbols are the same. This kind of trick is often employed to ensure that a signal stream contains a regular sequence of transitions, which then identify the symbol boundaries.

Although the symbol probabilities $\{P[x_i]\}$ are still all 1/3, the conditional probabilities $P[y_j \mid x_i]$ are not now all equal, and are best displayed by means of a table. A corresponding table of joint probabilities can then be drawn up, where

$$P[x_i, y_j] = P[x_i] . P[y_j \mid x_i]$$

$P[y_j \mid x_i]$ $x_i \backslash y_j$	$-A$	0	$+A$
$-A$	0	1/2	1/2
0	1/2	0	1/2
$+A$	1/2	1/2	0

$P[x_i, y_j]$ $x_i \backslash y_j$	$-A$	0	$+A$
$-A$	0	1/6	1/6
0	1/6	0	1/6
$+A$	1/6	1/6	0

Figure 15.2 shows these conditional probabilities on a *transition* probability diagram.

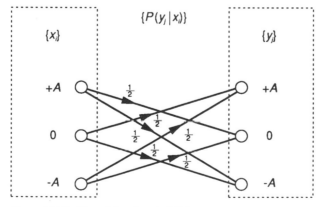

Figure 15.2 Transitional probabilities

The effect of dependence can now be seen in the joint probabilities. If these were measured for a practical case, and also the symbol probabilities $\{P[x_i]\}$, then the conditional probabilities could be calculated and the dependence between each pair of symbols assessed. ●

15.1.2 Calculation of the autocorrelation function

Now we can pursue this example still further, and use it to calculate the autocorrelation function for this sequence of signal samples. Expressing equation 15.3 in a form suitable for discrete-amplitude signals:

$$R_X(k) = E[X \cdot Y]$$

$$= \sum_x \sum_y x \cdot y P[x] \cdot P[y \mid x] \tag{15.8}$$

Now for *independence* between samples, $P[y \mid x] = P[y]$ so

$$R_X(k) = E[X] \cdot E[Y] = (E[X])^2 \tag{15.9}$$

Also, when $k = 0$, then $Y = X$ and $P[Y \mid X] = 1$, so

$$R_X(0) = E[X^2] \tag{15.10}$$

● **Example 15.4** Calculate the autocorrelation function for the two signals described in examples 15.2 and 15.3.

The autocorrelation function is calculated separately for each value of k.

From equation 15.10, the value at $k = 0$ will be the same for each case, ie. $R_X(0) = 2A^2/3$.

Now take $k = 1$:

a) For the independent case (example 15.2), since the conditional probability $P[y_j \mid x_i] = P[y_j]$, then

$$R_X(1) = (E[X])^2 = 0$$

b) For the dependent case (example 15.3), we must work through the probabilities in the tables to execute the sum of equation 15.8. In this case, $R_X(1) = -A^2/3$.

Now take $k = 2$:

a) For no dependence between y_j and x_i: $R_X(2) = 0$.
b) For the dependent case, the transitional probabilities are given in the table.

$$P[y_j \mid x_i] \quad x_i \backslash y_j \qquad -A \quad 0 \quad +A$$

$x_i \backslash y_j$	$-A$	0	$+A$
$-A$	1/2	1/4	1/4
0	1/4	1/2	1/4
$+A$	1/4	1/4	1/2

Evaluating equation 15.8, $R_X(2) = A^2/6$.

Subsequent values are calculated by extending this simple arithmetic.

Hence for the signal where each sample is *independent* of those adjacent to it, the autocorrelation function is

$$\{R_X(k)\} = \{2A^2/3, 0, 0, 0, 0 \ldots\}$$

When two similar adjacent signal values are prohibited, the autocorrelation function is

$$\{R_X(k)\} = \{2A^2/3, -A^2/3, A^2/6, -A^2/12, A^2/24, -A^2/48, \ldots, 0, \ldots\} \qquad \bullet$$

This simple example serves to show the importance of dependence, when calculating the autocorrelation function of a signal. In general there are two distinct regions to be considered:

a) Where adjacent values are *dependent*
b) Where adjacent values are *independent*.

Using this principle, it is possible to calculate the autocorrelation function for many different random signals, provided that the necessary conditional probabilities are known. Although these exercises give plenty of scope for ingenious algebra, they are not helpful for our immediate purpose, which is to define and then to use the autocorrelation function in order to understand the effect of filtering operations on a random signal.

So we now return to our main discussion of the properties and use of the autocorrelation function.

15.1.3 Properties of the autocorrelation function

Given a random process which generates a sequence of random variables $\{X(n)\}$, the autocorrelation function is defined as

$$R_X(k) = E[X(n) \cdot X(n-k)] \tag{15.11}$$

When $k = 0$, then the two random variables are identical, $X(n-k)$ is totally dependent upon $X(k)$ and correlation is at a maximum. As k increases, the correlation between the two random variables decreases, until at some value of k, $X(n-k)$ is independent of $X(n)$, and they may be said to be uncorrelated.

The major properties of the autocorrelation function are therefore:

$$R_X(0) = E[X^2] \tag{15.12}$$

$$R_X(k) \rightarrow (E[X])^2 \qquad k \rightarrow \infty \tag{15.13}$$

$$R_X(k) = R_X(-k) \tag{15.14}$$

since $E[X(n) \cdot X(n-k)] = E[X(n) \cdot X(n+k)]$.

A slightly different form of this function is the *autocovariance*, which is

$$C_X(k) = R_X(k) - (E[X])^2 \tag{15.15}$$

and

$$C_X(0) = V[X] \tag{15.16}$$

● **Example 15.5** A certain discrete-time random signal is described by a sequence of Gaussian random variables $\{X(n)\}$. It is known to have a mean value of $2 \cdot 0$ V and mean-square-value of $8 \cdot 0$ V^2.

The autocorrelation function is outlined by the limits:

$$R_X(0) = 8 \cdot 0 \text{ V}^2$$
$$R_X(k) \rightarrow 4 \cdot 0 \text{ V}^2 \quad \text{as} \quad k \rightarrow \infty$$

If adjacent samples are independent, then $R_X(k) = 4 \cdot 0$ for all $k \geqslant 1$.

If adjacent samples are independent only for $k \geqslant 5$ say, then

$$R_X(k) = 4 \cdot 0 \text{ V}^2 \qquad k \geqslant 5$$
$$8 \cdot 0 \text{ V}^2 \geqslant |R_X(k)| \qquad 1 \leqslant k < 5$$

Also, the autocovariance $C_X(k) = R_X(k) - 4 \cdot 0 \text{ V}^2$.   ●

A signal which varies dramatically from one sample to the next will automatically have a narrow autocorrelation function, since adjacent sample values are largely independent. On the other hand, a smoothed signal, which in the limit approaches a constant value, has an autocorrelation function which is spread widely and which changes slowly as k is increased from zero.

15.2 Averaging random signals

We have established that the autocorrelation function describes the sequential properties of a random signal, so we can now use it to determine the behaviour of averaged signals. The type of signal that we are considering here is one which has an identifiable average value A, and which is contaminated by some random disturbance or fluctuation $g(n)$. We considered such signals in chapter 10, although at that stage the disturbances were modelled by sinusoidal components. Thus

$$x(n) = A + g(n) \tag{15.17}$$

The problem then is to estimate the value of A given the signal $x(n)$, that is to estimate $E[X]$. An intuitive approach is to add together subsequent samples from the random signal $x(n)$, which will enhance the mean value while the random component samples will add destructively. So we set out to evaluate this effect with more precision. The first step is to discover the result of adding together two arbitrary random signals.

15.2.1 Adding two random signals

Consider the general case of two random signals, represented by random variables X and Y. These are added together to form a third random variable Z. Using the properties of expected value, mean-square-value and variance described in chapter 14:

$$Z = X + Y \tag{15.18}$$

$$E[Z] = E[X] + E[Y] \tag{15.19}$$

$$\begin{aligned} V[Z] &= E[Z^2] - (E[Z])^2 \\ &= E[(X+Y)^2] - (E[X] + E[Y])^2 \\ &= V[X] + V[Y] + 2(E[X.Y] - E[X].E[Y]) \end{aligned} \tag{15.20}$$

Now take the particular case where X and Y are different terms in the same sequence:

$$X \equiv X(n) \quad \text{and} \quad Y \equiv X(n-k) \tag{15.21}$$

$$E[Z] = 2E[X] \tag{15.22}$$

$$\begin{aligned} V[Z] &= 2V[X] + 2C_X(k) \\ &= 2[C_X(0) + C_X(k)] \end{aligned} \tag{15.23}$$

So, when adding two random signals, the mean values add. The variances add too, when the samples are *independent*, but include the additional term of the *autocovariance* when they are not.

● **Example 15.6** As an exercise, add together pairs of samples from the sequence described in example 15.4.
Since $E[X] = 0$, the mean of the sum is also zero.
From example 15.4,

$$\{C_X(k)\} = \{2A^2/3, -A^2/3, A^2/6, -A^2/12, A^2/24, -A^2/48, ..., 0, ...\}$$

Add adjacent samples $\quad z(n) = x(n) + x(n-1)$

then $\quad V[Z] = 2.2A^2/3 - 2.A^2/3 = 2A^2/3$

Add alternate samples $\quad z(n) = x(n) + x(n-2)$

then $\quad V[X] = 2.2A^2/3 + 2.A^2/6 = 5A^2/3$

Notice that when adjacent samples are added, the variance is reduced by the negative correlation between adjacent samples. ●

15.2.2 Radar signal averaging

In section 14.4.3, we estimated the probability of detection and the false-alarm probability for radar pulses, assuming that the target returns only one pulse at a time. In most systems, several pulses are received from each target contact.

Consider a rotating antenna, which generates a radio beam of width $\theta°$ and rotates at R rpm. A point target is therefore illuminated for a time of $\theta/(6R)$ sec (see figure 15.3).

If the transmitted pulses occur at a rate r pulses per second, then a total of M pulses is received from the target as the antenna sweeps past it:

$$M = \frac{r\theta}{6R} \tag{15.24}$$

These pulses may be added together to increase the detector signal-to-noise ratio, and figure 15.3 shows schematically how this is done. Each incoming sample value from the radar receiver is added to the accumulated total of previous values at this range. Let $x(n, j)$ represent the receiver output at the nth range cell, and for the jth transmitted pulse while the target is still illuminated. The processing algorithm is then simply

$$y(n) = \sum_{j=1}^{M} x(n, j) \tag{15.25}$$

Assuming that the target strength is constant over this time, then its effective magnitude is increased by M times.

Since the interval between pulses $1/r$ sec is much greater than the reciprocal of the receiver bandwidth, then the successive noise contributions are independent. If the received pulses are coherently demodulated (section 14.4.3), the noise is Gaussian, and its variance is increased also by M times

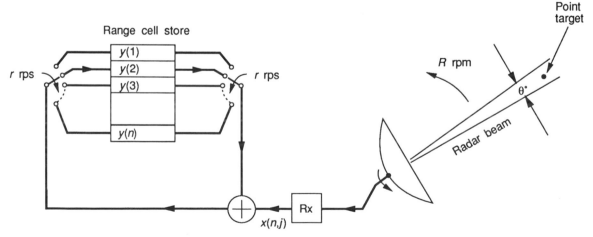

Figure 15.3 Radar rotation and averaging

due to the adding of successive pulses. The signal-to-noise ratio after adding SNR_O is related to the signal-to-noise ratio before adding SNR_I by

$$SNR_O = M \cdot SNR_I \tag{15.26}$$

Simple addition of successive pulses therefore provides a *processing gain* in the signal-to-noise ratio of the signal. An improved SNR allows much better detection to be accomplished.

● **Example 15.7**   A certain radar coherent receiver, operates at a signal-to-noise ratio of 3 dB, on a particular target. The antenna has a beamwidth of $1 \cdot 5°$ and rotates at 15 rpm. The pulse repetition rate is 1200 per second. Calculate the improvement in performance when pulses are added together.

Consider first the initial performance. Reference to figure 14.21 shows that, with an SNR of 3 dB, a detection probability of about $0 \cdot 5$ is just achievable at a false alarm probability of $0 \cdot 1$. Alternatively, a false-alarm probability of 10^{-3} is matched with a detection probability of about $0 \cdot 03$!

Applying equation 15.24 we find that $M - 20$. The improvement in SNR is therefore 13 dB, so that the processed SNR is now 16 dB. Here, a false-alarm probability of 10^{-6} may be achieved with a detection probability of $0 \cdot 95$, which is more realistic.   ●

15.3   Measurements in noise

15.3.1   The principle

There is a large class of measurement problems where the final result is a zero-frequency signal contaminated by white noise. The optimum process to separate signal from noise is then an *averager* or *smoother*. Lest it be thought that this case applies only to original signals which are zero-frequency, figure 15.4 gives the outline of a typical synchronous signal measurement.

Here we have a sinusoidal signal of frequency ω_0, which is passed through some kind of transmission medium, and emerges with scaled amplitude a. The object of the measurement is to determine the coefficient a. This is not a difficult operation *per se*, but becomes so when a is extremely small, or when there is sufficient noise within the transmission medium to mask the true signal.

Examples of such a measurement cover a wide range of applications. The network may be a filter circuit where we wish to measure its response in the *stopband*, a radio channel containing fading and interference, or a controlled mechanical system like a steerable vehicle. Materials measurements often fall within this class of problem; for instance when

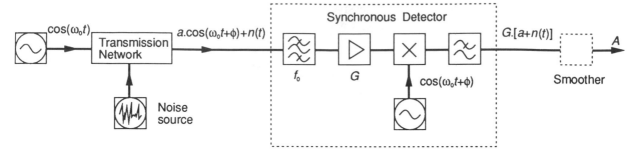

Figure 15.4 Signal transmission measurement

measuring the sound-insulating properties of a certain material, the sound pressure on the far side of the insulating material is very small but must be measured.

Measurement of small signals is inherently difficult since the high-gain amplifier required in order to observe the signals at all contributes electronic noise which may swamp the true signal. Example 14.2 shows how to calculate resistor noise, but the subsequent transistor adds to this noise power.

The synchronous detector in figure 15.4 follows normal heterodyning practice, which is more fully covered in chapter 4. It turns the input signal at frequency ω_0 into a zero-frequency signal, but with a magnitude proportional to that of the original signal.

This is a particular case of the general class of signal detectors, which are based on cross-correlation between the incoming signal and a local replica of the expected signal. A form of input signal more general than that of equation 15.17 is therefore:

$$x(n) = s(n) + g(n) \tag{15.27}$$

$s(n)$ may be a sinusoidal signal, a square wave or a pulse sequence. $g(n)$ is the general disturbance or system noise.

In order to detect the signal, we attempt to estimate the *cross-correlation* between $x(n)$ and a noise-free replica of the signal, $w(n)$. The received signal is designed to be a scaled version of the local reference, so $s(n) = a . w(n)$. The cross-correlation function is defined by

$$\begin{aligned} R_{XW}(k) &= E[X(n) . W(n-k)] \\ &= E[(S(n) + G(n)) . W(n-k)] \\ &= R_{SW}(k) + R_{GW}(k) \end{aligned} \tag{15.28}$$

Evaluating these terms, we note that

$$R_{SW}(k) = a . R_W(k) \tag{15.29}$$

and

$$R_{GW}(k) = 0 \tag{15.30}$$

since $W(n)$ and $G(n)$ are independent and $g(n)$ is zero mean.

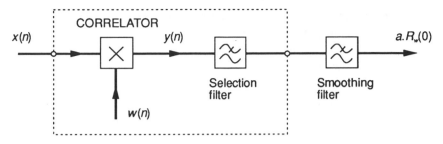

Figure 15.5 Cross-correlation detector

Now, since the local function $w(n)$ is known and chosen beforehand, it follows that its autocorrelation function is also known, and the unknown coefficient a has consequently been measured. In practice, the ensemble average implied by equation 15.28 is replaced by a time-average operation, and the system is as indicated generally in figure 15.5.

In a measuring instrument of this type, a signal $y(n)$ is generated, where

$$y(n) = w(n) \cdot s(n) + w(n) \cdot g(n) \tag{15.31}$$

The first of these terms is deterministic, while the second is random. For this measurement system to operate, the chosen signal $w(n)$ must be periodic and is therefore represented by a set of harmonically related frequencies (chapter 3 and appendix A2):

$$w(n) = \sum_{r=-M}^{M} \mathbf{C}_r \cdot \exp \mathrm{j}(\omega_0 T_s rn) \tag{15.32}$$

The product of $w(n)$ and $s(n)$ therefore generates many harmonic frequencies, including zero-frequency (chapter 4), and the selection lowpass filter is set to eliminate all of them except the zero-frequency component, which corresponds to the autocorrelation function of the signal (equation 15.29) at $k = 0$.

The product of $w(n)$ and $g(n)$ also generates a full set of harmonics, each of which carries the random signal $g(n)$, but again they are all eliminated by the selection lowpass filter except the one at zero-frequency—the bandwidth of $g(n)$ is much less than ω_0.

The output signal from this process is now in the form of equation 15.17 with which we commenced this discussion, and is processed by the smoothing filter in order to reduce the noise which masks the desired constant value.

Such signal processors are used under a variety of different names. If the signal is sinusoidal or a square-wave, then the device is often called a *phase-sensitive detector (PSD)*. If the signal is a periodic pulse sequence, called a *pseudo-random sequence (PRS)* or a *pseudo-noise sequence (PN)*, then the device is usually called a *correlator*. A *Boxcar Detector* is an adaptation of the sampling oscilloscope principle mentioned in section 6.1.2, but with additional averaging; it is capable of measuring signals which are completely immersed in noise. The *spectrum analyser* is another example, varying the local oscillator frequency in the synchronous detector of figure 15.4, to identify the frequency spectrum of the input signal. Section 16.4.3 gives a deeper analysis.

All these examples of commercial instruments employ some means of demodulation to convert the incoming noisy signal into a constant value plus noise, which must then be smoothed or averaged to reduce the noise content.

In order to apply the concepts of signal addition developed in section 15.2.1, we assume here that the output signal of the synchronous detector is sampled, and is $\{x(n)\}$. Continuous-time signals can also be dealt with by a similar argument, but the discrete-time case is the easier to comprehend.

15.3.2 Moving-average processor

All signal processing, in order to reduce the effect of noise, averages the signal samples together to enhance the signal relative to the noise. Figure 15.6 shows the signal flow-graph for a simple moving average filter, which corresponds with the discrete-time integrator or running-average process of section 10.3.3.

The input signal to the averager is $\{x(n)\}$, which has a mean value A and a noise component of variance N, so the input signal-to-noise ratio SNR_I is A^2/N.

The output signal is $\{y(n)\}$, and

$$y(n) = \frac{1}{M} \sum_{r=0}^{M-1} x(n-r) \tag{15.33}$$

Consequently

$$E[Y] = \frac{1}{M} \sum_{r=0}^{M-1} E[X]$$

$$= A \tag{15.34}$$

Assuming that the input signal samples are independent, then the output

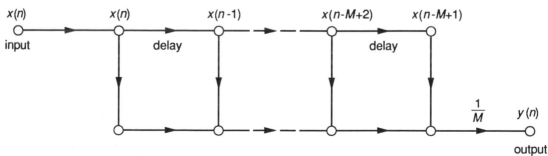

Figure 15.6 Signal flow-graph for moving-average filter

variance is just the sum of the input variances scaled by $1/M^2$:

$$V[Y] = \frac{1}{M^2} \sum_{r=0}^{M-1} V[X]$$

$$= N/M \qquad\qquad (15.35)$$

The output signal-to-noise ratio is therefore SNR_O, which is $M . A^2/N$, and the signal-to-noise ratio has improved by M times. The factor is sometimes known as the *processing gain*. While there is no theoretical limit to the degree of improvement which may be made, there remains a practical limit, since the signal must be stationary for MT_S sec while these sample values are averaged together.

If adjacent input samples are not independent, then the processing gain is reduced, and the calculation is more involved (see section 16.2.3).

A classical, though exotic, example of this technique is for the detection of weak radio signals from deep space, where averaging times of hours or even days, are employed.

● **Example 15.8**   The output signal from a certain measurement consists of a zero-frequency value together with Gaussian noise. The SNR is 1. For how long must this signal be averaged, in order to improve the SNR to 100? The sampling rate is 4 kHz.

An improvement of 100 times is required, so $M = 100$.

Since the sampling time interval is $0 \cdot 25$ ms, the averaging time interval is therefore 25 ms.   ●

15.3.3   PDF of the averaged signal

Calculation of the exact pdf of the averaged signal is complicated, but we can see the general principle by considering the addition of just two random variables, X and Y, to form a third random variable Z:

$$Z = X + Y \qquad\qquad (15.36)$$

Now the probability that the output random variable lies in the range $[z \leqslant Z \leqslant (z + \delta z)]$ is the probability that the two input variables lie in the ranges $[x \leqslant X \leqslant (x + \delta x)]$ and $[y \leqslant Y \leqslant (y + \delta y)]$ respectively. However, any given value of the output random variable Z may be generated from any one of an infinite number of joint pairs $\{X, Y\}$. Combining this information:

$$f_Z(z) . \delta z = \sum_x \sum_y f_{XY}(x, y) . \delta x . \delta y \qquad\qquad (15.37)$$

Now if the two input random variables are independent, the joint pdf may be replaced by the product of the two individual pdfs (see equation 15.7). Further, the two input variables in this problem are constrained in the sense that $y = z - x$, so that we may eliminate the variable y in the equation:

$$f_Z(z) \cdot \delta z = \sum_x f_X(x) \cdot f_Y(z - x) \cdot \delta x \cdot (\delta z - \delta x) \qquad (15.38)$$

Taking the limit, we see that this is a *convolution integral*:

$$f_Z(z) = \int_{-\infty}^{\infty} f_X(x) \cdot f_Y(z - x) \cdot dx \qquad (15.39)$$

or

$$f_Z(\ ) = f_X(\ ) * f_Y(\ ) \qquad (15.40)$$

So, when two independent random signals are added together, the pdf of the sum is the convolution of the two individual pdfs. In the case of averaging two independent samples from the same signal source, then the pdf is convolved with itself. In passing it is worth pointing out that the cumbersome convolution operation is often simplified by taking the Fourier Transform of each pdf, forming the *characteristic functions*. Convolution is then effected by multiplying together these two transforms, and then taking the inverse transform.

However, the main reason for investigating this property at all in this discussion is to show that the resulting pdf tends to a Gaussian shape. When a large number of signal samples (M) are added, as in the moving-average filter, the pdf of the output signal is the result of convolving the original pdf with itself, M times. Since the convolution of two Gaussian shapes yields a further Gaussian function, it is seen that the general trend is to generate a Gaussian pdf. This principle is formally established (although with some difficulty) in the *Central Limit Theorem*.

Figure 15.7 illustrates the Central Limit Theorem by showing the result of convolving a uniform pdf with itself 1, 2 and 3 times. The full-line curves show the result of each successive convolution, while the broken-line curves show the corresponding Gaussian function, which has the same variance as the convolved pdf. Note that when only three convolutions have taken place, the convolved pdf becomes very similar to the Gaussian pdf.

Having now established this principle, it is normally satisfactory to assume that the pdf resulting from averaging a number of independent noise samples is approximately Gaussian. Thus, for an input signal with mean value A and variance N, the residual output noise has variance N/M, and the output signal-to-noise ratio is $M \cdot A^2/N$.

Suppose that we require the measurement to be within a tolerance of $\pm \varepsilon A$ for a certain percentage of the time. We often express the result of such a measurement, as the probability that the measurement lies within this pair of limits P_M. Figure 15.8 shows the pdf of the averaged signal and these limits.

$$P_M = 1 - 2 \cdot Q[\varepsilon \sqrt{(SNR)}] \qquad (15.41)$$

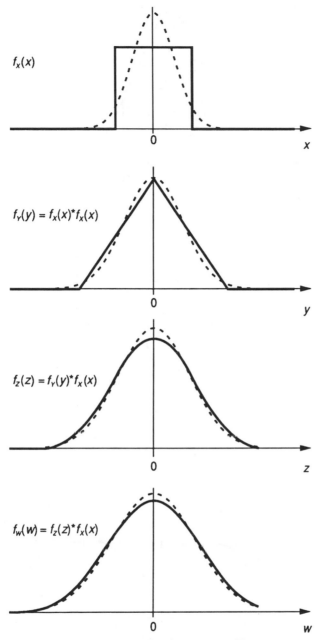

$f_x(x)$

$f_Y(y) = f_x(x)^* f_x(x)$

$f_Z(z) = f_Y(y)^* f_x(x)$

$f_W(w) = f_Z(z)^* f_x(x)$

Figure 15.7 Illustration of the Central Limit Theorem

Figure 15.9 illustrates this relationship in a graphical manner, which can be used as a quick-look chart to assess processing requirements. The chart indicates how precision (tolerance) can be traded for uncertainty (probability), at a given signal-to-noise ratio.

● **Example 15.9** Evaluate the performance of the averaging process of example 15.8.

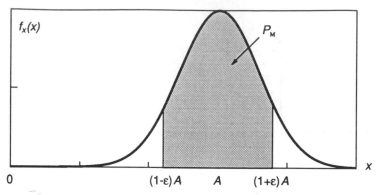

Figure 15.8 Result of an averaging process

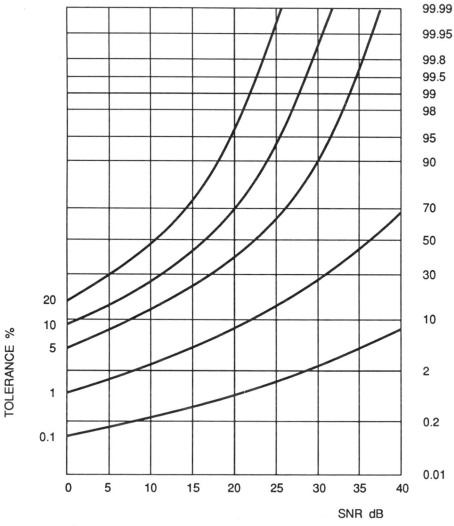

Figure 15.9 Probability that a measurement lies within a certain tolerance band

Before averaging, the SNR = 1, or 0 dB.

A measurement tolerance of $\pm 10\%$ is achieved for about 10% of the time.

After averaging, the SNR = 100 or 20 dB.

A measurement tolerance of $\pm 10\%$ is now achieved for about 70% of the time.   ●

● **Example 15.10**   A certain zero-frequency signal is expected to be of the order of 1 mV, but is contaminated with noise of power $0 \cdot 25 \ \mu V^2$. How many samples must be averaged together to give a value within 5% tolerance, with $99 \cdot 5\%$ probability?

The SNR is $1/0 \cdot 25$ or 6 dB.

Referring to figure 15.9, a 5% tolerance with $99 \cdot 5\%$ probability, requires an SNR of about 35 dB.

The processing gain is therefore to be 29 dB, or a ratio of 794. This is the number of samples which must be averaged.

Note that the number would be greater if successive signal samples were dependent or correlated.   ●

15.3.4   Recursive processor

In chapter 10 we encountered the recursive averager, as well as the straightforward integrator or moving-average processor. Figure 15.10 shows the signal flow-graph for such an averager, whose difference equation is

$$y(n) = (1 - \alpha) \cdot x(n) + \alpha \cdot y(n - 1) \tag{15.42}$$

The process is recursive because each output sample value $y(n)$ depends upon the previous output sample $y(n - 1)$, as well as on the current input sample $x(n)$. Remember that for the output to be a stable or converging sequence, the coefficient must be constrained by $|\alpha| < 1$.

Even if the input signal sequence has no dependence between adjacent sample values, the output sequence will clearly show correlation between

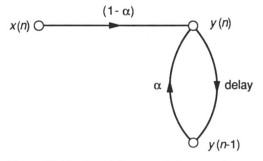

Figure 15.10 Signal flow-graph for recursive averager

adjacent samples. In other words, the autocorrelation function will be broadened, corresponding to a signal which fluctuates less than the input signal. We now wish to calculate the processing gain of this process.

Consider first the mean values of input and output, by taking the expectation of variables in equation 15.42:

$$E[Y(n)] = (1 - \alpha) . E[X(n)] + \alpha . E[Y(n-1)] \qquad (15.43)$$

Hence $\qquad E[Y] = E[X] \qquad (15.44)$

The zero-frequency gain of this process is therefore unity.

In order to find the output variance with a random input signal, we first find the output autocorrelation function. By definition,

$$\begin{aligned}
R_Y(k) &= E[Y(n) . Y(n-k)] \\
&= E[\{(1-\alpha) . X(n) + \alpha . Y(n-1)\} . Y(n-k)] \\
&= (1-\alpha) . R_{XY}(k) + \alpha . R_Y(k-1) \qquad (15.45)
\end{aligned}$$

Now $R_{XY}(k)$ is a *cross-correlation function*, and describes the dependence of $Y(n-k)$ on $X(n)$. Since $Y(n-k)$ occurs *before* $X(n)$, there clearly can be no dependence at all, if adjacent input samples are independent, so

$$R_{XY}(k) = E[X] . E[Y] \qquad k \neq 0 \qquad (15.46)$$

Note that this restricted interpretation of the cross-correlation function is relevant only to this particular type of problem; the cross-correlation function is a valuable random-signal system descriptor in its own right (section 16.2.4).

So, returning to equation 15.45, we can also express the principle in terms of variances:

$$C_Y(k) = \alpha . C_Y(k-1) \qquad (15.47)$$

We have therefore a recursive equation for successive values in the output signal autocovariance function. If the signal variance $C_Y(0)$ is known, then the subsequent values follow from it. So

$$C_Y(k) = \alpha^k . C_Y(0) \qquad (15.48)$$

$$R_Y(0) = E[\{(1-\alpha) . X(n) + \alpha . Y(n-1)\}^2] \qquad (15.49)$$
$$= (1-\alpha)^2 . R_X(0) + \alpha^2 . R_Y(0) + 2\alpha(1-\alpha) . R_{XY}(1)$$

We have already discussed the cross-correlation function (equation 15.46), and since $E[X] = E[Y] = 0$ for a noise signal, we can simplify this equation too:

$$C_Y(0) = \frac{1-\alpha}{1+\alpha} C_X(0) \qquad (15.50)$$

The input signal-to-noise ratio SNR_I is A^2/N, while the output SNR_O is given by

$$\text{SNR}_O = \frac{A^2(1 + \alpha)}{N(1 - \alpha)} \tag{15.51}$$

The processing gain is therefore

$$\frac{\text{SNR}_O}{\text{SNR}_I} = \frac{1 + \alpha}{1 - \alpha} \tag{15.52}$$

If $\alpha < 1$, then the processing gain is greater than unity, and can be very large if $\alpha \rightarrow 1$. For reference, the SNR improvement is plotted against values of the coefficient, in figure 15.11. Thus, for $\alpha = 0 \cdot 999$, the processing gain or signal-to-noise ratio improvement is 35 dB.

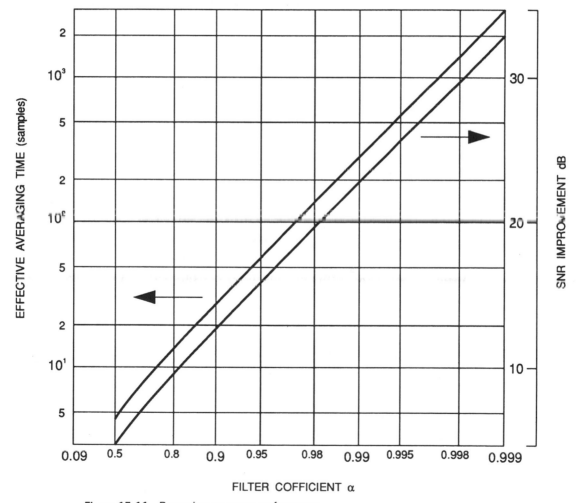

Figure 15.11 Recursive averager performance

There are two penalties to be paid for this performance:

a) As $\alpha \to 1$, so the arithmetic precision of the digital filter must be increased, leading to longer wordlengths in order to represent the coefficient to sufficient decimal places.

b) The input process must be stationary over a long time interval.

In order to estimate the second of these effects, we note that the autocorrelation function of the output noise is exponential in form. The correlation between two sample values which are k intervals apart is proportional to α^k. So if we take the point $k = m$, where the correlation has dropped to say 5% of its initial value, then we have an estimate of how many signal sample values are effectively included in the averaging operation. Since $\alpha^m = 0 \cdot 05$, then

$$m = -3 \cdot 0 / \ln(\alpha) \qquad (15.53)$$

This approximate measure is also included in the graph of Figure 15.11.

● **Example 15.11** Compare the performance of the moving-average process of example 15.9, with the recursive averager.

The desired processing gain is 100, or 20 dB.

This corresponds, from figure 15.11, to $\alpha = 0 \cdot 98$, and an average of effectively 150 samples. The moving-average process uses only 100 samples. ●

The recursive averager therefore uses more signal samples to carry out the average to a certain degree of noise reduction, but of course the samples in the distant past are weighted less than those in the recent past (see section 10.3.2). The major attraction about the recursive averager is that it is a simpler algorithm to implement than the moving-average process. It also lends itself neatly to cases where the statistics of the signal are changing slowly with time, because of its exponential weighting of signal values in the distant past. Both processes are used in commercial instruments.

STUDY QUESTIONS

1 Determine the autocorrelation function of a square-wave signal, which is sampled at intervals T_S. The signal amplitude is $\pm A$ and the period is $2N \cdot T_S$.

2 A certain three-level signal has allowed amplitudes of $\{0, A, 2A\}$, and successive sample values are independent. Determine the joint probability for each of the possible pairs of adjacent signal values.

$$P[X = 0] = 0 \cdot 25 \qquad P[X = A] = 0 \cdot 5 \qquad P[X = 2A] = 0 \cdot 25$$

Note that the transitional probabilities are equal to $\{P[Y]\}$.

3 A certain three-level signal has allowed amplitudes of $\{0, A, 2A\}$, but is now constrained so that no two successive signal values can be the same. Draw out the first few tables of conditional probabilities for pairs of values separated by k intervals, and hence calculate the autocorrelation function of this signal.

$$P[X = 0] = \tfrac{1}{3} \qquad P[X = A] = \tfrac{1}{3} \qquad P[X = 2A] = \tfrac{1}{3}$$

4 The autocorrelation function of a certain discrete-time signal is $R_X(k)$ and

$$R_X(k) = 5(1 + 2 \cdot \exp - |k|) \, V^2$$

What is the mean power of this signal, and what is its mean value?

5 Considering the signal in question 4, what separation between sample values will give an autocovariance less than $0 \cdot 2 \, V^2$?

6 Calculate the mean and variance of the signal formed by adding adjacent samples of the signal in question 4.

7 The random signal of question 4 is now added to an independent Gaussian signal having a mean value of 2 V and a noise variance of 4 V^2. What is the mean and the variance of the result?

8 The antenna of a certain radar system has a beam width of 2° and rotates at 10 rpm. Determine the pulse-repetition-frequency which will improve detection by $14 \cdot 77$ dB.

9 A sampled thermocouple measurement yields a zero-frequency signal and Gaussian noise, with an SNR of 7 dB. Determine the number of stages in a moving-average processor which will increase the SNR to better than 15 dB. What is the probability that this measurement will be in error by more than 20%?

10 A certain zero-frequency signal which is contaminated with Gaussian noise is to be measured with an error of less than 1% for 99% of the time. The signal value is about 10 mV; what value of noise can be tolerated if a moving-averager with 1000 stages has been used?

11 Given the same conditions as equation 10, what is the value of the recursive-averager coefficient which will produce a similar result?

16 Random Signals: Power Spectral Density and Linear Filtering

OBJECTIVES

To introduce power spectral density as a measure of the frequency content of a random signal.

To show how linear filtering operations may be analysed using power spectral density, autocorrelation function and crosscorrelation function.

COVERAGE

Commencing with a definition of power spectral density, its properties are explored and then applied to linear systems. Considering the power spectral density of the signal at the output of a linear system leads to a definition of *equivalent noise bandwidth*. A moving-average (FIR) linear filter is then analysed for non-white input noise, using both power spectral density and autocorrelation function.

Discussion of cross properties between input and output signals leads to a technique for measuring the system unit-pulse response using a PRBS.

Particular examples of using autocorrelation function and power spectral density include a further discussion of the averaging process, sampling and quantisation noise, data transmission, matched filtering, and measurement of the power spectral density itself.

Analysing random signals can be a difficult exercise, replete with unwieldy algebra. We have seen in earlier chapters, however, that measurements of amplitude variation can be accomplished with reasonable effort, and that even the autocorrelation function allows further insight into time variations of random signals without too much trouble.

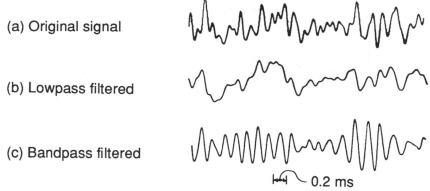

(a) Original signal

(b) Lowpass filtered

(c) Bandpass filtered

0.2 ms

Figure 16.1 Filtering a random signal

We now wish to delve deeper into the influence of linear filters on random signals, which will involve a frequency description appropriate to random signals, and so we are entering a more complicated arena. With deterministic signals, frequency responses and spectra involve integrals and summations of cosinusoidal functions, and these are now to be combined with the properties of randomness. Our introduction to this topic will stick to fairly simple and basic problems, but as we move towards more practical cases the algebra quickly becomes insoluble and practical system design is often done by numerical calculation and simulation.

We aim to emphasise the general principles and to give a flavour of the topic. However, the results can be applied immediately to practical measurements on random signals.

Fortunately, in most situations involving random signals, we require only a calculation of the *signal-to-noise ratio (SNR)* after processing, rather than a complete description of the output signal power spectral density. The SNR often can be calculated quite simply.

Figure 16.1 shows the result of filtering a specific random signal. Lowpass filtering [signal(b)] causes the short-term fluctuations of the signal to be slower, as one might expect. Bandpass filtering [signal(c)] causes the signal to take on the broad characteristic of the bandpass filter frequency, since other frequency components are reduced in magnitude. Filtering a random signal therefore has a similar qualitative effect to filtering a deterministic signal, so presumably a feasible frequency description of a random signal exists.

16.1 Power spectral density

The power spectral density is a frequency description of a random signal or process, and we define it before exploring its properties.

Since we can talk about random signals only in the context of *average* properties, any spectral description must be an averaged measure. We already have the autocorrelation function which is an *expected* or time-averaged function, so we will deduce a frequency description starting from this point.

Recall that the *Discrete-Fourier-Transform (DFT)* $\mathbf{X}(\omega)$ is a measure of the frequency properties of a discrete-time signal $x(n)$ (see sections 7.2.3 and 13.1.2):

$$\mathbf{X}(\omega) = \sum_{n=-\infty}^{\infty} x(n) . \exp j(-\omega T_s n) \qquad (16.1)$$

The autocorrelation function $R_X(k)$ is used to define the power spectral density $S_X(\omega)$, via the DFT:

$$S_X(\omega) = \sum_{k=-\infty}^{\infty} R_X(k) . \exp j(-\omega T_s k) \qquad (16.2)$$

Since $R_X(k)$ is even and real, this expression simplifies to

$$S_X(\omega) = R_X(0) + 2 . \sum_{k=1}^{\infty} R_X(k) . \cos(\omega T_s k) \qquad (16.3)$$

From this equation we draw the following conclusions:

a) $S_X(\omega)$ is also real.
b) $S_X(\omega)$ is periodic in ω, with period ω_S; as befits the spectral description of any discrete-time signal (section 6.3.1).
c) $S_X(\omega)$ has the units of *power*, since the autocorrelation function also has units of power.

So we see that $S_X(\omega)$ merits the description *power spectrum*, but why use the term *density*? Since $S_X(\omega)$ is the DFT of the autocorrelation function $R_X(k)$, it follows that the inverse DFT yields $R_X(k)$ from a knowledge of the spectrum $S_X(\omega)$, as in equation 13.7. Hence, setting $k = 0$:

$$E[X^2] = R_X(0) = \frac{1}{2\pi} . \int_{-\pi}^{\pi} S_X(\omega) . d(\omega T_S) \qquad (16.4)$$

The integral of $S_X(\omega)$ therefore gives the total power in the random signal, which justifies the term *density* as applied to this function. We shall hence refer to the quantity $S_X(\omega)$, as defined in equation 16.2 or 16.4, as the *Power Spectral Density*.

A couple of examples will illustrate how the function behaves.

● **Example 16.1** The classical *noise* signal consists of a sequence of samples which are independent of one another.

The autocorrelation function for a noise signal of power N is therefore

$$R_X(k) = \{N, 0, 0, 0, 0, ...\} \qquad k \geqslant 0$$

Using equation 16.3 $\qquad S_X(\omega) = N$

The power spectral density is therefore constant with frequency, and such a signal is known as *white noise*, since all spectral components have equal weighting, like white light. ●

● **Example 16.2** Consider a noise signal whose adjacent samples are dependent in some way, and whose autocorrelation function is given by

$$R_X(k) = \{N, N/2, 0, 0, 0, ...\} \qquad k \geqslant 0$$

Applying equation 16.3 again, we see that

$$S_X(\omega) = N \cdot [1 + \cos(\omega T_S)]$$

Notice that dependence between adjacent samples has changed the power spectral density considerably, and has concentrated power in the region near to zero frequency—see figure 16.2, which shows the power spectral density under its corresponding autocorrelation function.

Changing the autocorrelation function so that adjacent samples have negative correlation, alters the form of the power spectral density again. Thus, suppose that

$$R_X(k) = \{N, - N/2, 0, 0, 0, ...\} \qquad k \geqslant 0$$

$$S_X(\omega) = N \cdot [1 - \cos(\omega T_S)]$$

Power is now concentrated in the region near to the Nyquist frequency. These functions are also illustrated in figure 16.2, and show that as the autocorrelation function is widened from a single non-zero value, so the power spectral density is restricted in frequency. There is evidence here of the *reciprocal spreading* principle.

Figure 16.2 also shows typical signal sequences for these three cases. Notice that positive correlation gives a small degree of smoothing, while negative correlation tends to make the sequence values alternate in sign. These features correspond to the general spectral properties of lowpass and highpass signals. ●

Before we leave this definition phase, consider the case of a signal with a mean value A. Then we may express the autocorrelation function in terms of the *autocovariance function* $C_X(k)$:

$$R_X(k) = C_X(k) + A^2 \tag{16.5}$$

Consequently, from equation 16.3, the power spectral density is

$$S_X(\omega) = C_X(0) + 2 \cdot \sum_{k=1}^{\infty} C_X(k) \cdot \cos(\omega T_S k) + A^2 \cdot \delta(\omega) \tag{16.6}$$

where $\delta(\omega)$ is the impulse function.

Thus, we may always treat the mean value of a random signal separately from the fluctuation part of the signal, using the autocovariance function

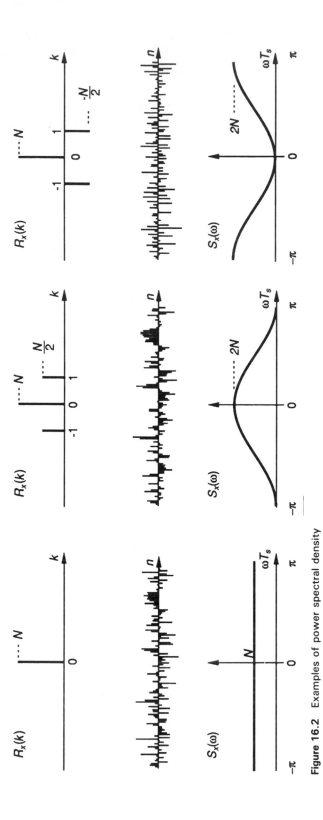

Figure 16.2 Examples of power spectral density

rather than the autocorrelation function. Having said that, in the majority of cases, random signals have zero mean anyway, so both autocorrelation function and autocovariance function amount to the same thing.

16.2   Linear system relationships

16.2.1   Input and output signals

The signal spectral density $S_X(\omega)$ evidently gives the power around the frequency ω for a random signal, and so is useful for showing where filtering must be carried out. The next question is how to calculate the corresponding spectral function at the output of a linear filter which has a frequency response $\mathbf{H}(\omega)$?

We note first that although $S_X(\omega)$ is a real function of ω, the frequency response $\mathbf{H}(\omega)$ is a complex function of ω, and includes phase. Recall that the frequency response $\mathbf{H}(\omega)$ is defined as the response of the filter to a phasor input (section 3.3). Thus

Input	$x(n) = \exp\ \mathrm{j}(\omega T_S n)$	(16.7a)
Output	$y(n) = \mathbf{H}(\omega).\exp\ \mathrm{j}(\omega T_S n)$	(16.7b)

So $\mathbf{H}(\omega)$ defines the complex output/input signal ratio for a phasor with an angular frequency of ω. When considering a random signal input, the only function we have which relates in any way to the frequency variable ω is the power spectral density. How then does $\mathbf{H}(\omega)$ relate to the *output-power/input-power* ratio?

Since $|\mathbf{H}(\omega)|$ is the ratio of the rms values of output and input phasors, their powers will be related by $|\mathbf{H}(\omega)|^2$. This property is also demonstrated by considering the nature of the power spectral density in more detail.

First we isolate a small region of the frequency axis, $\delta\omega$, at a frequency ω. The elementary input signal defined by this region has a power of $S_X(\omega).\delta\omega$, and if $\delta\omega$ is small, corresponds to a phasor of frequency ω and expected rms magnitude of $\sqrt{[S_X(\omega).\delta\omega]}$. After being processed by the filter, the expected rms magnitude becomes $|\mathbf{H}(\omega)|.\sqrt{[S_X(\omega).\delta\omega]}$, which corresponds to $\sqrt{[S_Y(\omega).\delta\omega]}$.

Equating these two interpretations, we may write

$$S_Y(\omega) = S_X(\omega).|\mathbf{H}(\omega)|^2 \tag{16.8}$$

The output-signal power spectral density is therefore simply linked with the input-signal power spectral density, and in a fairly obvious manner.

● **Example 16.3**   Consider a simple digital filter, which has the unit-pulse response $g(k)$:

$$\{g(k)\} = \{1, 2, 1, 0, 0, 0, ...\} \qquad k \geqslant 0$$

Then the frequency response $\mathbf{H}(\omega)$, is found by using the DFT:

$$\mathbf{H}(\omega) = \sum_{k=0}^{\infty} g(k) \cdot \exp \mathrm{j}(-\omega T_S k)$$

$$= 2[1 + \cos(\omega T_S)] \cdot \exp \mathrm{j}(-\omega T_S)$$

If the input signal is random white noise, then $S_X(\omega) = N$.

$$S_Y(\omega) = 4N \cdot [1 + \cos(\omega T_S)]^2$$

$$= 2N \cdot [3 + 4 \cdot \cos(\omega T_S) + \cos(2\omega T_S)] \qquad \bullet$$

We may also use equation 16.8 to deduce a complementary relationship for the time-domain, by taking the Inverse-DFT. Remembering that

$$|\mathbf{H}(\omega)|^2 = \mathbf{H}(\omega) \cdot \mathbf{H}^*(\omega) \tag{16.9}$$

we may write in shorthand form:

$$R_Y(k) = R_X(k) * g(k) * g(-k) \tag{16.10}$$

where $*$ denotes convolution and $g(k)$ is the unit-pulse response.

This general relationship is helpful in establishing conceptual relationships, but often is difficult to apply because of the multiple convolutions, although they can be carried out numerically. So we shall now investigate a simpler way of calculating noise powers by using an *equivalent bandwidth*, and then examine one case in more detail, by applying these ideas to the moving-average filter.

16.2.2 Equivalent noise bandwidth

When the input signal is white, or at least has a constant power spectral density N over the *bandwidth* of the signal process, we may combine equations 16.4 and 16.8 to calculate the output-signal power $E[Y^2]$:

$$E[Y^2] = \frac{N}{2\pi} \cdot \int_{-\pi}^{\pi} |\mathbf{H}(\omega)|^2 \cdot \mathrm{d}(\omega T_S) \tag{16.11}$$

This calculation might be simplified by referring to the *equivalent noise bandwidth* ω_N, the principle of which is illustrated in figure 16.3. Assume for simplicity that the filter or system has unity gain at zero frequency. Then the *equivalent noise bandwidth* ω_N defines a response of constant magnitude over the frequency range of $\pm \omega_N$, which, when the input signal is white noise, gives the same output-signal power as the system response $\mathbf{H}(\omega)$.

Thus using equation 16.11 for this equivalent filter, we may also express the output-signal noise power as

$$E[Y^2] = \frac{N \cdot 2\omega_N T_S}{2\pi}$$

$$= 2 \cdot N f_N T_S \tag{16.12}$$

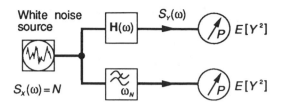

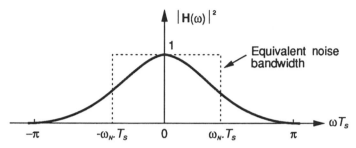

Figure 16.3 Equivalent noise bandwidth

Combining this statement with equation 16.11, we have a definition for ω_N:

$$\omega_N = \frac{\omega_S}{4\pi} \cdot \int_{-\pi}^{\pi} |\mathbf{H}(\omega)|^2 \cdot d(\omega T_S) \tag{16.13}$$

So, if the equivalent noise bandwidth is known for a certain filter, then the output noise power can easily be obtained. Averaging operations which were discussed in chapter 15 can now be understood as restricting the bandwidth of the input noise, and so enhancing the SNR by bandwidth reduction.

As an example of such a calculation, consider the recursive averager of section 15.3.4, for which the integral of equation 16.13 has a simple solution:

$$y(n) = (1 - \alpha) \cdot x(n) + \alpha \cdot y(n - 1) \tag{16.14}$$

We analyse this difference equation to find the frequency response $\mathbf{H}(\omega)$, and then evaluate equation 16.13 to find the equivalent noise bandwidth ω_N:

$$\omega_N = \frac{\omega_S(1 - \alpha)}{2(1 + \alpha)} \tag{16.15}$$

Note that ω_N is not the 3 dB bandwidth of this filter; compare this equation with equation 9.14 in example 9.5. However, for high-order filters the two bandwidths are generally similar.

● **Example 16.4** A certain measured signal is sampled at 2 kHz. The signal is at zero frequency, but is in the presence of random noise which is bandlimited to $\pm 1 \cdot 0$ kHz. The signal-to-noise power ratio is thought to be 7 or 5 dB.

Now a linear filter is to be used in order to enhance the SNR to 30 dB. Let it have unity gain at zero frequency.

The ratio of improvement is to be 25 dB, or a ratio of 316.

The noise must be reduced by a factor of 316.

From equation 16.12, the required equivalent noise bandwidth is therefore $2000/(316 \times 2)$ or $3 \cdot 2$ Hz.

Solving equation 16.15, we deduce that the recursive averager coefficient α is to be $0 \cdot 9936$.

Notice that this result is exactly the same as that which is shown by the graph in figure 15.11, and deduced by the autocorrelation function argument in section 15.3.4. We are merely looking at the problem from the frequency domain point of view instead of the time domain, and sometimes this is more convenient to do.   ●

An interesting and useful by-product of the *equivalent noise bandwidth* concept is that the effective sampling rate of such a signal is approximately $2f_N$ Hz, where $f_N = \omega_N/2\pi$. Although the signal may be sampled at a greater rate than this, $2f_N$ Hz represents the rate at which *independent* samples occur.

16.2.3   A moving-average filter

A general moving-average filter with M coefficients is described by the difference equation:

$$y(n) = \sum_{r=0}^{M-1} a(r) \cdot x(n-r) \tag{16.16}$$

As well as using this difference equation to calculate the frequency response of the filter and thence to find the output-signal power spectral density, we can also use it to calculate the output-signal autocorrelation function. Thus

$$R_Y(k) = E[Y(n) \cdot Y(n-k)] \tag{16.17}$$

$$= E\left[\sum_{r=0}^{M-1} a(r) \cdot X(n-r) \cdot \sum_{s=0}^{M-1} a(s) \cdot X(n-s-k) \right]$$

$$= \sum_{r=0}^{M-1} \sum_{s=0}^{M-1} a(r) \cdot a(s) \cdot R_X(r-s-k) \tag{16.18}$$

Notice that this equation is really a more detailed version of equation 16.10. However it is not in the best form, since for a given delay k, terms having the same value of $(r-s)$ must be found and combined. In fact, r and s are not independent variables. So, let $(r-s) = n$, whence $s = (r-n)$.

Now in order to see how these summations go, look at figure 16.4. Here, the two axes represent the independent indices r, s. At a certain positive

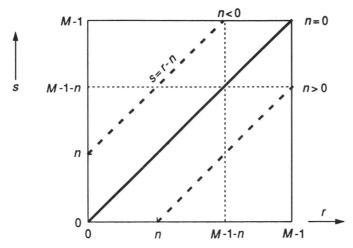

Figure 16.4 Range of indices for summation

value of n, s can run only over the range $0 \rightarrow (M - 1 - n)$ before the coefficients $a(r)$ are exhausted. Alternatively, for a negative value of n, s is restricted to the range $|n| \rightarrow (M - 1)$.

Then the equation may be re-written:

$$R_Y(k) = \sum_{s=0}^{M-1} a^2(s) . R_X(k) + \sum_{n=1}^{M-1} \sum_{s=0}^{M-1-n} a(s+n) . a(s) . R_X(n-k)$$

$$+ \sum_{n=-1}^{-(M-1)} \sum_{s=|n|}^{M-1} a(s+n) . a(s) . R_X(n-k) \tag{16.19}$$

While this equation is still difficult to understand, it does enable the output-signal autocorrelation function of the filter to be calculated from a knowledge of the input-signal autocorrelation function. This is an important advance over previous discussions, where white noise has been the only type of input signal that we have been able to accommodate.

Evaluating the double-summation by hand is cumbersome, although it is ideally suited to machine computation. However we can use the principle in order to discover the output-noise power when averaging a non-white noise signal, which is a much simpler operation.

An M-point averager has all the coefficients equal, so that referring to equation 16.16:

$$a(r) = 1/M \qquad 0 \leqslant r \leqslant (M-1) \tag{16.20}$$

We could now calculate the output-signal autocorrelation function for a moving-average filter, using equation 16.19. More to the point, we can calculate the output-signal noise power when the input-signal is not white noise, by setting $k = 0$. Noting that there are $2(M - n)$ terms to be added

355

for each value of n,

$$E[Y^2] = R_Y(0) = \frac{1}{M} \cdot R_X(0) + \frac{2}{M} \sum_{n=1}^{M-1} \left(1 - \frac{n}{M}\right) \cdot R_X(n) \qquad (16.21)$$

The first term on the right-hand side of the equation represents the output-signal power if the input signal is white noise, that is with an autocorrelation function $R_X(k)$ which is non-zero only at $k = 0$. The summation represents an *additional* term, which makes the averaging process less effective.

This result should be compared with the simplified discussion in section 15.2.1, which gave birth to equation 15.23 for a white-noise input.

When the number of averager points M is large and $R_X(k)$ has vanished well before $k = M$, the expression simplifies again to

$$E[Y^2] = R_Y(0) \simeq \frac{1}{M} \cdot R_X(0) + \frac{2}{M} \sum_{n=1}^{M-1} R_X(n) \qquad (16.22)$$

● **Example 16.5** Consider an averager with $M = 8$, and let the input signal be the one whose autocorrelation function is

$$\{R_X(k)\} = \{4, 3, 2, 1, 0, 0, 0 \dots\} \qquad k \geqslant 0$$

The input-signal power is $R_X(0)$ which is 4 V^2.

If the signal was white noise, then the output-signal power would be $R_X(0)/M$ or 0·5 V^2.

However, the input-signal is not white, and equation 16.21 must be evaluated in order to calculate the output-signal power. The additional term due to the summation is 1·19, giving a total power of 1·69 V^2.

This numerical case emphasises the necessity of finding out whether the input-signal to an averager has independent samples or not. If some correlation exists between adjacent sample values, then the averaging effect of the filter is reduced. In this case it would be necessary to extend the averager to something like 30 points, in order to achieve an output signal power approaching 0·5 V^2.

Figure 16.5a illustrates these relationships, showing the input-signal autocorrelation function $R_X(k)$ and the output-signal autocorrelation function $R_Y(k)$. The smoothing action of the averager is clearly seen.

A graph is also shown in figure 16.5b, which gives the output-signal power versus the averager parameter M, and shows how the approximate relationship of equation 16.22 becomes valid in this case, for $M > 20$. ●

We can also interpret this operation by using the power spectral density of the signal. Compare the output-signal power expression of equation 16.22 with the expression for $S_X(\omega)$ in equation 16.3. Then we see that

$$E[Y^2] = R_Y(0) \simeq S_X(0)/M \qquad (16.23)$$

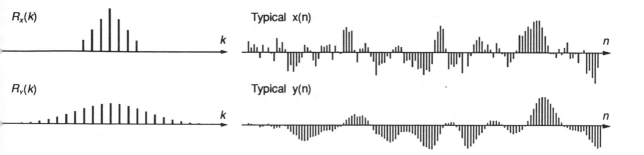

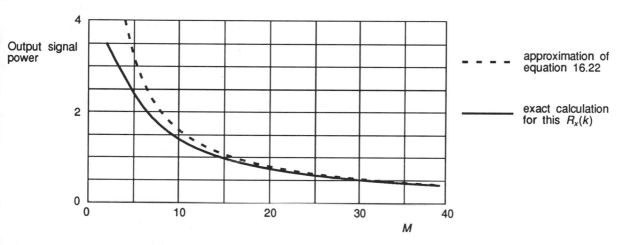

(a) **Autocorrelation functions and typical signals for $M = 8$**

Output signal power

- - - - approximation of equation 16.22

——— exact calculation for this $R_x(k)$

(b) **Output power versus M for a given input-signal autocorrelation function**

Figure 16.5 Averaging a non-white noise signal

In many cases, particularly those where we are concerned with the output-noise power of the averager, and the input-noise has been obtained by filtering white noise, then $S_X(0)$ can easily be obtained. So, if the input-noise has an equivalent noise bandwidth of ω_N, the effective input-noise power is

$$E[X^2] \simeq 2f_N T_S . S_X(0) \tag{16.24}$$

If the noise power is to be reduced by a factor P, then

$$P = 2f_N T_S . S_X(0) . \frac{M}{S_X(0)}$$

$$= 2 . f_N T_S . M \quad \text{or} \quad 2Mf_N/f_S \tag{16.25}$$

For white-noise, the output-power from an averager would be reduced by M times, so this process is less effective by a factor $2f_N/f_S$, which may be explained by an argument based on the sampling theorem.

A noise source with an equivalent noise bandwidth of f_N Hz defines via the sampling theorem, a sequence which has *independent* values at intervals of $1/2f_N$ sec, even though the sampling rate f_s is higher than this. The effective sampling rate, for independent samples, is therefore reduced by $2f_N/f_s$.

Equation 16.25 evaluates the number of moving-averager stages M which will reduce the noise power by a factor P. A recursive averager on the other hand has an equivalent noise bandwidth described by equation 16.15, and for the non-white noise case, using a similar argument, the factor P is given approximately by $2f_N(1 - \alpha)/f_s(1 + \alpha)$. The coefficient α is therefore

$$\alpha = \frac{M - 1}{M + 1} \qquad\qquad (16.26)$$

● **Example 16.6** Let us re-examine the averaging of the non-white signal in example 16.5, using the power spectral density approach. Then the input signal has an autocorrelation function:

$$\{R_X(k)\} = \{4, 3, 2, 1, 0, 0, 0 \ldots\} \qquad k \geqslant 0$$

The autocorrelation function falls to zero when $k = 4$.

Applying equation 16.3, the power spectral density is

$$S_X(\omega) = 4 + 6 . \cos(\omega T_S) + 4 . \cos(2\omega T_S) + 2 . \cos(3\omega T_S)$$

and $\quad S_X(0) = 16 \text{ V}^2$

In example 16.5, averaging with $M = 8$ was calculated. Applying equation 16.23 we obtain a figure for the output-signal power of $2 \cdot 0 \text{ V}^2$, compared with the exact figure of $1 \cdot 69 \text{ V}^2$ calculated from the autocorrelation function and equation 16.21.

However, we note that $R_X(k)$ falls to zero only at $k = 4$, which is not sufficiently removed from $k = M$, so the approximation implied in equation 16.22 is not a good one.

Setting $M = 20$ say, gives an output-signal power of $0 \cdot 8 \text{ V}^2$ which is close to the calculated value shown in figure 16.5b.   ●

Figure 16.6 gives a more general view of this technique for calculating output power. The approximation is seen to be valid when the bandwidth of the noise-reducing filter is much less than the bandwidth of the source. In the case of the moving-averager, the equivalent noise bandwidth is $1/(2T_S M)$ Hz.

● **Example 16.7** As a further illustration of the use of power spectral density in calculating the output-signal from a smoothing filter, consider two sources:

Source A is white noise, power 4 V^2 and

$$\{R_N(k)\} = \{4, 0, 0, \ldots\} \qquad k \geqslant 0$$

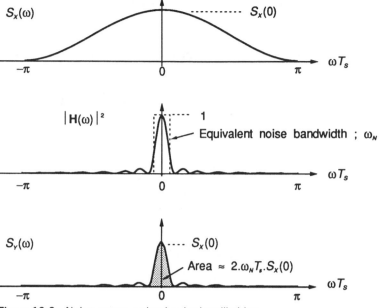

Figure 16.6 Noise power reduction by bandlimiting

Source B is non-white noise power 4 V^2 and

$$\{R_N(k)\} = \{4, 3, 2, 1, 0, 0, ...\} \qquad k \geqslant 0$$

Determine the number of stages M in a moving-averager, which will reduce the output-noise power to $0 \cdot 1$ V^2.

For source A, which is white noise, the output-noise power is $R_N(0)/M$ and so $M = 40$.

For source B, which is non-white noise, the output noise power is approx $S_N(0)/M$, and hence using the value of $S_N(0)$ from example 16.6, $M = 160$. Notice that the autocorrelation function drops to zero when $k = 4$, which is much less than the value of M, so this estimate should be accurate.

Using equation 16.26 we may also calculate the recursive-averager coefficient that is required in the two cases:

Source A: $\alpha = 0 \cdot 975$ Source B: $\alpha = 0 \cdot 99375$ ●

16.2.4 Crosscorrelation and the PRBS

Some interesting system properties can be found by comparing output and input signals. Consider the system shown in figure 16.7. The system is described in terms of its unit-pulse response $g(n)$, or its frequency response $\mathbf{H}(\omega)$. Thus

$$y(n) = \sum_{r=0}^{M-1} g(r) \cdot x(n-r) \qquad (16.27)$$

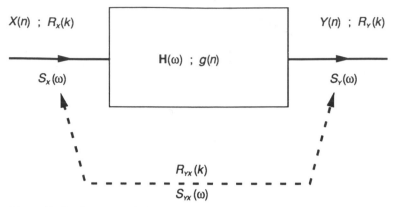

Figure 16.7   Cross-correlating signals in a system

The output and input signals can be related by the *crosscorrelation function*:

$$R_{YX}(k) = E[Y(n) . X(n - k)] \tag{16.28}$$

$$= E\left[\sum_{r=0}^{M-1} g(r) . X(n - r) . X(n - k)\right]$$

Therefore

$$R_{YX}(k) = \sum_{r=0}^{M-1} g(r) . R_X(k - r) \tag{16.29}$$

or $R_{YX}(k) = g(k) * R_X(k)$ $\tag{16.30}$

Thus, the crosscorrelation function between output-signal and input-signal is the convolution of the unit-pulse response of the system and the autocorrelation function of the input signal. Since the crosscorrelation function can easily be measured, the unit-pulse response of a system can be determined from it, provided that the autocorrelation function of the input signal is like a unit-pulse.

Recall from example 16.1 that the autocorrelation function of white noise is a unit-pulse, so white noise can be used as a test signal in order to determine the unit-pulse response of a system.

This kind of testing can also be viewed from the frequency domain, by taking the DFT of equation 16.29, whence

$$\mathbf{S}_{YX}(\omega) = S_X(\omega) . \mathbf{H}(\omega) \tag{16.31}$$

Notice that the *cross-spectrum* $\mathbf{S}_{YX}(\omega)$ is complex, whereas a power spectral density is real. Observe too that if the input-signal is white noise, then $S_X(\omega) = N$, and $\mathbf{S}_{YX}(\omega) = N . \mathbf{H}(\omega)$.

This technique is important in those situations where the unit-pulse response of a given system is extremely small in magnitude, and where a large magnitude pulse drive cannot be applied. Frequently it is necessary to measure the characteristics of a system while it is being controlled, for

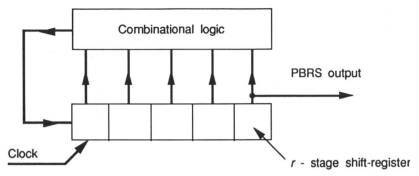

Figure 16.8 PRBS generator

instance a chemical plant, a flying vehicle like an aeroplane or a rocket; and a noise test can be carried out at low level while the overall system is still functioning in the usual way.

Instead of a pure white noise source, a binary signal sequence is commonly used. This is the *pseudo random binary sequence (PRBS)*, and it is generated from a simple shift-register logic circuit as shown in figure 16.8.

A shift-register is a serial connection of binary storage elements. When all stages are clocked simultaneously, the stored pattern is shifted one place to the right, and a new binary digit is input to the extreme lefthand stage. The pattern held by the register is known as its *state*.

The PRBS generator generates the next input digit from a linear calculation based on the present state of the shift-register. Linear operations on binary digits are carried out by *exclusive-OR* gates.

Let this register have r stages, which therefore describe 2^r different states. The *maximum-length* sequence has R steps in it, and

$$R = 2^r - 1 \qquad (16.32)$$

One state, the all-zero state, is not allowed since that would generate no feedback and the sequence would then stop. The algorithm which calculates the feedback digit must be chosen carefully in order to achieve this maximum-length sequence, but the nature of this design choice is not our concern here. We are just going to use the signal properties of the sequence. A 31-bit sequence generated from a 5-stage register is

1001011001111100001101110101010000

It turns out that the sequence has extremely good randomness properties. The balance between ones and zeros is almost exact. In fact there will be 2^{r-1} ones and $(2^{r-1} - 1)$ zeros in the sequence. The probability of finding a *one* for example is therefore almost exactly $0 \cdot 5$ if R is large. The probability of obtaining a group of, say, k ones is $0 \cdot 5^k$, which is one measure of randomness.

When calculating the autocorrelation function of this signal, we consider

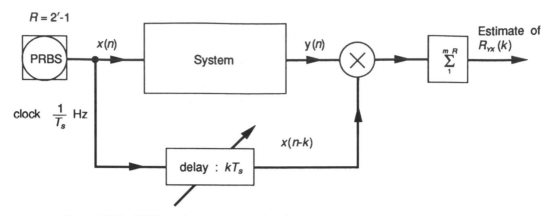

$R = 2^r - 1$

clock $\frac{1}{T_s}$ Hz

delay : kT_s

Figure 16.9   PRBS testing to obtain unit-pulse response

the signal values to be $(+1, -1)$ instead of $(0, 1)$, so that

$$R_X(0) = 1 \quad \text{and} \quad R_X(k) = -1/R \qquad k \neq 0 \tag{16.33}$$

This is almost the ideal autocorrelation function for testing, since $1/R$ is very small. So figure 16.9 shows the measurement process for determining the unit-pulse response of a discrete-time system.

The averager operates over an integral number of sequence lengths, to give the exact crosscorrelation function. Note that we are replacing an *expectation* operation with a time average, and hence are assuming that the signal is *ergodic* (see section 14.2.1). Noise generated within the process is reduced by the averager, according to the principles which we have discussed in section 15.3 at some length.

Although we have developed the theory in the discrete-time domain for convenience, it still applies to the world of continuous-time signals, with small modifications. We shall explore briefly the interface between these two worlds in section 16.3 and show how functions like the power spectral density must be interpreted slightly differently in these two contexts.

16.3   Sampling and analog-to-digital conversion

Figure 16.10 shows the sub-system that we are to consider. An analog or continuous-time random-signal $x(t)$ is sampled and digitised to give the discrete-time signal $x(n)$, assuming unity 'gain' in the converter. Before sampling, $x(t)$ is bandlimited with an anti-alias filter, which we assume to have the ideal bandwidth of $f_S/2$ Hz.

We now study the progression of the signal through these stages, in terms of the autocorrelation functions and power spectral densities, and make special mention of the signal-to-noise power ratio SNR, by considering a sinusoidal signal in the presence of noise.

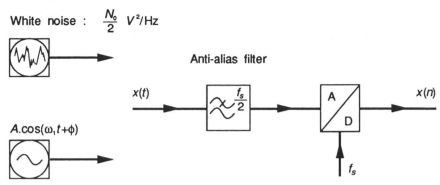

White noise : $\dfrac{N_0}{2}$ V^2/Hz

$A.\cos(\omega, t + \phi)$

Figure 16.10 A sampling sub-system

16.3.1 Continuous-time signals

If $x(t)$ is random, its autocorrelation function will be defined as before, but on a continuous-time basis:

$$R_X(\tau) = E[X(t) . X(t - \tau)] \tag{16.34}$$

This form of autocorrelation function has familiar properties:

$$R_X(0) = E[X^2] \qquad R_X(\tau \to \infty) \to (E[X])^2$$

However, since the signal is not sampled, it exists for all values of τ.

The power spectral density is found in a similar fashion to equation 16.2, except that an integral rather than a summation relationship is used:

$$S_X(\omega) = \int_{-\infty}^{\infty} R_X(\tau) . \exp \mathrm{j}(-\omega\tau) \, \mathrm{d}\tau \tag{16.35}$$

Again, since the signal is not sampled, the power spectral density is not periodic. The other major point to note is that this $S_X(\omega)$ for a continuous-time signal is not normalised, but has dimensions of V^2-sec. Practical units of power spectral density are usually V^2/Hz, and the function is normally plotted along the frequency axis, ie. $S_X(f)$.

$$R_X(\tau) = \frac{1}{2\pi} \int_{-\infty}^{\infty} S_X(\omega) . \exp \mathrm{j}(\omega\tau) \, \mathrm{d}\omega$$

$$= \int_{-\infty}^{\infty} S_X(f) . \exp \mathrm{j}(\omega\tau) \, \mathrm{d}f \tag{16.36}$$

$$R_X(0) = E[X^2] = \frac{1}{2\pi} \int_{-\infty}^{\infty} S_X(\omega) \, \mathrm{d}\omega$$

$$= \int_{-\infty}^{\infty} S_X(f) \, \mathrm{d}f \tag{16.37}$$

● **Example 16.8** Consider the case where the input signal is white noise. This has a uniform power spectral density which is $N_0/2$ V^2/Hz.

363

The factor of 2 is important. Although the definition of $S_X(f)$ is *double-sided* and extends over both positive and negative frequency axes, any measured value (N_0) is for a *single-sided* model of positive frequencies only.

Applying equation 16.36,

$$R_N(\tau) = (N_0/2) \cdot \int_{-\infty}^{\infty} \exp j(\omega\tau) \, df$$

$$= (N_0/2) \cdot \delta(\tau)$$

where $\delta(\ )$ is the usual delta function.

Notice that the power in this signal, $R_N(0)$, is infinite. Clearly this is an unrealistic model. In practice, such a white noise signal is always bandlimited, so that it has finite power. See next section. ●

● **Example 16.9** Consider now a sinusoidal signal:

$$x(t) = A \cdot \cos(\omega_1 t + \varphi).$$

Since this signal consists of two phasors, its power spectral density is given by

$$S_X(f) = (A^2/4) \cdot \delta(f - f_1) + (A^2/4) \cdot \delta(f + f_1)$$

The autocorrelation function follows from equation 16.32:

$$R_X(\tau) = (A^2/2) \cdot \cos(\omega_1 \tau)$$

Notice that both these functions are independent of the initial phase of the original signal. Figure 16.11 summarises their properties. ●

If we had both the noise and the sinusoidal signal present, the signal-to-noise ratio could not be defined, because the noise power is apparently infinite.

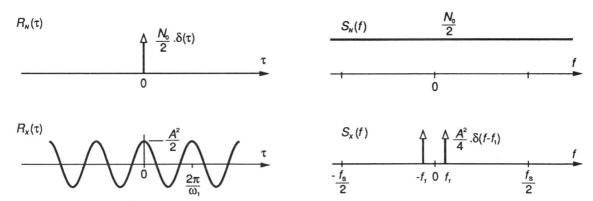

Figure 16.11 Descriptions of noise and sinusoidal signals

16.3.2 Anti-alias filter

Let us assume an ideal anti-alias filter, of bandwidth $\pm f_s/2$ Hz, and unity gain. Then

$$|H(f)| = 1 \qquad -f_s/2 \leqslant f \leqslant f_s/2 \tag{16.38}$$

After this filter, any sinusoidal signal components will be unaltered, but any white noise signal will become bandlimited. Applying equation 16.37, we see that the noise power N after the filter is

$$N = \int_{-f_s/2}^{f_s/2} (N_0/2) \, \mathrm{d}f$$
$$= N_0 f_s/2 \tag{16.39}$$

Hence $\quad R_N(\tau) = (N_0 f_s/2) \cdot \mathrm{sinc}(f_s \tau) \tag{16.40}$

where $\mathrm{sinc}(x) = \sin(\pi x)/(\pi x)$.

Assuming that the noise and the sinusoidal signal are applied simultaneously, then the signal-to-noise ratio at this point is

$$\mathrm{SNR} = \frac{A^2}{N_0 f_s} \tag{16.41}$$

16.3.3 After sampling

Consider now the effect of sampling this signal plus noise. The nature of the signals has changed considerably, but we shall show that the SNR has not.

Thus, the signal is now

$$x(n) = A \cdot \cos(\omega_1 T_s n + \varphi) \tag{16.42}$$

and

$$R_X(k) = A^2 \cdot \mathrm{avg} \, [\cos(\omega_1 T_s n + \varphi) \cdot \cos(\omega_1 T_s(n-k) + \varphi)]$$
$$= (A^2/2) \cdot \cos(\omega_1 T_s k) \tag{16.43}$$

since the signal is deterministic.

Hence

$$S_X(f) = (A^2/4) \cdot \delta(f - f_1) + (A^2/4) \cdot \delta(f + f_1) \tag{16.44}$$

The noise is now described by taking the sampled form of the autocorrelation function in equation 16.40:

$$R_N(k) = (N_0 f_s/2) \cdot \mathrm{sinc}(k)$$
$$R_N(k) = (N_0 f_s/2) \qquad k = 0$$
$$= 0 \qquad\qquad k \neq 0 \tag{16.45}$$

Thence

$$S_N(\omega) = (N_0 f_s/2) \qquad -\omega_s/2 \leqslant \omega \leqslant \omega_s/2 \tag{16.46}$$

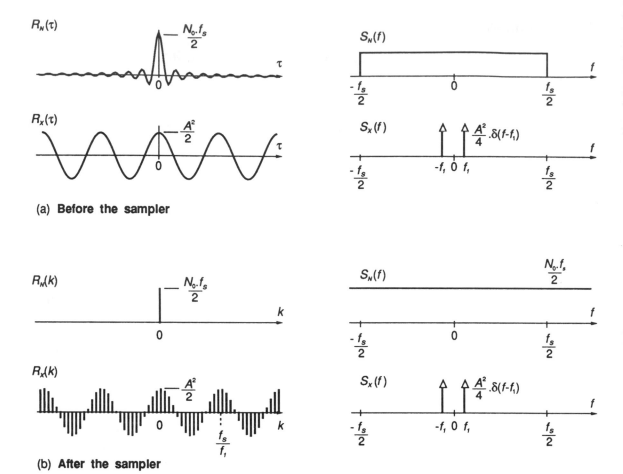

(a) **Before the sampler**

(b) **After the sampler**

Figure 16.12   Signal properties around a sampler

So the SNR is unchanged from what it was before sampling. These functions are sketched in figure 16.12, to compare conditions on both sides of the sampler.

● **Example 16.10**   A certain voice-frequency signal is sinusoidal and has a measured rms magnitude of 2 V. It is contaminated by white noise having a power spectral density of $N_0 = 10^{-8}$ V^2/Hz, and is to be digitised at a sampling rate of 8 kHz. Calculate the SNR after the A-D converter, assuming an ideal anti-alias filter.

Now from equation 16.45, we can calculate the effective noise power after the A-D converter:

$$R_N(0) = (10^{-8} \times 8 \times 10^3)/2 = 4 \times 10^{-5}$$

The rms signal magnitude is 2, so the SNR at this point is

$$\text{SNR} = (2^2)/(4 \times 10^{-5}) = 10^5 \text{ or } 50 \text{ dB} \quad ●$$

16.3.4 Quantisation noise and sampling rate

So far in this discussion, we have ignored the effect of quantisation noise (see section 6.4), which worsens the signal-to-noise ratio of the signal.

Consider now a binary analog-to-digital converter with b-bit words, operating over an input-signal range of $\pm V_M$. Then the quantum step between adjacent digital signal levels is q, and the corresponding quantisation noise power is N_Q:

$$q = 2V_M \cdot 2^{-b} \tag{16.47}$$

$$N_Q = q^2/12 \tag{16.48}$$

Provided that the input signal is not constant, successive quantisation errors are independent, so the autocorrelation function $R_Q(k)$ and power spectral density $S_Q(\omega)$ of this noise are

$$\begin{aligned} R_Q(k) &= N_Q \qquad k = 0 \\ &= 0 \qquad k \neq 0 \end{aligned} \tag{16.49}$$

$$S_Q(\omega) = N_Q \tag{16.50}$$

Clearly then the power spectral density of the quantisation noise is independent of the sampling frequency f_S.

Since the quantisation noise is not related to the input-noise N, the net noise power after digitisation is $(N + N_Q)$, and the SNR is modified appropriately.

● **Example 16.11** The A-D converter used in example 16.10 has an input-signal range of ± 3 V, and generates 8-bit words. Determine the effect of quantising noise on the SNR of this signal.

Using equation 16.48 we first calculate the quantisation-noise power

$$N_Q = (2 \times 3 \times 2^{-8})^2/12 = 4 \cdot 6 \times 10^{-5}$$

So the total noise power is

$$N + N_Q = 4 \times 10^{-5} + 4 \cdot 6 \times 10^{-5} = 8 \cdot 6 \times 10^{-5}$$

The new SNR is therefore

$$\text{SNR} = 2^2/8 \cdot 6 \times 10^{-5} = 4 \cdot 65 \times 10^4 = 46 \cdot 7 \text{ dB}$$

Compared with the SNR before taking quantisation noise into account, a loss of $3 \cdot 3$ dB has occurred, which could be serious, depending upon the application.   ●

Quantisation noise can be reduced by using a non-linear transfer characteristic in the A-D converter, but this of course distorts the signal itself. An alternative technique is to sample at a rate which is much greater

367

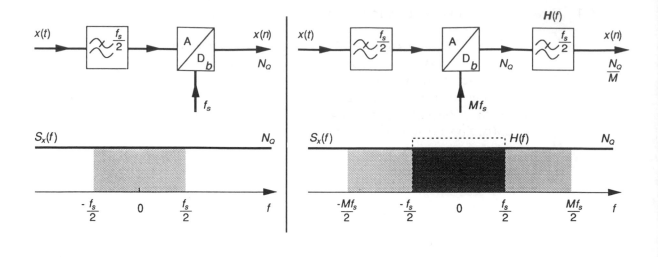

(a) **Normal sampling**　　　　　　　(b) **Oversampling by** M **times**

Figure 16.13　Oversampling and quantisation noise

than that stipulated by the sampling theorem. Thus, consider the two cases depicted in figure 16.13.

In normal sampling, case (a), the quantisation noise, is as calculated above. However, in case (b) the sampling rate is increased to Mf_S Hz. Now although the power spectral density $S_Q(\omega)$ has not changed, not all the available frequency range from 0 to $Mf_S/2$ Hz is of interest to us. The signal occupies only the range 0 to $f_S/2$ Hz.

A digital lowpass filter $\mathbf{H}(f)$ is therefore included, with bandwidth $f_S/2$, and the reduced quantisation noise power is calculated by re-writing equation 16.4 with f as the variable instead of ωT_S:

$$\mathbf{N}_{Q2} = \frac{1}{Mf_S} \int_{-Mf_S/2}^{Mf_S/2} S_Q(f) \cdot | \mathbf{H}(f) |^2 \cdot \mathrm{d}f$$

$$= \frac{1}{Mf_S} \int_{-f_S/2}^{f_S/2} N_Q \cdot \mathrm{d}f$$

$$= N_Q/ M \qquad\qquad (16.51)$$

A substantial reduction in quantisation-noise power has therefore been achieved, although at the expense of a higher sampling rate for the A-D converter, and some digital signal processing after the converter. For audio frequency signals, such as for voice processing and transmission, or for line-signal processing in telecommunications, such penalties can be paid in order to increase the apparent dynamic range of the converter. An 8-bit converter can be made to behave as though it was a 10-bit converter for instance.

In order to evaluate the improvement in terms of effective A-D converter resolution, recall the calculation of the quantum interval q (equation 16.47). Thus, if the actual number of bits in the converter is b, and the equivalent number of bits corresponding to the quantisation noise is b_Q, then

$$2^{-b_Q} = 2^{-b}/\sqrt{M}$$

or $\quad b_Q - b = \frac{1}{2}.\log_2(M)$ (16.52)

An improvement of 1 bit in the effective performance of the A-D converter can be achieved by making $M = 4$, while an improvement of 2 bits requires $M = 16$,

● **Example 16.12** Following on from example 16.11, let us calculate what could be achieved by oversampling.

Let $M = 4$, then the new sampling frequency is 32 kHz, but the converter has 9-bit performance.

Applying equation 16.51, we calculate the new quantisation-noise power:

$$N_{Q2} = 4 \cdot 6 \times 10^{-5}/4 = 1 \cdot 15 \times 10^{-5}$$

The total noise power is

$$N + N_{Q2} = 4 \times 10^{-5} + 1 \cdot 15 \times 10^{-5} = 5 \cdot 15 \times 10^{-5}$$

The SNR is therefore $(2)^2/5 \cdot 15 \times 10^{-5}$ or $48 \cdot 9$ dB, an improvement of $2 \cdot 2$ dB.

For $M = 16$, the new sampling rate would be 128 kHz, but performance is equivalent to 10 bits.

The SNR is now $49 \cdot 7$ dB, and the effect of quantisation-noise has almost been removed. ●

An example of this technique in commercial use is a high-quality audio-frequency ADC, offering 18-bit performance at a sampling frequency of $44 \cdot 1$ kHz. 128 times oversampling is used, and the digital filter employs coefficients to 32-bit accuracy, with the arithmetic truncated to 24 bits.

16.4 Applications

16.4.1 Data transmission

In section 14.4.2 we discussed data transmission as an example of a situation where random signals occur, and where the random noise at the detector can produce wrong decisions on the data symbols. At that time we considered only the noise power at the point of decision, relative to the signal magnitude, and then calculated the probability of error from this ratio. We gave no thought to how these magnitudes would be determined in practice, but now we have enough information to discuss this question.

Figure 16.14 shows the arrangement of a typical data signal receiver for a baseband or unmodulated signal transmission.

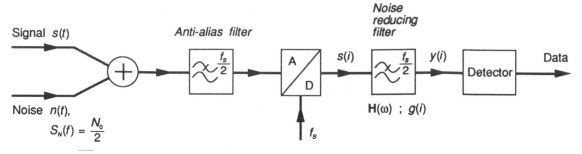

Figure 16.14 Data transmission baseband receiver

Let us assume that the signal element $s(i)$ is a polar signalling symbol, of length MT_S, that is

$$s(i) = A \qquad 0 \leqslant i \leqslant M - 1$$
$$\text{or} \qquad s(i) = 0 \qquad 0 \leqslant i \leqslant M - 1 \tag{16.53}$$

The noise input is white noise, power spectral density $N_0/2$ V^2/Hz, and the sampling rate is f_S Hz. Now we have found in section 16.3.3, that the power spectral density of the noise after the converter is $N = N_0 f_S/2$.

Since the noise is spread evenly right up to the Nyquist frequency, and the signal energy will be restricted to lower frequencies, we should be able to improve the signal-to-noise power ratio by lowpass filtering. Such a filter is shown in the diagram, and has a unit-pulse response $g(i)$ and frequency response $\mathbf{H}(\omega)$.

The noise power at the detector can be calculated from the unit-pulse response of the filter and the input-signal autocorrelation function, by calculating the output-signal autocorrelation function as in equation 16.10, and setting $k = 0$. A narrowband filter will reduce this output-noise power.

On the other hand, a narrowband filter will distort the signal, as we noted briefly in section 9.4.4. So we shall want to identify the maximum value of the output signal, which then causes the filtered signal to stand out from the noise by the greatest amount. The signal response is calculated via the convolution equation.

For simplicity, and to make the analytical point clearly, let us assume that the filter is a recursive lowpass filter as discussed in section 16.2.2. In practice, we are unlikely to use such a simple filter if noise is a problem, as will become clear after the analysis has been done.

The noise output power of the filter will be N_Y, say:

$$N_Y = N \cdot \left(\frac{1 - \alpha}{1 + \alpha} \right) \tag{16.54}$$

Since $\alpha < 1$, the noise power will be reduced by this filter, which is all to the good. However, the filter also distorts the input signal, and these two effects must carefully be balanced if an optimum SNR is to be achieved.

The unit-pulse response of this filter is $g(i) = (1 - \alpha) \cdot \alpha^i$, which is a

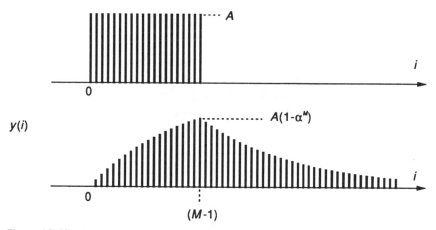

Figure 16.15 Input and output signals for recursive filter

decaying exponential. When a data symbol of magnitude A is applied to the filter, we apply the convolution relationship (section 7.2.2), and discover that the output grows exponentially, as seen in figure 16.15. Consequently the peak output occurs at the end of the symbol interval.

In general,

$$y(i) = \sum_{r=-\infty}^{\infty} g(r) . s(i - r)$$

At the end of the symbol,

$$y(M - 1) = \sum_{r=0}^{M-1} A . (1 - \alpha) . \alpha^{M-1-r}$$

$$= A . (1 - \alpha^M) \tag{16.55}$$

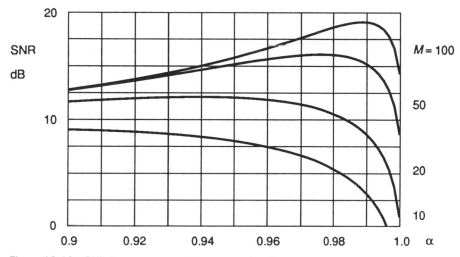

Figure 16.16 SNR improvement using a recursive filter

371

The maximum signal-to-noise ratio at the detector is therefore

$$\text{SNR}_{\text{max}} = \frac{A^2 \cdot (1 - \alpha^M)^2 \cdot (1 + \alpha)}{N \cdot (1 - \alpha)} \qquad (16.56)$$

The ratio $\text{SNR}_{\text{max}} \cdot N/A^2$ is plotted in figure 16.16 against the parameter α, for various values of M. A broad optimum occurs, which denotes the best conditions for symbol detection, and this general principle is illustrated in example 16.13. However this type of filter is generally unsuitable, because the long unit-pulse response causes successive symbols to interfere with each other (see figure 16.15).

● **Example 16.13** Consider a polar signalling system at 1 ksymbol/sec, where the sampling rate is 20 kHz. The input signal magnitude is 30 mV, and the noise power spectral density is measured as $N_0 = 3 \times 10^{-8}$ V^2/Hz.

The effective noise power after the anti-alias filter is

$$N = N_0 f_s/2 = 3 \times 10^{-8} \cdot 2 \times 10^4/2 = 3 \times 10^{-4} \text{ V}^2$$

Consequently A^2/N is $3 \cdot 0$.

Now each symbol represents 20 sampling instants, so $M = 20$.

Let the recursive filter have a coefficient $\alpha = 0 \cdot 95$. Then from equation 16.56,

$$\text{SNR}_{\text{max}} = 48 \cdot 2 \text{ or } 16 \cdot 8 \text{ dB}$$

The bit-error-rate is then

$$P_e = Q[\sqrt{(\text{SNR}_{\text{max}}/2)}] = Q[4 \cdot 9] = 5 \times 10^{-7} \qquad ●$$

This example shows how the information we have discussed can be used to calculate the bit-error rate for a simple data receiver. We have not yet tried to optimise the design, but an optimum kind of filter can be found, and we will derive it in the next section.

16.4.2 The matched filter, an optimum detector

The input signal symbol is $s(t)$, of length MT_S sec, and to this is added a white noise signal $n(t)$ having a power spectral density $N_0/2$ V^2/Hz. Signal-plus-noise passes through the anti-alias filter and is digitised. Quantisation noise is ignored.

Consider an arbitrary filter for signal processing, with a finite unit-pulse response $g(i)$ of length M, and one isolated input symbol $s(i)$. The output signal will consist of two components, $y_S(i)$ and $y_N(i)$.

Then since the input noise is white, the noise power at the output of the filter may be calculated from the output-signal autocorrelation function, equation 16.19. Since $R_X(0) = N$, and $R_X(k) = 0$ for $k \neq 0$:

$$E[Y_N^2] = N \cdot \sum_{r=0}^{M-1} g^2(r) \tag{16.57}$$

The output signal will be

$$y_S(i) = \sum_{r=0}^{M-1} g(r) \cdot s(i-r) \tag{16.58}$$

For best performance, we need to choose the set $\{g(r)\}$ so that the optimum signal-to-noise ratio, $\max[y_S^2(i)]/E[Y_N^2]$, is maximised. The formal mathematics for this operation is not difficult, but slightly cumbersome, so we shall content ourselves with an heuristic argument.

From the convolution expression in equation 16.58, we note that each input-signal sample is weighted by one coefficient in the unit-pulse response sequence.

Let $g(r) = s(M-1-r)$. This strategy has the effect that small signal-values are multiplied by small weights, and the noise that is present at these times is hence reduced. Consequently in general,

$$y_S(i) = \sum_{r=0}^{M-1} s(M-1-r) \cdot s(i-r)$$

$$= a \cdot R_S(i - M + 1) \tag{16.59}$$

where a is a scaling constant.

In shape therefore, the output signal looks like the delayed autocorrelation function of the input signal. Figure 16.17 shows some of the individual functions which are involved in the equations. The maximum value of the output signal occurs when $i = (M-1)$, and is $a \cdot R_S(0)$, so

$$\max[y_S(i)] = y_S(M-1) = \sum_{r=0}^{M-1} s^2(M-1-r)$$

$$= E/T_S \tag{16.60}$$

where E is the input-signal energy.

From equation 16.57, we see that since $g(r) = s(M-1-r)$, the noise power is now $E \cdot N_0 f_S / 2 T_S$; so the optimum SNR is

$$\text{SNR}_{\max} = \frac{2E}{N_0} \tag{16.61}$$

This classical result reminds us that the maximum SNR at the output of a *matched filter* depends only on the signal *energy* and the white-noise power spectral density. The limit is achieved by making the unit-pulse response of the filter equal to the time-reversed signal. It is then a *correlation* detector, and can be shown to be the best linear filter that is possible.

The matched filter is the one assumed when calculating the performance of any ideal pulse detection or transmission system.

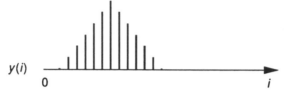

Figure 16.17 Illustration of the matched filter convolution

● **Example 16.14** Consider how a matched filter would improve the performance of the receiver in example 16.13.

In this case the matched filter becomes a simple moving-averager, with equal weights at each tap. Thus

$$g(r) = A \qquad 0 \leqslant r \leqslant M - 1$$

The pulse energy is given by

$$A^2 . MT_S = (0 \cdot 03)^2 \times 20/2 \times 10^4 = 9 \cdot 0 \times 10^{-7}$$

The optimum SNR is therefore

$$2E/N_0 = 2 \cdot 9 \times 10^{-7}/3 \times 10^{-8} = 60 \text{ or } 18 \text{ dB}$$

The bit-error-rate is therefore

$$Q[\sqrt{(60/2)}] = Q[5 \cdot 5] = 2 \cdot 2 \times 10^{-8}$$

An improvement of $1 \cdot 2$ dB in SNR has therefore been achieved, with the consequent reduction of the bit-error-rate by almost one order of magnitude. ●

The improvement in this simple case of a rectangular pulse is not impressive, but with more elaborate signals, having a peakier autocorrelation function, the improvements can be considerable.

16.4.3 Measuring the power spectral density

It may seem perverse to use the measurement of power spectral density as an example of its application! However, this measurement is frequently carried out, and nicely illustrates some of the problems that occur in the assessment of any random signal.

Spectrum analysers are commonly used in communication engineering to examine signal spectra, and to measure the noise level due to miscellaneous modulation processes, some intentional and some incidental. Radio-frequency oscillators, for instance, often exhibit *phase noise* due to jitter in the oscillation process, and this is evaluated by a spectrum analyser. Voice signals are analysed in order to determine their spectral content, and thence to encode or interpret the spoken sound. In control systems, the relationship of equation 16.8 is often used to measure the frequency response of the controlled system, and this requires the power spectral density of input and output signals to be measured. Mechanical systems are often examined for modes of vibration by taking the power spectral density of a signal from a velocity transducer.

Suppose that we have a continuous-time random signal $x(t)$, with power spectral density $S_X(f)$ V^2/Hz. In order to measure this function, we digitise the signal and then examine a small section of the spectrum with a bandpass filter. This filter has an equivalent noise bandwidth ω_B rad/sec, centred at ω_0 rad/sec. Figure 16.18 illustrates the situation for a simplified digital instrument.

After the converter, the signal has a power spectral density $S_X(\omega)$. The analysis filter frequency response is $\mathbf{H}(\omega)$, where

$$|\mathbf{H}(\omega)| = 1 \qquad \begin{aligned} (\omega_0 - \omega_B/2) &\leqslant \omega \leqslant (\omega_0 + \omega_B/2) \\ -(\omega_0 + \omega_B/2) &\leqslant -\omega \leqslant -(\omega_0 - \omega_B/2) \end{aligned} \qquad (16.62)$$

Hence

$$\begin{aligned} S_Y(\omega) = S_X(\omega) \qquad & (\omega_0 - \omega_B/2) \leqslant \omega \leqslant (\omega_0 + \omega_B/2) \\ & -(\omega_0 + \omega_B/2) \leqslant -\omega \leqslant -(\omega_0 - \omega_B/2) \\ = 0 \qquad & \text{elsewhere} \end{aligned} \qquad (16.63)$$

The signal power at the output of the filter is therefore

$$\begin{aligned} E[Y^2] = R_Y(0) &= \frac{1}{2\pi} \int_{-\pi}^{\pi} S_Y(\omega T_S) . \, \mathrm{d}(\omega T_S) \\ &\simeq \frac{\omega_B T_S}{\pi} . \, S_X(\omega_0 T_S) \end{aligned} \qquad (16.64)$$

Now, converting from the normalised frequency scale by dividing by T_S, and also using units of Hz, we obtain

$$E[Y^2] = 2B . S_X(f_0) \qquad (16.65)$$

where $\omega_B = 2\pi . B$.

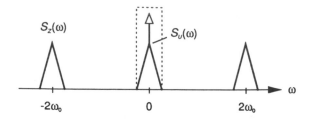

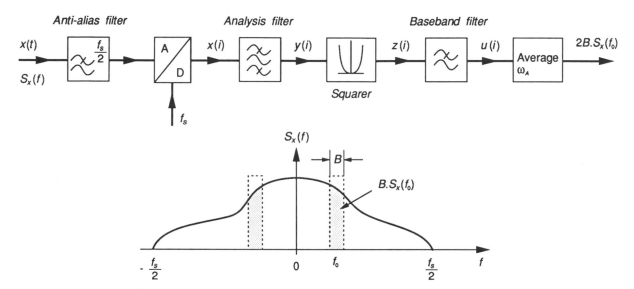

Figure 16.18 Measurement of power spectral density

The output power from this filter is therefore a scaled measure of the signal power spectral density at this particular frequency.

The whole spectrum can be evaluated by having a bank of filters each at a different frequency. In practice, the bank of filters is often implemented by the FFT (see chapter 13). An alternative approach is to sweep a single filter slowly across the frequency range, which is the principle of the conventional swept-frequency analyser. Practical analysers usually employ the heterodyne principle introduced in section 4.3.3, rather than physically changing the centre frequency of a bandpass filter.

● **Example 16.15**   A certain power spectral density measurement takes place at a sampling frequency of 10 kHz. A power of $0 \cdot 2$ V^2 is measured in a bandwidth of $0 \cdot 5$ kHz at a centre frequency of $2 \cdot 0$ kHz.

The true power spectral density of the input signal is therefore, from equation 16.65,

$$S_X(2 \text{ kHz}) = 0 \cdot 2 / (2 \times 0 \cdot 5 \times 10^3) = 0 \cdot 2 \text{ mV}^2/\text{Hz}$$

Note, the spectrum is sometimes expressed in rms form, in this case $1 \cdot 41 \times 10^{-2}$ V/$\sqrt{\text{Hz}}$.

The digital bandpass filter is centred on an angle of $\omega_0 T_S$ or $1\cdot 26$ rad, and its bandwidth corresponds to $\omega_B T_S$ or $0\cdot 31$ rad.　●

So, a simple bandpass filter enables the power spectral density of an input signal to be measured, provided that we can estimate the power contained in this bandwidth. The output signal of the filter is random and narrowband, and the analysis procedures outlined in chapters 14, 15, 16 apply. The proper analysis is complicated, so we shall merely apply some intelligent approximations in order to appreciate the nature of the solution and arrive at a usable result.

In order to estimate the power output of the filter, a square-law detector is used, giving a signal $z(i) = y^2(i)$. Since the random variable Y represents a narrowband process ($\omega_B \ll \omega_0$), we may write

$$Y(i) = A(i) . \cos[\omega_0 T_S i + \Phi(i)] \tag{16.66}$$

$A(i)$ is the *envelope* function of the modulated sine wave, which varies relatively slowly, and is a random variable possessing *Rayleigh* statistics (see section 14.4.3). The random variable $\Phi(i)$ is the phase function which also varies slowly, and has a uniform probability density function between the limits of $\pm \pi$ radians.

When this signal is squared, two terms arise at distinct frequencies:

$$Z(i) = A^2(i)/2 + [A^2(i)/2] . \cos[2(\omega_0 T_S i + \Phi(i))] \tag{16.67}$$

The first lowpass filter after the squarer removes the term at frequency $2\omega_0$, leaving the baseband term containing the squared envelope, $U(i)$. The amplitude statistics of this term can be calculated with some difficulty using the techniques of section 14.3, and yield the following results for a Gaussian signal Y with variance σ_Y^2:

$$E[U] = \sigma_Y^2 \qquad V[U] = \sigma_Y^4 \tag{16.68}$$

The ratio of mean-value-squared to variance (ie. the SNR) is therefore unity! In order to make sense of the measurement it is necessary to *smooth* or *average* the result. Now it can be shown that the power spectral density of this signal, in the region of zero frequency, is

$$S_U(f = 0) \simeq \sigma_Y^4 . \delta(f) + \sigma_Y^4/B \tag{16.69}$$

The first term represents the mean value of the signal, the estimate of power from the bandpass filter ($E[Y^2]$), while the second represents the fluctuation or disturbance which masks the true measurement. In order to reduce this disturbance, we can only reduce its bandwidth, using the smoothing filter shown in figure 16.18. So, if the smoothing filter has an equivalent noise bandwidth of W Hz or ω_A rad/sec, the disturbance is reduced to a power of $2\sigma_Y^4 W/B$, and the effective 'signal-to-noise ratio' as far as the measurement of power is concerned is

$$\mathrm{SNR}_{\mathrm{meas}} = B/(2W) \tag{16.70}$$

If $W \ll B$, then conditions are right for the central limit theorem to apply, and we may assume that the resulting disturbance or noise has approximately Gaussian statistics. Hence the estimates of error can follow exactly along the lines of section 15.3, with results from the graph in figure 15.9 applying.

● **Example 16.16**   Continuing example 16.15, suppose that the output signal from the bandpass filter is applied to a square-law detector and followed by a smoothing filter of equivalent noise bandwidth 5 Hz.

The effective SNR is therefore $500/10 = 50$ or 17 dB.

Referring to the graph in figure 15.9, we observe that the estimate of power spectral density will be within $\pm 20\%$ of the correct value for 90% of the time. It will yield a result which is within 1% of the true value for only one reading out of 10.

Other values can of course be deduced, either by calculation or by reference to the graph.   ●

When we try to discover for how long the output signal should be averaged in order to achieve this result, we run up against the problem that the input-noise to the averager is bandlimited and is not white. Implementation of equation 16.22 to find the output-noise power would be cumbersome, since we have first to calculate the autocorrelation function of the output-noise. However, using the power spectral density of the input signal, we may use equation 16.25 to obtain an approximate result.

In this case, the input-noise to the squarer is bandpass with an equivalent noise bandwidth of B Hz, so that the equivalent noise bandwidth of the lowpass input signal to the averager is $B/2$ Hz.

● **Example 16.17**   Without the smoothing filter, the effective SNR at the point of measurement in example 16.16 would have been unity, so the process gain or SNR improvement required of the averager is 17 dB, or a ratio of $P = 50$.

Now the bandwidth of the analysis filter in example 16.15 is 500 Hz, but using our baseband definition of equivalent noise bandwidth (section 16.2.1), $f_N = 250$ Hz.

Hence using equation 16.25,

$$M = (50 \times 10^4)/(2 \times 250) = 1000$$

The averaging time is therefore 100 ms since the sampling rate is 10 kHz.

The equivalent noise bandwidth of the averager is 5 Hz.

A recursive averager would give a similar result if according to equation 16.26:

$$\alpha = (1000 - 1)/(1000 + 1) = 0 \cdot 998   ●$$

STUDY QUESTIONS

1 Find the power spectral density of a discrete-time random signal which has an autocorrelation function $R_x(k)$:

$$\{R_x(k)\} = \{N, N/4, 0, 0 \ldots\} \qquad k \geqslant 0$$

2 The random signal of question 1 is now passed through a filter which has a unit-pulse response $g(i)$. Find the power spectral density of the filtered signal.

$$\{g(i)\} = \{2, 2, 0, 0, \ldots\} \qquad i \geqslant 0$$

3 Find the noise equivalent bandwidth of a filter which has the frequency response $H(\omega)$:

$$|H(\omega)| = (1/2) \cdot [1 + \cos(b\omega T_s)] \qquad |\omega| \leqslant \pi/bT_s$$
$$= 0 \qquad \text{elsewhere}$$

4 A discrete-time white-noise source having a power of $0 \cdot 1\ V^2$ is applied to the filter of question 3, which now has $b = 3$. What is the output-noise power?

5 A certain zero-frequency measured signal is sampled at 10 kHz. The signal is in the presence of white noise which is bandlimited to ± 5 kHz, and gives an SNR of 2 dB.

   A filter like that in question 3 is used to improve the SNR to 22 dB. Calculate the necessary noise equivalent bandwidth of the filter, and the value of coefficient b.

6 A certain noise signal has an autocorrelation function $R_N(k)$. The noise power is to be reduced by an equally-weighted moving-average filter having 16 stages. Calculate the output-noise power (a) under these conditions, and (b) assuming that the noise samples are independent.

$$\{R_N(k)\} = \{4, 3, 2, 1, 0, 0, 0 \ldots\} \qquad k \geqslant 0$$

7 Discuss one practical application for the crosscorrelation function.

8 The signal at the input to a line receiver is 200 mV rms at a frequency of $1 \cdot 5$ kHz, and has added noise with a double-sided power spectral density of $0 \cdot 5\ \mu V^2/Hz$.

   The receiver is to be implemented digitally at a sampling rate of $7 \cdot 5$ kHz, and an ideal anti-alias filter is used. Calculate the SNR after the signal has been digitised.

9 The A-D converter in question 8 operates over a range of ± 2 V. What is the minimum number of bits that must be used, if the quantising-noise power is to be less than the input-noise power?

10 Discuss the principle whereby an increased sampling rate may be used to reduce the amount of quantisation-noise power in a digital system. What sampling rate should be used to ensure that the quantisation-noise power in the system of question 8 is less than 20% of the input-noise power?

11 After analog-digital conversion at 8 kHz, a certain pulse signal $s(i)$ has a cosine shape:

$$s(i) = (1 - \cos(2\pi i/K)) \qquad 0 \leqslant i \leqslant K - 1$$

It is received in the presence of white noise having a double-sided power spectral density of 20 $\mu V^2/Hz$. Calculate the SNR at the threshold detector, after the signal has been processed by a matched filter. Take $K = 4$ for a simple calculation.

12 Calculate the SNR for the signal and noise of question 11, if an equal-weight moving-average filter is used instead of the matched filter. Note that the performance will be inferior.

13 A digital spectrum analyser has an analysis bandpass filter with an equivalent noise bandwidth of 100 Hz at a centre frequency of 800 Hz. The sampling rate is 10 kHz. The mean power of the signal at the output of the bandpass filter is measured as $0 \cdot 1$ V^2. What is the value of the original signal power spectral density at 800 Hz?

14 What averaging-filter bandwidth is necessary, in order to measure the result given in question 13 to an accuracy of 10% with 90% confidence? What averaging time is needed for a moving-average filter, and what coefficient α would be necessary for a recursive averager in order to achieve this result?

Appendices

OBJECTIVES

To summarise some of the mathematical concepts and techniques behind the development of signal processing in this book.

COVERAGE

A variety of topics is introduced, useful results are summarised, and some of the derivations are hinted at. More complete treatments can be obtained from the wide range of available textbooks, so these appendices just give the flavour of the subjects.

In particular, the Fourier series and transform and the Laplace, Z and Discrete-Fourier transforms are summarised. The emphasis here is on understanding how they are related and how they may be used, rather than on rigorously deriving their properties.

Appendix A1
Phasor Algebra and Trigonometric Identities

A1.1 Complex arithmetic

A *complex number* locates a point in a plane, and may be expressed in either *rectangular* or *polar* co-ordinates. The co-ordinate in the vertical direction is labelled with the operator j, and the plane is usually referred to as *complex*. Consider a point $\{a, b\}$, which in polar form is $\{r, \theta\}$ where

$$r = \sqrt{(a^2 + b^2)} \tag{A1.1}$$

$$\theta = \arctan(b/a) \tag{A1.2}$$

Then this point may be expressed succinctly by a complex number $\mathbf{A}$:

$$\mathbf{A} = a + jb \tag{A1.3}$$

or $\quad \mathbf{A} = r \cdot \exp j(\theta) \tag{A1.4}$

where $\quad \exp j(\theta) = \cos(\theta) + j \cdot \sin(\theta) \tag{A1.5}$

$\exp j(\theta)$ is then a vector in this complex plane, describing an angle θ and unit length. Note that

$$\exp j(\pi/2) = j \tag{A1.6}$$

The j operator therefore corresponds to a rotation of $\pi/2$ radians, and consequently $j^2 \equiv -1$.

A *conjugate* complex number $\mathbf{A}^*$ has the same radius or magnitude as the complex number $\mathbf{A}$, but a negative angle. It is defined thus:

$$\mathbf{A}^* = a - jb$$
$$= r \cdot \exp j(-\theta) \tag{A1.7}$$

Clearly $\quad \mathbf{A} \cdot \mathbf{A}^* = r^2 \tag{A1.8}$

Arithmetic with complex numbers may be done with either rectangular co-ordinates or polar co-ordinates, as convenient.

● **Example A1.1**   Add together two complex numbers **A** and **B**:

$\quad$ **A** = 3 + j6 $\qquad$ **B** = 5 − j8

Then $\quad$ **A** + **B** = (3 + 5) + j(6 − 8) = 8 − j2 $\quad$ ●

● **Example A1.2**   Multiply together the two complex numbers.
$\quad$ This is best done by first converting both numbers to polar form. Using a calculator,

$\quad$ **A** = 6·71 ∠ 1·11 $\qquad$ **B** = 9·43 ∠ − 1·01

Then $\quad$ **A** . **B** = 6·71 × 9·43 ∠ (1·11 − 1·01) = 63·28 ∠ 0·10

The magnitudes are multiplied, and the angles added. Division proceeds similarly, except that the magnitudes are now divided and the angles are subtracted. $\quad$ ●

Integrating or summing complex numbers follows normal procedures, and two important results are

$$\int_{-\pi}^{\pi} \exp j(\theta) \, d\theta = 0 \tag{A1.9}$$

Clearly, also

$$\int_{-\infty}^{\infty} \exp j(\theta) \, d\theta = 0 \tag{A1.9a}$$

$$\sum_{r=0}^{N-1} \exp j(r\theta) = \frac{1 - \exp j(N\theta)}{1 - \exp j(\theta)} \tag{A1.10}$$

Equation A1.10 is a widely used result, and is derived from the summation formula for a *Geometric series* having a constant ratio exp j(θ).

A1.2   Trigonometric identities

The conventional trigonometric identities can be derived from these complex number concepts. From equation A1.5, we can see that

$$\cos(\theta) = [\exp j(\theta) + \exp j(-\theta)]/2 \tag{A1.11}$$

$$\sin(\theta) = [\exp j(\theta) - \exp j(-\theta)]/2j \tag{A1.12}$$

Hence $\quad \cos^2(\theta) + \sin^2(\theta) = 1$ $\hspace{3cm}$ (A1.13)

and $\quad \cos^2(\theta) - \sin^2(\theta) = \cos(2\theta)$ $\hspace{2.3cm}$ (A1.14)

Now consider the product $\cos(\theta) . \cos(\varphi)$:

$$\cos(\theta) . \cos(\varphi)$$
$$= [\exp j(\theta) + \exp j(-\theta)] . [\exp j(\varphi) + \exp j(-\varphi)]/4$$
$$= [\exp j(\theta + \varphi) + \exp j(-\theta - \varphi) + \exp j(-\theta + \varphi) + \exp j(\theta - \varphi)]/4$$
$$= [\cos(\theta + \varphi) + \cos(\theta - \varphi)]/2 \tag{A1.15}$$

This result is widely used in modulation studies such as those in chapter 5, together with the following:

$$2 . \cos(\theta) . \cos(\varphi) = \cos(\theta - \varphi) + \cos(\theta + \varphi) \tag{A1.16}$$

$$2 . \sin(\theta) . \sin(\varphi) = \cos(\theta - \varphi) - \cos(\theta + \varphi) \tag{A1.17}$$

$$2 . \sin(\theta) . \cos(\varphi) = \sin(\theta - \varphi) + \sin(\theta + \varphi) \tag{A1.18}$$

From these results:

$$2 . \cos^2(\theta) = 1 + \cos(2\theta) \tag{A1.19}$$

$$2 . \sin^2(\theta) = 1 - \cos(2\theta) \tag{A1.20}$$

The following identities for additive angles can be derived in similar fashion:

$$\cos(\theta \pm \varphi) = \cos(\theta) . \cos(\varphi) \mp \sin(\theta) . \sin(\varphi) \tag{A1.21}$$

$$\sin(\theta \pm \varphi) = \sin(\theta) . \cos(\varphi) \pm \cos(\theta) . \sin(\varphi) \tag{A1.22}$$

whence

$$\cos(2\theta) = 2 . \cos^2(\theta) - 1 \tag{A1.23}$$

$$\sin(2\theta) = 2 . \sin(\theta) . \cos(\theta) \tag{A1.24}$$

A1.3 Phasors

A *phasor* is a rotating vector in a complex plane, a complex function of time. We may express it thus:

$$\mathbf{x}(t) = \mathbf{A} . \exp j(\omega t) \tag{A1.25}$$

It consists of two distinct parts:

- The *unit phasor* $\exp j(\omega t)$ having an angle which is steadily increasing with time, and hence rotating at constant velocity with a rate of ω rads/sec.
- The *complex magnitude* $\mathbf{A}$, which defines the actual magnitude and starting angle of this phasor.

The average power in this complex phasor signal will be given by

$$\text{avg}[\mathbf{x}(t) . \mathbf{x}^*(t)] = |\mathbf{A}|^2 \tag{A1.26}$$

If all phasors in a given situation have the same frequency and are hence based on the same unit phasor, then arithmetic involving these phasors need be concerned only with their complex magnitudes. Chapter 3 explores these applications.

Where phasors have different frequencies, then these frequencies may be expressed relative to some reference frequency, and their non-linear combination will yield new frequencies, as discussed in chapter 5.

Appendix A2
Fourier Series

A2.1 Purpose

To represent a periodic signal $x(t)$ by a set of frequency components or complex phasors. Each frequency is a multiple of a *fundamental* frequency ω_0 (or f_0) which corresponds with the period of the original periodic signal. A periodic bandlimited time waveform is then described completely by a knowledge of the fundamental frequency and the set of complex coefficients which are the complex magnitudes of the individual phasors. See chapter 4 for the application of this principle.

Periodic signals are in the class of *power* signals—see section 2.3.

A2.2 Proposal

$$x(t) = \sum_{n=-\infty}^{\infty} \mathbf{c}(n) \cdot \exp \ j(n\omega_0 t) \tag{A2.1}$$

Hence

$$\mathbf{c}(n) = \frac{1}{2\pi} \int_{-\pi}^{\pi} x(t) \cdot \exp \ j(-n\omega_0 t) \cdot d(\omega_0 t) \tag{A2.2}$$

where the signal period is T sec, and $\omega_0 = 2\pi/T$.

A2.3 Proof

In order to verify the inversion formula, equation A2.2, consider the following average:

$$\text{avg}\,[x(t) \cdot \exp \ j(-m\omega_0 t)]$$

$$= \frac{1}{2\pi} \int_{-\pi}^{\pi} \exp \ j(-m\omega_0 t) \cdot x(t) \cdot d(\omega_0 t)$$

$$= \frac{1}{2\pi} \int_{-\pi}^{\pi} \exp \ j(-m\omega_0 t) \cdot \sum_{n=-\infty}^{\infty} \mathbf{c}(n) \cdot \exp \ j(n\omega_0 t) \cdot d(\omega_0 t)$$

$$= \frac{1}{2\pi} \sum_{n=-\infty}^{\infty} \mathbf{c}(n) \int_{-\pi}^{\pi} \exp \mathrm{j}[(n-m)\omega_0 t] \, . \, \mathrm{d}(\omega_0 t)$$

Now from equation A1.9 we see that this integral is zero except when $n = m$, when it equals 2π; whence the rhs of the equation becomes $\mathbf{c}(m)$. The lhs of the equation corresponds with the rhs of equation A2.2, so this inversion relation is therefore demonstrated.

A2.4 Properties

$y(t)$ and $x(t)$ are real periodic functions of time, and are described by the sets of Fourier coefficients $\{\mathbf{b}(n)\}$ and $\{\mathbf{c}(n)\}$ respectively. Then

Units If $x(t)$ has units of volts, then $|\mathbf{c}(n)|$ also has units of volts.

Symmetry If $x(t)$ is real, then $\mathbf{c}(-n) = \mathbf{c}^*(n)$ ⠀⠀⠀⠀⠀⠀⠀⠀(A2.3)

Time shift If $y(t) = x(t - \tau)$, then $\mathbf{b}(n) = \mathbf{c}(n) \, . \exp \mathrm{j}(-n\omega_0\tau)$ ⠀⠀(A2.4)

Parseval's relation Signal power may be calculated from either the time waveform, or from the set of Fourier coefficients, as the energy per period:

$$P = \frac{1}{T} \int_0^T x^2(t) \, \mathrm{d}t = \sum_{n=-\infty}^{\infty} |\mathbf{c}(n)|^2 \tag{A2.5}$$

Convolution If $x(t)$ is a periodic signal which is applied to a system having a frequency response $\mathbf{H}(\omega)$ and impulse response $h(t)$, then the output signal $y(t)$ is given by

$$y(t) = x(t) * h(t) = \sum_{n=-\infty}^{\infty} \mathbf{c}(n) \, . \, \mathbf{H}(n\omega_0) \, . \exp \mathrm{j}(n\omega_0 t) \tag{A2.6}$$

Table A2.1   *Some Fourier series coefficients*

Square wave:	$c(n) = \dfrac{2}{\pi n} \cdot \sin(n\pi/2)$ (A2.7)
Rectangular pulse:	$c(n) = \dfrac{1}{\pi n} \cdot \sin(n\pi t_w/T)$ (A2.8)
Triangular wave:	$c(n) = \dfrac{4}{n^2 \pi^2}$ n odd (A2.9)
Ramp:	$c(n) = \dfrac{-1}{\pi n} \cdot \cos(n\pi)$ (A2.10)

● **Example A2.1**   Consider a signal with period T sec, and which is

$$x(t) = \exp(-at) \qquad 0 \leqslant t \leqslant T$$

Then from equation A2.2, where $\omega_0 = 2\pi/T$,

$$c(n) = \frac{1}{T} \int_0^T \exp - (at + jn\omega_0 t) \, . \, \mathrm{d}t$$

$$= \frac{1}{(a + jn\omega_0)T} (1 - \exp - (aT + jn\omega_0 T))$$

$$\simeq \frac{1}{aT + jn2\pi} \qquad \text{if } \exp(-aT) \ll 1 \qquad (A2.11) \quad ●$$

Appendix A3
Fourier Transform

A3.1 Purpose

To represent an aperiodic pulse signal in the frequency domain, extending the complex phasors used for the Fourier series. The Fourier transform applies to the class of signals which have finite energy.

A3.2 Proposal

$$x(t) = \int_{-\infty}^{\infty} \mathbf{X}(f) . \exp \mathrm{j}(\omega t) \, \mathrm{d}f \qquad (A3.1)$$

and

$$\mathbf{X}(f) = \int_{-\infty}^{\infty} x(t) . \exp \mathrm{j}(-\omega t) \, \mathrm{d}t \qquad (A3.2)$$

A3.3 Proof

We may derive the Fourier transform relationships from the Fourier series equations A2.1 and A2.2, by taking the limiting case where $T \rightarrow \infty$, and hence $\omega_0 \rightarrow 0$. Now for a bandlimited periodic signal we have, from equation A2.1,

$$x(t) = \sum_{n=-N}^{N} \mathbf{c}(n) . \exp \mathrm{j}(n\omega_0 t) \qquad (A3.3)$$

However as $T \rightarrow \infty$, so the number of phasors in the summation becomes infinite, and their spacing (ω_0) tends to zero. Consequently the Fourier coefficient $\mathbf{c}(n)$ also tends to zero, from equation A2.2. We define a new frequency function, which is preserved in this limiting process:

$$\mathbf{X}(f) = \frac{\mathbf{c}(n)}{f_0} \qquad (A3.4)$$

Substituting into equation A2.2 and pressing to the limit we arrive at equation A3.2 above. Since the upper limit of n now tends to infinity, evidently the product nf_0 becomes a continuous frequency variable f, and equation A3.3 becomes of the form of equation A3.1.

A3.4 Properties

Units If $x(t)$ has units of volts, then $|X(f)|$ has units of volt-sec or volt/Hz. This is a *density* function.

Symmetry For a real signal $\quad X(-f) = X^*(f)$ (A3.5)

Scaling If $y(t) = x(at)$ then $\quad Y(f) = (1/a) . X(f/a)$ (A3.6)

Time shift If $y(t) = x(t - \tau)$ then $\quad Y(f) = X(f) . \exp \mathrm{j}(-\omega\tau)$ (A3.7)

Frequency shift If $Y(f) = X(f - f_0)$ then $\quad y(t) = x(t) . \exp \mathrm{j}(\omega_0\tau)$
(A3.8)

Parseval's relation Calculation of signal energy can be done in either the time or the frequency domains:

$$E = \int_{-\infty}^{\infty} |x(t)|^2 \, \mathrm{d}t = \int_{-\infty}^{\infty} |X(f)|^2 \, \mathrm{d}f$$ (A3.9)

Convolution If $w(t) = x(t) . y(t)$ then $\quad W(f) = X(f) * Y(f)$ (A3.10)

If $W(f) = X(f) . Y(f)$ then $\quad w(t) = x(t) * y(t)$ (A3.11)

A3.5 Dictionary

Table A3.1 *Some Fourier transforms*

Rectangular pulse:		NB. $\mathrm{sinc}(u) = \sin(\pi u)/(\pi u)$ $X(f) = \mathrm{sinc}(fT)$ (A3.12)
Triangular pulse:		$X(f) = \mathrm{sinc}^2(fT)$ (A3.13)
Impulse: $x(t) = \delta(t)$		$X(f) = 1$ (A3.14)
Sinusoid: $x(t) = \cos(\omega_0 t)$	$X(f) = \tfrac{1}{2} . \delta(f - f_0) + \tfrac{1}{2} . \delta(f + f_0)$	(A3.15)
$x(t) = \sin(\omega_0 t)$	$X(f) = (-\mathrm{j}/2) . \delta(f - f_0) +$ $(\mathrm{j}/2) . \delta(f + f_0)$	(A3.16)

● **Example A3.1** Consider the pulse signal:

$$x(t) = \exp(-at) \qquad t \geqslant 0$$
$$= 0 \qquad\qquad t < 0$$

Then from equation A3.2,

$$\mathbf{X}(f) = \int_0^\infty \exp(-at - j\omega t) \, dt$$

$$= \frac{1}{a + j2\pi f} \qquad\qquad\qquad (A3.17)$$

Note the dimensions of sec^{-1}, and compare with equation A2.11 which reflects the discrete nature of the frequency scale in that case. ●

Appendix A4
Laplace Transform

A4.1 Purpose

The Laplace transform is a generalisation of the Fourier transform, extending those basic ideas to use *complex-frequency*, $\mathbf{s} = \sigma + j\omega$. Essentially it deals with transient signals having finite energy, although periodic signals can be represented as for the Fourier transform.

The complex frequency plane, or *s*-plane as it is usually referred to, enables the attributes of *poles* and *zeros* to be employed in the description of signals. However, the greatest use for the Laplace transform is perhaps to describe a linear system performance in terms of its impulse response and its complex-frequency response.

A4.2 Proposal

$$\mathbf{X}(s) = \int_{-\infty}^{\infty} x(t) \cdot \exp(-st) \, dt \tag{A4.1}$$

where $s = \sigma + j\omega$.

$$x(t) = \frac{1}{2\pi j} \int_{\sigma - j\infty}^{\sigma + j\infty} \mathbf{X}(s) \cdot \exp(st) \, ds \tag{A4.2}$$

Notice that the inversion formula, equation A4.2, involves a *contour integration* in the complex *s*-plane, and requires a knowledge of functions of a complex variable before it can successfully be accomplished. The region of convergence must first be located.

Since the Laplace transform $\mathbf{X}(s)$ normally consists of a ratio of rational polynomials, $\mathbf{N}(s)$ in the numerator and $\mathbf{D}(s)$ in the denominator, inversion is usually done by splitting $\mathbf{X}(s)$ up into partial fractions, so that each term is in the form of one of the standard entries in the dictionary.

A4.3 Proof

No proof is offered, except to point out the similarities between equations A4.1 and A4.2, and the corresponding Fourier transform definitions A3.1

and A3.2. In the Fourier transform equations, $j\omega$ has been replaced by the complex variable s.

A4.4 Properties

Units   If $x(t)$ has units of volts, $|\mathbf{X}(s)|$ has units of volts-sec.

Symmetry   If $x(t)$ is real, then $\mathbf{X}(s^*) = \mathbf{X}^*(s)$.   (A4.3)

Scaling   If $y(t) = x(at)$ then $\mathbf{Y}(s) = (1/|a|) \cdot \mathbf{X}(s/a)$ (A4.4)

Stability   All poles must lie in the left-half-plane, if the signal $x(t)$ is not to diverge.

Time shift   If $y(t) = x(t - \tau)$ then $\mathbf{Y}(s) = \mathbf{X}(s) \cdot \exp(-s\tau)$ (A4.5)

Frequency shift   If $y(t) = x(t) \cdot \exp(s_0 t)$ then $\mathbf{Y}(s) = \mathbf{X}(s - s_0)$ (A4.6)

Frequency response   If a system has an impulse response $h(t)$, a corresponding Laplace transform $\mathbf{H}(s)$, and a frequency response $\mathbf{H}(\omega)$, then

$$\mathbf{H}(\omega) = \mathbf{H}(s)\big|_{s=j\omega}$$ (A4.7)

Convolution   If $v(t) = x(t) * y(t)$ then $\mathbf{V}(s) = \mathbf{X}(s) \cdot \mathbf{Y}(s)$ (A4.8)

A4.5 Dictionary

Note: $u(t)$ is the *unit step*

$$u(t) = 1 \quad t \geqslant 0$$
$$= 0 \quad t < 0$$

Table A4.1   *Some Laplace transforms*

$x(t)$	$\mathbf{X}(s)$	
$\delta(t)$	1	(A4.9)
$u(t)$	$1/s$	(A4.10)
$u(t) \cdot \exp(-\alpha t)$	$\dfrac{1}{(s + \alpha)}$	(A4.11)
$u(t) \cdot \exp(-\alpha t) \cdot \cos(\omega_0 t)$	$\dfrac{s + \alpha}{(s + \alpha)^2 + \omega_0^2}$	(A4.12)
$u(t) \cdot \exp(-\alpha t) \cdot \sin(\omega_0 t)$	$\dfrac{\omega_0}{(s + \alpha)^2 + \omega_0^2}$	(A4.13)

● **Example A4.1**   Consider the transform of the transient signal:

$$x(t) = \exp(-at) \qquad t \geqslant 0$$
$$ = 0 \qquad\qquad t < 0$$

Then using equation A4.1,

$$\mathbf{X}(s) = \int_0^\infty \exp(-at - st)\ dt$$

$$= \frac{1}{s + a} \tag{A4.14}$$

This equation is identical to the Fourier transform equation A3.17, except that $j\omega$ has been replaced by s (see equation A4.7).   ●

Appendix A5
Discrete Fourier
Transform

A5.1 Purpose

To represent a discrete-time signal $\{x(n)\}$ by a function $\mathbf{X}(f)$ (or $\mathbf{X}(\omega)$) in the frequency domain. It is a simple extension of the Fourier transform concept, but relies heavily upon the Fourier series although time and frequency domains are interchanged.

The relationship was first introduced in chapter 7, and has been examined in more detail and applied in chapter 13. Signals suitable for analysis by the DFT are intrinsically in the class of *energy* signals.

A5.2 Proposal

$$x(n) = \frac{1}{2\pi} \int_{-\pi}^{\pi} \mathbf{X}(\omega) \cdot \exp\ j(\omega T_S n)\ d(\omega T_S) \tag{A5.1}$$

and

$$\mathbf{X}(\omega) = \sum_{n=-\infty}^{\infty} x(n) \cdot \exp\ j(-\omega T_S n) \tag{A5.2}$$

A5.3 Proof

Equation A5.1 follows directly from the Fourier transform relationship A3.1, given that

    a) The continuous-time variable t is replaced by the discrete-time variable nT_S.

    b) The frequency function $\mathbf{X}(\omega)$ is periodic in ω_S since the time signal is sampled at rate f_S.

Equation A5.2 follows from equation A3.2, and is a direct consequence of the signal having values only at discrete intervals of time.

A5.4 Properties

Units If $x(n)$ has units of volts, then $|\mathbf{X}(\omega)|$ also has units of volts.

Symmetry If $x(n)$ is real, then $\mathbf{X}(-\omega) = \mathbf{X}^*(\omega)$ (A5.3)

Periodicity From equation A5.2 it follows that $\mathbf{X}(\omega)$ is periodic with period ω_S

Time shift If $y(n) = x(n-m)$ then $\mathbf{Y}(\omega) = \mathbf{X}(\omega) . \exp j(-\omega T_S m)$

      (A5.4)

Frequency shift If $\mathbf{Y}(\omega) = \mathbf{X}(\omega - \omega_0)$ then $y(n) = x(n) . \exp j(\omega_0 T_S n)$

      (A5.5)

Parseval's relation

$$\text{Energy } E = \sum_{n=-\infty}^{\infty} x^2(n) = \frac{1}{2\pi} \int_{-\pi}^{\pi} |\mathbf{X}(\omega)|^2 \, d(\omega T_S) \qquad (A5.6)$$

Note: this equation implies that $\{x(n)\}$ is an *energy* signal.

Convolution If $\mathbf{W}(\omega) = \mathbf{X}(\omega) . \mathbf{Y}(\omega)$ then $w(n) = x(n) * y(n)$ (A5.7)

If $w(n) = x(n) . y(n)$ then $\mathbf{W}(\omega) = \mathbf{X}(\omega) * \mathbf{Y}(\omega)$ (A5.8)

Note: since the spectra are periodic, equation A5.8 results in a *circular convolution* (section 13.4.3). However, this is often not significantly different from aperiodic convolution if the spectral functions are lowpass.

A5.5 Dictionary

Since the spectral function $\mathbf{X}(\omega)$ is obtained by a summation, there are not many signals that can be given analytical treatment. The DFT is normally used in its numerical form. One simple case is the rectangular pulse, which is used in the analysis of window functions (section 13.3)

Unit-pulse $x(q) = 1$

                $x(n) = 0$ $n \neq q$

$\mathbf{X}(\omega) = \exp j(-\omega T_S q)$ (A5.9)

Rectangular pulse $x(n) = 1$ $0 \leqslant n \leqslant M-1$

                      $= 0$ $0 > n \geqslant M$

$$\mathbf{X}(\omega) = \exp j(-(M-1)\omega T_S/2) . \frac{\sin(M\omega T_S/2)}{\sin(\omega T_S/2)} \qquad (A5.10)$$

Phasor $x(n) = \exp j(\omega_0 T_S n)$

$\mathbf{X}(\omega) = \delta(\omega - \omega_0)$ (A5.11)

● **Example A5.1** Consider a signal which is the sampled version of that employed in previous examples, A2.1, A3.1, A4.1:

$$x(n) = \exp(-aT_sn) \qquad n \geqslant 0$$

Then, using equation A5.2, we see that

$$\mathbf{X}(\omega) = \sum_{n=0}^{\infty} \exp - (aT_sn + j\omega T_sn)$$

Now, noting that this is a geometric series, and using the summation formula in equation A1.10, we conclude:

$$\mathbf{X}(\omega) = \frac{1}{1 - \exp - (aT_S + j\omega T_S)}$$

Notice the exponential term, which is characteristic of DFTs.   ●

A5.6   Discrete-frequency finite-DFT

This is the computable version of the DFT, which is discussed in some detail in chapter 13.

$$x(n) = \frac{1}{N} \sum_{k=0}^{N-1} \mathbf{X}(k) . \mathbf{W}^{kn} \tag{A5.12}$$

$$\mathbf{X}(k) = \sum_{n=0}^{N-1} x(n) . \mathbf{W}^{-kn} \tag{A5.13}$$

where $\mathbf{W} = \exp j(2\pi/N)$.

Note that the $1/N$ factor appears sometimes in the forward transformation like equation A5.13, instead of in the inverse, equation A5.12. There is no essential difference, it is just a matter of scaling.

Signal and spectrum are now represented by numerical vectors, $\{x(n)\}$ and $\{\mathbf{X}(k)\}$ respectively, and both are effectively periodic in N intervals. Consequently, the properties are now as follows:

Units   If $x(n)$ has units of volts, then $|\mathbf{X}(k)|$ also has units of volts.

Symmetry   If $x(n)$ is real, then $\mathbf{X}(N-k) = \mathbf{X}^*(k)$ \hfill (A5.14)

Periodicity $\mathbf{X}(k + N) = \mathbf{X}(k)$ \hfill (A5.15)

Time shift   If $y(n) = x(n-m)$ then $\mathbf{Y}(k) = \mathbf{X}(k) . \mathbf{W}^{-km}$ \hfill (A5.16)

Frequency shift   If $\mathbf{Y}(k) = \mathbf{X}(k-q)$ then $y(n) = x(n) . \mathbf{W}^{nq}$ \hfill (A5.17)

Parseval's relation

$$\text{Energy } E = \sum_{n=0}^{N-1} |\mathbf{x}(n)|^2 = \frac{1}{N} \sum_{k=0}^{N-1} |\mathbf{X}(k)|^2 \tag{A5.18}$$

Convolution   If $\mathbf{V}(k) = \mathbf{X}(k) \cdot \mathbf{Y}(k)$ then $v(n) = x(n) * y(n)$ (A5.19)

If $v(n) = x(n) \cdot y(n)$ then $\mathbf{V}(k) = \mathbf{X}(k) * \mathbf{Y}(k)$ (A5.20)

Note: since all these vectors are periodic, both convolutions are *circular*.

A5.6.1   Dictionary

Unit-pulse $x(q) = 1$
$$x(n) = 0 \qquad n \neq q$$

$$\mathbf{X}(k) = \mathbf{W}^{-kq} \tag{A5.21}$$

Rectangular pulse $x(n) = 1 \qquad 0 \leqslant n \leqslant M - 1$
$$\qquad\qquad\qquad\qquad = 0 \qquad M \leqslant n \leqslant N - 1$$

$$\mathbf{X}(k) = \mathbf{W}^{-(M-1)k/2} \cdot \frac{\sin(\pi k M / N)}{\sin(\pi k / N)} \tag{A5.22}$$

Phasor $x(n) = \exp \mathrm{j}(2\pi n q / N) \equiv \mathbf{W}^{nq}$

$$\mathbf{X}(k) = \mathbf{W}^{-(N-1)(k-q)/2} \cdot \frac{\sin[\pi(k-q)]}{\sin[\pi(k-q)/N]} \tag{A5.23}$$

● **Example A5.2**   Consider now the analysis of the signal from example A5.1, according to this discrete-frequency finite-DFT. Then

$$x(n) = \exp(-aT_S n) \qquad 0 \leqslant n \leqslant N - 1$$

From equation A5.13,

$$\mathbf{X}(k) = \sum_{n=0}^{N-1} \exp - \left(aT_S n + \mathrm{j} \frac{2\pi}{N} kn \right)$$

This is a finite geometric series, so using the summation formula of equation A1.10:

$$\mathbf{X}(k) = \frac{1 - \exp - (aT_S N + \mathrm{j}2\pi k)}{1 - \exp - (aT_S + \mathrm{j}2\pi k / N)} \tag{A5.24}$$

Notice the numerator term, which is a direct consequence of truncating the exponential decay of $x(n)$.   ●

Appendix A6
Z-transform

A6.1 Purpose

To transform a discrete-time signal $\{x(n)\}$ to the complex z-plane, and hence provide a means for avoiding convolution. For discrete-time signals, the Z-transform mirrors the Laplace transform for continuous-time signals. Since a complex plane domain is used, poles and zeros emerge as useful descriptors of the prototype signal.

Chapter 8 discusses the Z-transform and its application, we aim here merely to summarise certain of its properties.

A6.2 Proposal

A definition is
$$\mathbf{X}(z) = \sum_{n=0}^{\infty} x(n) . z^{-n} \tag{A6.1}$$

This is the *one-sided* transform and assumes that signals do not exist before $n = 0$ (or $t = 0$). Inverting it to obtain a signal sequence involves integration over a complex plane, which is expressed formally:

$$x(n) = \frac{1}{2\pi j} \oint_C \mathbf{X}(z) . z^{n-1} \, dz \tag{A6.2}$$

The integral is a *contour integral* and can only be evaluated from a detailed knowledge of functions of a complex variable, which is beyond the scope of this introductory treatise.

A more common way of inverting the Z-transform is by long division. In most cases the transform $\mathbf{X}(z)$ consists of two rational polynomials, $\mathbf{N}(z)$ in the numerator and $\mathbf{D}(z)$ in the denominator. Dividing these through puts $\mathbf{X}(z)$ in the form of a continuing polynomial in z^{-n}, from which the values of $\{x(n)\}$ can immediately be identified (equation A6.1).

A6.3 Proof

The transform equation A6.1 is presented as a proposition, and no proof is advanced. Chapter 8 discusses its properties in more detail, and shows

399

that a number of useful applications follow immediately.

It is possible to advance an argument for its derivation starting from the Laplace transform.

A6.4 Properties

Units If $x(n)$ is in volts, then $|X(z)|$ is also in volts.

Variable z is a complex variable, portrayed on the z-plane.

Stability The signal is convergent if all poles lie within the unit circle.

Symmetry If $x(n)$ is real then $X(z^*) = X^*(z)$ (A6.3)

Time shift If $y(n) = x(n - m)$ then $Y(z) = X(z) . z^{-m}$ (A6.4)

Frequency spectrum If a signal sequence $\{x(n)\}$ has a Z-transform $X(z)$ and a Fourier transform (frequency spectrum) $X(\omega)$, then

$$X(\omega) = X(z)|_{z = \exp j(\omega T_s)}$$ (A6.5)

Convolution If $v(n) = x(n) * y(n)$ then $V(z) = X(z) . Y(z)$ (A6.6)

A6.5 Dictionary

Table A6.1 *Some Z-transforms*

$x(n)$	$X(n)$	
c^n	$\dfrac{z}{z - c}$	(A6.7)
n	$\dfrac{z}{(z - 1)^2}$	(A6.8)
$a^n . \cos(\theta n)$	$\dfrac{z[z - a . \cos(\theta)]}{z^2 - 2a . \cos(\theta) . z + a^2}$	(A6.9)
$a^n . \sin(\theta n)$	$\dfrac{a . \sin(\theta) . z}{z^2 - 2a . \cos(\theta) . z + a^2}$	(A6.10)

● **Example A6.1** Consider the signal sequence $\{x(n)\}$,

$$x(n) = \exp(- aT_s n) \qquad n \geqslant 0$$

From equation A6.1, we see that

$$X(z) = \sum_{n=0}^{\infty} \exp(- aT_s n) . z^{-n}$$

This too is a geometric series, with a common ratio of $z^{-1} . \exp(- aT_s)$.

Hence, using the summation formula of equation A1.10,

$$\mathbf{X}(z) = \frac{1}{1 - z^{-1} . \exp(-aT_S)}$$

$$= \frac{z}{z - \exp(-aT_S)} \qquad\qquad (A6.11)$$

Substituting $z = \exp \mathrm{j}(\omega T_S)$ in this expression yields the frequency spectrum of example A5.1. $\bullet$

Appendix A7
Decibels

The *decibel* is a logarithmic measure of power ratio. There are many cases where a signal magnitude can vary over a large range of values, and by using a logarithmic scale, the *decibel* compresses this large range of numbers into a more manageable scale.

For instance, consider figure A7.1, which shows the set-up for measuring the amplification of a high-gain amplifier. Suppose that the amplifier has a gain of 2 000 000. In order to test its gain we must apply an input signal of the order of 1 μV, which is difficult to measure, so we use a resistive attenuator circuit to produce the required input signal from a signal source of 1 V. The input signal to the amplifier is then 0·000 001 V. By expressing these voltages in terms of logarithmic ratios, we reduce the range of values.

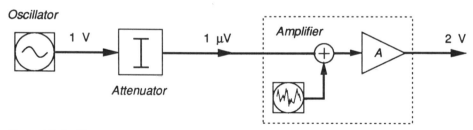

Figure A7.1 Measurement of amplifier gain

A7.1 Power ratio

The *decibel* value of a power ratio R is defined for a signal power P, relative to a reference power P_r:

$$R = 10 \cdot \log_{10}\left(\frac{P}{P_r}\right) \text{ dB} \tag{A7.1}$$

The reference power P_r is sometimes taken to be a fixed value. Thus

dBm   refers to power measured relative to 1 mW
dBW   refers to power measured relative to 1 W

Often, the reference power is taken as some interfering or *noise* power, and

Table A7.1 *Decibel values of power ratio*

P/P_r	1	2	3	10	20	30	100	1000	10 000	100 000	
R	0	3	4·8	10	13	14·8	20	30	40	50	dB

the signal power is expressed as the ratio of signal-to-noise power (SNR). Thus, if the signal power is S V^2, and the noise power is N V^2, then

$$\text{SNR} = 10 \cdot \log_{10}\left(\frac{S}{N}\right) \text{ dB} \tag{A7.2}$$

This situation occurs in figure A7.1, where in addition to the legitimate input signal at frequency f Hz, there will also be a *noise* signal, generated by the first transistor in the amplifier circuit. The output of the amplifier will therefore consist of signal plus noise, and SNR is a measure of signal quality.

If the SNR $= 0$ dB, then signal and noise have equal power. If SNR $= 10$ dB then the ratio is $10:1$, while if the SNR $= -10$ dB the ratio is $1:10$. For a reasonable quality audio system, SNR > 40 dB, while for telephone conversations SNR $= 10$ dB is still usable.

A7.2 Voltage ratio

Signal level is normally measured in terms of voltage (or current), but this measure is easily related to the power ratio. Recall that in signal processing we routinely refer to the squared signal as *power*, even in the absence of a definable resistance. Thus for a signal x, the power ratio may be expressed as

$$R = 10 \cdot \log_{10}\left(\frac{x^2}{x_r^2}\right)$$

$$R = 20 \cdot \log_{10}\left(\frac{x}{x_r}\right) \text{ dB} \tag{A7.3}$$

A voltage ratio may therefore be expressed in decibels by using 20 as the multiplying constant.

For example, in the case of the amplifier in figure A7.1, the input voltage

Table A7.2 *Decibel values of voltage ratio*

V/V_r	1	2	3	10	20	30	100	1000	10 000	100 000	
R	0	6	9·6	20	26	29·6	40	60	80	100	dB

is V_1, the output voltage is V_2, and the voltage gain may be written as

$$A = 20 \cdot \log_{10}\left(\frac{V_2}{V_1}\right) \text{ dB} \qquad (A7.4)$$

In this case, $A = 2\,000\,000$ or 123 dB.

Notice that because the decibel is a logarithmic measure, the gains of cascaded stages may be added in dB. Referring again to figure A7.1, the attenuator has a voltage gain of -120 dB and the amplifier has a voltage gain of 123 dB, so the overall gain is 3 dB.

Signal-to-noise ratio can also be expressed as the ratio of *rms* voltages. For instance, in the example above, the input signal to the amplifier is 1 μV rms. Suppose that the noise generated in the amplifier input stage is 250 nV rms. The SNR is therefore 12 dB.

STUDY QUESTIONS

1 Express the following power ratios in dB: 2, 100, 200, 20000.

2 Express the following voltage ratios in dB: 4, 10, 40, 400.

3 A certain SNR is 30 dB. What are the corresponding voltage and power ratios?

4 A certain coaxial cable has an attenuation of 10 dB/km at 10 MHz, and 30 dB/km at 100 MHz. Two equal-magnitude signals, at 10 MHz and 100 MHz, are launched into such a cable. What is their voltage ratio after a distance of 10 km?

Appendix A8
Probability
Q-function

The Q[ ] function describes the area under the tail of a Gaussian probability density function, and is defined as

$$Q[y] = \frac{1}{\sqrt{(2\pi)}} \int_y^\infty \exp(-x^2/2)\, dx \qquad (A8.1)$$

If $y > 4$ then

$$Q[y] \simeq \frac{1}{y\sqrt{(2\pi)}} \exp(-y^2/2) \qquad (A8.2)$$

For a Gaussian random variable with variance σ^2, the probability that a certain level a is exceeded is

$$P[X \geqslant a] = Q[a/\sigma] \qquad (A8.3)$$

The table which follows is calculated according to equations A8.1 and A8.2 as appropriate. Figure A8.1 gives a graphical view of the function.

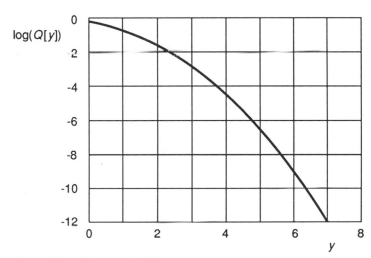

Figure A8.1 The Q probability function

405

y	0	0·2	0·4	0·6	0·8
0	0·5000	0·4920	0·4841	0·4761	0·4681
0·1	0·4602	0·4523	0·4444	0·4365	0·4286
0·2	0·4208	0·4130	0·4052	0·3975	0·3898
0·3	0·3821	0·3745	0·3670	0·3595	0·3520
0·4	0·3446	0·3373	0·3300	0·3228	0·3157
0·5	0·3086	0·3016	0·2947	0·2878	0·2810
0·6	0·2743	0·2677	0·2612	0·2547	0·2483
0·7	0·2420	0·2358	0·2297	0·2237	0·2178
0·8	0·2119	0·2062	0·2005	0·1950	0·1895
0·9	0·1841	0·1789	0·1737	0·1686	0·1636
1·0	0·1587	0·1539	0·1492	0·1446	0·1401
1·1	0·1357	0·1314	0·1272	0·1231	0·1191
1·2	0·1151	0·1113	0·1075	0·1039	0·1003
1·3	0·0969	0·0935	0·0902	0·0870	0·0838
1·4	0·0808	0·0779	0·0750	0·0722	0·0695
1·5	0·0668	0·0643	0·0618	0·0594	0·0571
1·6	0·0548	0·0527	0·0505	0·0485	0·0465
1·7	0·0446	0·0427	0·0410	0·0392	0·0376
1·8	0·0360	0·0344	0·0329	0·0315	0·0301
1·9	0·0287	0·0274	0·0262	0·0250	0·0239
2·0	0·228	0·0217	0·0207	0·0197	0·0188
2·1	0·0179	0·0170	0·0162	0·0154	0·0146
2·2	0·0139	0·0132	0·0126	0·0119	0·0113
2·3	0·0107	0·0102	0·0096	0·0091	0·0087
2·4	0·0082	0·0078	0·0073	0·0070	0·0066
2·5	0·0062	0·0059	0·0055	0·0052	0·0049
2·6	$4·67e-3$	$4·40e-3$	$4·15e-3$	$3·91e-3$	$3·68e-3$
2·7	$3·47e-3$	$3·27e-3$	$3·08e-3$	$2·89e-3$	$2·72e-3$
2·8	$2·56e-3$	$2·40e-3$	$2·26e-3$	$2·12e-3$	$1·99e-3$
2·9	$1·87e-3$	$1·75e-3$	$1·64e-3$	$1·54e-3$	$1·44e-3$
3·0	$1·35e-3$	$1·27e-3$	$1·19e-3$	$1·11e-3$	$1·04e-3$
3·1	$9·70e-4$	$9·06e-4$	$8·47e-4$	$7·91e-4$	$7·38e-4$
3·2	$6·89e-4$	$6·43e-4$	$6·00e-4$	$5·59e-4$	$5·21e-4$
3·3	$4·85e-4$	$4·52e-4$	$4·21e-4$	$3·92e-4$	$3·64e-4$
3·4	$3·39e-4$	$3·15e-4$	$2·93e-4$	$2·72e-4$	$2·53e-4$
3·5	$2·34e-4$	$2·18e-4$	$2·02e-4$	$1·87e-4$	$1·74e-4$
3·6	$1·61e-4$	$1·49e-4$	$1·38e-4$	$1·28e-4$	$1·18e-4$
3·7	$1·10e-4$	$1·01e-4$	$9·38e-5$	$8·67e-5$	$8·02e-5$
3·8	$7·41e-5$	$6·85e-5$	$6·33e-5$	$5·85e-5$	$5·40e-5$
3·9	$4·99e-5$	$4·61e-5$	$4·25e-5$	$3·93e-5$	$3·62e-5$

y	0	0·2	0·4	0·6	0·8
4·0	3·35e − 5	3·07e − 5	2·82e − 5	2·59e − 5	2·37e − 5
4·1	2·18e − 5	2·00e − 5	1·83e − 5	1·67e − 5	1·53e − 5
4·2	1·40e − 5	1·28e − 5	1·17e − 5	1·07e − 5	9·81e − 6
4·3	8·96e − 6	8·18e − 6	7·47e − 6	6·82e − 6	6·22e − 6
4·4	5·67e − 6	5·17e − 6	4·71e − 6	4·29e − 6	3·90e − 6
4·5	3·55e − 6	3·23e − 6	2·94e − 6	2·67e − 6	2·43e − 6
4·6	2·20e − 6	2·00e − 6	1·82e − 6	1·65e − 6	1·49e − 6
4·7	1·36e − 6	1·23e − 6	1·11e − 6	1·01e − 6	9·12e − 7
4·8	8·25e − 7	7·46e − 7	6·75e − 7	6·10e − 7	5·51e − 7
4·9	4·98e − 7	4·49e − 7	4·06e − 7	3·66e − 7	3·30e − 7
5·0	2·97e − 7	2·68e − 7	2·41e − 7	2·17e − 7	1·96e − 7
5·1	1·76e − 7	1·58e − 7	1·42e − 7	1·28e − 7	1·15e − 7
5·2	1·03e − 7	9·25e − 8	8·30e − 8	7·45e − 8	6·68e − 8
5·3	5·98e − 8	5·36e − 8	4·80e − 8	4·30e − 8	3·85e − 8
5·4	3·44e − 8	3·08e − 8	2·75e − 8	2·46e − 8	2·19e − 8
5·5	1·96e − 8	1·75e − 8	1·56e − 8	1·39e − 8	1·24e − 8
5·6	1·10e − 8	9·83e − 9	8·76e − 9	7·79e − 9	6·93e − 9
5·7	6·16e − 9	5·48e − 9	4·87e − 9	4·33e − 9	3·84e − 9
5·8	3·41e − 9	3·02e − 9	2·68e − 9	2·38e − 9	2·11e − 9
5·9	1·87e − 9	1·65e − 9	1·46e − 9	1·29e − 9	1·15e − 9
6·0	1·01e − 9	8·95e − 10	7·91e − 10	6·98e − 10	6·16e − 10
6·1	5·44e − 10	4·80e − 10	4·23e − 10	3·73e − 10	3·29e − 10
6·2	2·89e − 10	2·55e − 10	2·24e − 10	1·97e − 10	1·73e − 10
6·3	1·52e − 10	1·34e − 10	1·18e − 10	1·03e − 10	9·06e − 11
6·4	7·95e − 11	6·97e − 11	6·11e − 11	5·36e − 11	4·69e − 11
6·5	4·11e − 11	3·59e − 11	3·14e − 11	2·75e − 11	2·40e − 11
6·6	2·10e − 11	1·84e − 11	1·60e − 11	1·40e − 11	1·22e − 11
6·7	1·06e − 11	9·28e − 12	8·09e − 12	7·04e − 12	6·13e − 12
6·8	5·34e − 12	4·65e − 12	4·04e − 12	3·51e − 12	3·05e − 12
6·9	2·65e − 12	2·30e − 12	2·00e − 12	1·74e − 12	1·51e − 12

Note: 5·66e − 3 denotes $5·66 \times 10^{-3}$.

$Q[y]$	10^{-3}	10^{-4}	10^{-5}	10^{-6}	10^{-7}	10^{-8}	10^{-9}	10^{-10}
y	3·10	3·70	4·27	4·76	5·22	5·61	6·00	6·36

Appendix A9
Probability Functions of Two Variables

A9.1 Continuous-valued variables

A9.1.1 Joint probability

Consider two random variables X and Y, whose values fall on a continuous magnitude scale. Then we may define a *joint probability density function* $f_{XY}(x, y)$, which expresses the probability that two conditions are met:

$$f_{XY}(x, y) . \delta x . \delta y$$
$$= \text{Prob}[[x \leqslant X \leqslant (x + \delta x)] \quad \text{and} \quad [y \leqslant Y \leqslant (y + \delta y)]] \qquad (A9.1)$$

It is thus similar to the single variable pdf, and possesses many similar properties. For instance:

$$\int_{-\infty}^{\infty} \int_{-\infty}^{\infty} f_{XY}(x, y) \, dx \, dy = 1 \qquad (A9.2)$$

$$\int_{-\infty}^{\infty} f_{XY}(x, y) \, dx = f_Y(y) \qquad (A9.3)$$

$$\int_{-\infty}^{\infty} f_{XY}(x, y) \, dy = f_X(x) \qquad (A9.4)$$

Two-variable moments may also be defined. The general definition is

$$m_{rs} = \int_{-\infty}^{\infty} \int_{-\infty}^{\infty} x^r . y^s . f_{XY}(x, y) \, dx \, dy \qquad (A9.5)$$

Particular moments are therefore:

$$m_{10} = E[X] \qquad m_{01} = E[Y] \qquad (A9.6)$$
$$m_{20} = E[X^2] \qquad m_{02} = E[Y^2] \qquad (A9.7)$$
$$m_{11} = E[X . Y] \qquad (A9.8)$$

A9.1.2 Conditional probability

In the above discussion, we have assumed that the random variables X and Y can be chosen freely. In many circumstances the choice is constrained, and in particular when one of the two variables is *dependent* upon the other the probabilities are said to be *conditional*.

Conditional probability density functions may be defined thus:

$$f_{XY}(y \mid x) . \delta y = \text{Prob}[[y \leqslant Y \leqslant (y + \delta y)] \text{ and } [X = x]] \qquad (A9.9)$$

$$f_{XY}(x \mid y) . \delta x = \text{Prob}[[x \leqslant X \leqslant (x + dx)] \text{ and } [Y = y]] \qquad (A9.10)$$

Calculation of the conditional probability density functions proceeds according to a relationship known as *Bayes' Theorem*:

$$f_{XY}(x, y) = f_Y(y) . f_{XY}(x \mid y) \qquad (A9.11)$$

$$f_{XY}(x, y) = f_X(x) . f_{XY}(y \mid x) \qquad (A9.12)$$

It follows that

$$\int_{-\infty}^{\infty} f_{XY}(x \mid y) \, dx = 1 \qquad (A9.13)$$

$$\int_{-\infty}^{\infty} f_{XY}(y \mid x) \, dy = 1 \qquad (A9.14)$$

A9.2 Discrete-valued variables

The random variable X is now restricted to a finite set of values $\{x_i\}$, and Y is similarly restricted to a set $\{y_j\}$. Probability density functions are now replaced by probabilities, for instance the joint probability $P[x_i, y_j]$. A similar set of relationships exists to those above, except that now the integrals are finite summations. Hence we may summarise a number of properties thus:

$$\sum_i \sum_j P[x_i, y_j] = 1 \qquad (A9.15)$$

$$\sum_j P[x_i, y_j] = P[x_i] \qquad (A9.16)$$

$$\sum_i P[x_i, y_j] = P[y_j] \qquad (A9.17)$$

$$P[x_i, y_j] = P[x_i] . P[y_j \mid x_i] \qquad (A9.18)$$

$$P[x_i, y_j] = P[y_j] . P[x_i \mid y_i] \qquad (A9.19)$$

$$\sum_i P[x_i \mid y_j] = 1 \qquad (A9.20)$$

$$\sum_j P[y_j \mid x_i] = 1 \qquad (A9.21)$$

General moment:

$$m_{rs} = \sum_r \sum_s x_i^r \cdot y_j^s \cdot P[x_i, y_j] \tag{A9.22}$$

● **Example A9.1** The following statistics refer to a binary-valued signal source, where the two variables x_i and y_j are an adjacent pair in the signal sequence.

x_i	$P[x_i]$		$P[x_i, y_j]$	x_i 0	1		y_j	$P[y_j]$		$P[y_j \mid x_i]$	x_i 0	1
0	0·3	y_j 0		0·2	0·4		0	0·6	y_j 0		0·67	0·57
1	0·7	1		0·1	0·3		1	0·4	1		0·33	0·43

The formulae A9.15 to A9.21 can now be verified by inspection. For instance, global summations of the probabilities $P[x_i]$, $P[y_j]$ and $P[x_i, y_j]$ yield unity in each case.

Summing the columns of the $P[y_j \mid x_i]$ table also yields unity for each column, as per equation A9.21.

Calculation of the conditional probability, equation A9.18, can also be verified.   ●

Reading List

There is a vast number of books on Signal Processing, and the following list is a personal view of a few of these.

Signal properties, transforms and linear systems

M L MEADE & C R DILLON, *Signals and Systems* (Van Nostrand 1986)
A clear and readable small book, which fills in some of the gaps in my treatment of the subject. Contains additional mathematical derivations of common signal properties.

P A LYNN & W FUERST, *Digital Signal Processing* (Wiley 1989)
Covers the classical material, but with the practical support of a number of computer programs which allow the reader to explore the use of the concepts.

A V OPPENHEIM, A S WILSKY with I T YOUNG, *Signals and Systems* (Prentice-Hall 1983)
A reference book, which deals thoroughly with the fundamental concepts of signal processing, but contains numerous examples for teaching the subject systematically.

Signal processing in practice

C S WILLIAMS, *Designing Digital Filters* (Prentice-Hall 1986)
An excellent tutorial guide to the techniques available for filter design.

A BATEMAN & W YATES, *Digital Signal Processing Design* (Pitman 1988)
One of the few books which relates signal processing to specific DSP hardware designs. A very readable and helpful practical book.

D J DEFATTA & J I LUCAS, *Digital Signal Processing* (Wiley 1988)
Presents the classical theory of signal processing including random signals, from a systems point of view. Includes practical filter designs and overall system designs for signal detection, etc.

A V OPPENHEIM & R W SHAFER, *Digital Signal Processing* (Prentice-Hall 1975)

A classical treatment of signal processing, ranging widely over the fundamental issues, including advanced applications and implementation difficulties like finite arithmetic precision.

Random signals

M SCHWARTZ & L SHAW, *Signal Processing—Discrete Spectral Analysis, Detection and Estimation* (McGraw-Hill 1975)

Very clear development of the topic of random signals, leading through to extremely valuable and practical applications.

D J DEFATTA & J I LUCAS, *Digital Signal Processing* (Wiley 1988)

Presents the classical theory of signal processing including random signals, from a systems point of view. Includes practical filter designs and overall system designs for signal detection, etc.

A PAPOULIS, *Probability, Random Variables and Stochastic Processes* (McGraw-Hill 1965)

Gives a classical foundation to the topic, with rigorous derivations. Not an easy read, but valuable as a reference source.

Numerical Solutions
to Study Questions

Introduction to Information
1 $11 \cdot 3$ bits

Chapter 1
4 Lowpass, bandstop

Chapter 2
4 0; 1250
   5; 75
   2; $208 \cdot 5$
   $0 \cdot 36$; $0 \cdot 9$
5 $3 \cdot 0$, $2 \cdot 77$, $2 \cdot 12$, $1 \cdot 15$, 0, $-1 \cdot 15$, $-2 \cdot 12$, $-2 \cdot 77$
   $3 \cdot 0$, 0, $-3 \cdot 0$, 0, $3 \cdot 0$, 0, $-3 \cdot 0$, 0
6 avg $= 0$, $P = 4 \cdot 5$, $E = 72$ mJ
7 $0 \cdot 444$ V^2, $0 \cdot 0668$

Chapter 3
2 $6 \cdot \exp \mathrm{j}(0 \cdot 4) = 5 \cdot 53 + \mathrm{j}2 \cdot 34$
3 $6 \cdot \exp \mathrm{j}(0 \cdot 16n + 0 \cdot 4)$; $6 \cdot \exp \mathrm{j}(0 \cdot 1n + 0 \cdot 4)$; $6 \cdot \exp \mathrm{j}(0 \cdot 4)$
4 Complex magnitude is $(a/2) \cdot \exp \mathrm{j}(-\theta - \pi/2)$
6 $z(t) = 5 \cdot 88 \cdot \cos(500t + 130 \cdot 8°)$
7 $y(t) = 2 \cdot 7 \cdot \cos(\omega t - 33°)$
10 Ratio is $0 \cdot 06 : 1$

Chapter 4
2 *a*) 0 V, $17 \cdot 0$ V^2, $4 \cdot 12$ V
   b) $5 \cdot 0$ V, $29 \cdot 5$ V^2, $5 \cdot 43$ V
3 $2 \cdot 48$ V^2, $1 \cdot 57$ V
4 Filter A output: $y(t) = 0 \cdot 2\sqrt{2} \cdot \cos(2\pi 0 \cdot 6t)$, t in ms
   $P(\mathrm{A}) = 0 \cdot 04$ V^2, $P(\mathrm{B}) = 0 \cdot 34$ V^2, $P(\mathrm{C}) = 1 \cdot 69$ V^2, $P(\mathrm{D}) = 0 \cdot 41$ V^2
5 *a*) $y(t) = 21 \cdot 2 \cdot \cos(100t - 45°) + 10 \cdot 0 \cdot \cos(150t - 56 \cdot 3°)$
   b) $y(t) = 30 \cdot 0 + 8 \cdot 05 \cdot \cos(200t - 63 \cdot 4°)$
6 *a*) $x(n) = 5 \cdot 0 \cdot \cos(0 \cdot 2n) + 3 \cdot 0 \cdot \cos(0 \cdot 3n)$
   b) $x(n) = 5 \cdot 0 + 3 \cdot 0 \cdot \cos(0 \cdot 4n)$

7 a) $y(n) = 14 \cdot 56 \cdot \cos(0 \cdot 2n - 23 \cdot 9°) + 7 \cdot 98 \cdot \cos(0 \cdot 3n - 33 \cdot 4°)$
   b) $y(n) = 16 \cdot 7 + 6 \cdot 70 \cdot \cos(0 \cdot 4n - 37 \cdot 5°)$
8 a) $y(t) = 1 \cdot 5$; b) $y(t) = 0$
9 0, 800, 1000, 2000, 2800 rad/sec

Chapter 5
2 a) $0 \cdot 54$ V^2; b) $0 \cdot 56$ V^2; c) $0 \cdot 25$ V^2
3 Envelope max = $1 \cdot 3$, min = $0 \cdot 7$
   Signal power = $0 \cdot 52$ V^2 Sidebands are $1 \cdot 197$ and $1 \cdot 203$ MHz
4 a) $T_S = 0 \cdot 167$ μs and number of calculations is 60 000
   b) $T_S = 66 \cdot 7$ μs and number of calculations is 150
5 $\omega(t) = 2000 - 25 \cdot \cos(10t)$ rad/sec
6 15 kHz (or $62 \cdot 8$ Mrad/sec, $\pm 94 \cdot 2$ krad/sec); $P = 4 \cdot 5$ V^2
7 $\mathbf{A}(t) = \exp j[5 \cdot \cos(\omega_m t)]$

Chapter 6
3 $f_S = 25$ kHz; 10 kHz, 20 kHz
5 $7 \cdot 6 < f_S < 11$ kHz
6 a) 9   11   15   13   8   6   6   7
   b) 35   45   60   51   32   26   23   29
7 a) $0 \cdot 03$ $0 \cdot 0$ $0 \cdot 03$ $0 \cdot 07$ $0 \cdot 0$ $-0 \cdot 07$ $0 \cdot 03$ $-0 \cdot 03$
   b) $0 \cdot 0$ $0 \cdot 01$ $-0 \cdot 01$ $0 \cdot 0$ $0 \cdot 0$ $0 \cdot 01$ $0 \cdot 01$ $0 \cdot 0$
10 a) 9 bits; b) 13 bits

Chapter 7
1 a) Linear, b) Non-linear, c) Non-linear, d) Linear
2 $\{g(n)\} = \{\frac{1}{2}, 1, 1, 1, 1, 1, 1 \dots\}$ Integrator
3 $\{y(n)\} = \{\frac{1}{2}, 2, 5, 7, 6, 4, 3, 3, 3, 3, 3 \dots\}$
4 $\{y(n)\} = \{1, 1, 2, -4, -2, 0, 2, 0, 0 \dots\}$
   Note the accurate slope estimate.
5 a) $\mathbf{H}(\omega) = j2[2 \sin(\omega T_S) - \sin(2\omega T_S)] \cdot \exp j(-2\omega T_S)$
   b) $\mathbf{H}(\omega) = j/[2 \cdot \tan(\omega T_S/2)]$
6 a) $2T_S$, b) Additional phase is $0 \cdot 24°$/Hz

Chapter 8
1 $\{y(n)\} = \{0, 1, 1, 1, 1, 0, -2, -2, -2, 1 \dots\}$
2 a) $X(z) = z^{-1} + 2z^{-2} + 3z^{-3} + 4z^{-4} + 4z^{-5} + 2z^{-6} - 2z^{-8} - z^{-9} \dots$
       $Y(z) = z^{-1} + z^{-2} + z^{-3} + z^{-4} - 2z^{-6} - 2z^{-7} - 2z^{-8} + z^{-9} \dots$
   b) $X(z) = z/(z - 0 \cdot 9)$
   c) $X(z) = z/(z - 1 \cdot 1)$
   d) $X(z) = z/[z - 0 \cdot 8 \cdot \exp j(\pi/8)]$
   e) $X(z) = 1$
3 See table 8.1
4 $\{g(n)\} = \{2, -1, 0, 0, 0, 0 \dots\}$ $H'(z) = 2 - z^{-1}$
   $\{y(n)\} = \{0, 2, 3, 4, 5, 4, 0, -2, -4 \dots\}$

5 P1: $H(\omega) = 1 - \exp \mathrm{j}(-\omega T_S)$ P4: $H(\omega) = 2 - \exp \mathrm{j}(-\omega T_S)$

6 a) $H'(z) = 2 - 3z^{-1} + z^{-2}$
      $H(0) = 0$; $|H(\omega_S/4)| = 3 \cdot 16$; $|H(\omega_S/2)| = 6$

    b) $H'(z) = 1$

    c) $H'(z) = (1 - z^{-1})/(3 - z^{-1})$
      $H(0) = 0$; $|H(\omega_S/4)| = 0 \cdot 82$; $|H(\omega_S/2)| = 0 \cdot 5$

Chapter 9

1 $y(n) = x(n) - x(n - 2)$ $|H(\omega_S/4)| = 2$

2 $y(n) = x(n) + 1 \cdot 338 x(n - 1) + x(n - 2)$
    At 0 kHz: $|H(\omega)| = 3 \cdot 34$. At 6 kHz: $|H(\omega)| = 0 \cdot 66$.

3 $y(n) = x(n) + 2 \cdot 176 x(n - 1) + 3 \cdot 176 x(n - 2)$
           $+ 3 \cdot 176 x(n - 3) + 2 \cdot 176 x(n - 4) + x(n - 5)$

4 $y(n) = x(n) - 0 \cdot 9 y(n - 1)$
    At $\omega = 0$: $|H(\omega)| = 0 \cdot 53$. At $\omega = \omega_S/2$: $|H(\omega)| = 10 \cdot 0$.

5 Ratio before the filter is $4 \cdot 0$ or 12 dB
    Ratio after the filter is $44 \cdot 8$ or 33 dB.

6 $y(n) = x(n) + 1 \cdot 592 y(n - 1) - 0 \cdot 968 y(n - 2)$
    At 0 kHz: $|H(\omega)| = 2 \cdot 66$. At 100 Hz: $|H(\omega)| \simeq 53$.
    At 500 Hz: $|H(\omega)| = 0 \cdot 28$.

Chapter 10

2 $4 \cdot 5$ or $6 \cdot 5$ dB; $T = 12 \cdot 57 n$ ms where n is integer

4 $\geqslant 1 \cdot 33$ sec

5 $\geqslant 3111$

6 $\pm 1 \cdot 0$ V

7 $RC - 10$ ms; eg. $R = 100$ kΩ, $C = 0 \cdot 1$ μF

8 305 times or $24 \cdot 8$ dB

9 $\leqslant 1 \cdot 49$ kHz; $1 \cdot 33$ sec

10 $15 \cdot 7$ times or $12 \cdot 0$ dB

Chapter 11

1 Gain error is $1 \cdot 5\%$; $C = 561$ pF

2 $\tau = 0 \cdot 56$ ms; Gain at 5 kHz is $17 \cdot 4$

3 Error is $36 \cdot 3\%$ for $N = 3$; Error is $10 \cdot 0\%$ for $N = 2$

4 Design: $\tau \geqslant 0 \cdot 317$ ms; eg. $R = 10$ kΩ, $C = 31 \cdot 7$ nF
          $A \geqslant 250$
    Phase error at 100 Hz is $1 \cdot 15°$

5 Gain is 50; Error is $5 \cdot 7\%$ and $31°$

Chapter 12

2 $x_I(t) = 2 \cdot 5 \cdot \cos(300t + 2)$; $x_Q(t) = 2 \cdot 5 \cdot \sin(300t + 2)$

3 Magnitude is $2 \cdot 5$; Angle is 2 rad

4 $x_I(n) = 2 \cdot 5 \cdot \cos(0 \cdot 5n + 2)$; $x_Q(n) = 2 \cdot 5 \cdot \sin(0 \cdot 5n + 2)$;
    Mag. $2 \cdot 5$; Angle 2 rad

5 Magnitude is $\sqrt{[41 + 40 . \cos(200t)]}$

7 $x_I(8) = 2 \cdot 400$; $x_Q(8) = -0 \cdot 699$; Mag. $2 \cdot 5$; Angle $0 \cdot 283$ rad
   Shifting coefficients are: $\cos(\theta) = 0 \cdot 707$; $\sin(\theta) = -0 \cdot 707$
   After shifting: $y_I(8) = 1 \cdot 203$; $y_Q(8) = -2 \cdot 191$; Mag. $2 \cdot 5$; Angle $-1 \cdot 069$ rad

8

Angle	$+45°$	$-45°$	$+135°$	$-135°$
y_I	$0 \cdot 707$	$0 \cdot 707$	$-0 \cdot 707$	$-0 \cdot 707$
y_Q	$0 \cdot 707$	$-0 \cdot 707$	$0 \cdot 707$	$-0 \cdot 707$

9 $0 \cdot 89$ to $8 \cdot 9$ kHz   **10** $\delta \leqslant 0 \cdot 063$ rad or $3 \cdot 6°$

Chapter 13

1

Frequency	-1150	-1100	-900	-850	850	900	1100	1150 rad/s
Complex magnitude	j	$2 \cdot 5$	$2 \cdot 5$	$-$j	j	$2 \cdot 5$	$2 \cdot 5$	$-$j

Periodic frequency is 50 rad/sec

2 $(T/2) . \mathrm{sinc}\,[(f + f_0)T] + (T/2) . \mathrm{sinc}\,[(f - f_0)T]$

3 $$\mathbf{X}(\omega) = \frac{\exp \mathrm{j}[(-M-1)(\omega - \omega_0)T_S/2] . \sin[(\omega - \omega_0)MT_S/2]}{\sin[(\omega - \omega_0)T_S/2]}$$

4 $x(1) = 0 \cdot 5$, $x(-1) = 0 \cdot 5$, $x(n) = 0$ elsewhere

5 $a)$ $(1/2) . \mathbf{W}(\omega - \omega_1) + (1/2) . \mathbf{W}(\omega + \omega_1)$
   $b)$ $(-\mathrm{j}/2) . \mathbf{W}(\omega - \omega_1) + (\mathrm{j}/2) . \mathbf{W}(\omega + \omega_1)$

6

k	0	1	2	3	4	5	6	7
magnitude	$3 \cdot 0$	$2 \cdot 80$	$2 \cdot 24$	$1 \cdot 47$	$1 \cdot 0$	$1 \cdot 47$	$2 \cdot 24$	$2 \cdot 80$
angle (rad)	0	$-1 \cdot 32$	$-2 \cdot 68$	$2 \cdot 07$	0	$-2 \cdot 07$	$2 \cdot 68$	$1 \cdot 32$

7 Improvement is of the order of 114   **9** $1 \cdot 29$; $28 \cdot 8$ dB
8 $35 \cdot 5$ **10** $3 \cdot 2$ dB; $26 \cdot 3$ dB

Chapter 14

3 $1 \cdot 22$; $1 \cdot 31$; 0   **8** $0 \cdot 159$
4 $1 \cdot 0$; $1/9$ **9** $5 \cdot 0$ V^2; $0 \cdot 691$
5 $0 \cdot 556$; $0 \cdot 469$ **10** $6 \cdot 6 \times 10^{-2}$
6 $1 \cdot 5$; $3 \cdot 15$ **11** $72 \cdot 8\%$
7 $0 \cdot 014$

Chapter 15

1 $R_X(k) = A^2(1 - 2k/N)$ $0 \leqslant k \leqslant N$
   $R_X(k) = A^2(2k/N - 3)$ $N \leqslant k \leqslant 2N$

2 $P[X = 0, Y = 0] = 1/16$ $P[0, A] = 1/8$ $P[0, 2A] = 1/16$
   $P[A, 0] = 1/8$ $P[A, A] = 1/4$ $P[A, 2A] = 1/8$
   $P[2A, 0] = 1/16$ $P[2A, A] = 1/8$ $P[2A, 2A] = 1/16$

3 $\{R_X(k)\} = \{5A^2/3, 2A^2/3, 7A^2/6, 11A^2/12, \ldots A^2, \ldots\}$

4 $15 \cdot 0$ V^2 $2 \cdot 24$ V   **5** $k \geqslant 4$

6	4·48 V	27·4 V^2	7	4·24 V	14·0 V^2
8	prf is 900 Hz		9	$M = 7$	Prob = 0·23
10	1·5 μV^2		11	$\alpha = 0·998$	

Chapter 16

1 $S_X(\omega) = N[1 + \frac{1}{2}.\cos(\omega T_S)]$
2 $S_Y(\omega) = 8N[(5/4) + (3/2)\cos(\omega T_S) + (1/4)\cos(2\omega T_S)]$

3	$(3\omega_S)/(16b)$ rad/sec	9	5 bits
4	0·0125 V^2	10	30 kHz
5	50 Hz; $b = 37·5$	11	37·5 or 15·7 dB
6	0·92 V^2; 0·25 V^2	12	25 or 14 dB
		13	0·5 mV^2/Hz
8	10·67 or 10·3 dB	14	0·2 Hz; 2·5 sec; $\alpha = 0·999996$

Appendix 7

1	3, 20, 23, 43 dB	2	12, 20, 32, 52 dB
3	31·6, 1000	4	10^{10} or 200 dB

Index